学习 OpenCV 4
基于 Python 的算法实战

肖铃 / 著

电子工业出版社
Publishing House of Electronics Industry
北京·BEIJING

内 容 简 介

本书使用 Python 语言，以 OpenCV 源码结构中的模块为章节，系统地介绍了 OpenCV 在图像处理和计算机视觉领域的应用。

本书第 1 章主要介绍 OpenCV 的源码，以及开发环境的搭建和使用源码方式的编译技巧；第 2～14 章主要介绍 OpenCV 的各个模块，包括模块导读，以及模块中重点算法使用案例的讲解，并在章节的最后提供了相应的进阶知识。

本书以 OpenCV 源码结构的方式讲解，一方面可以让读者深入了解 OpenCV 软件及相应的源码，另一方面可以加深读者对软件的透彻理解。初学者可以参考本书前 6 章系统地学习图像处理应用，特别是第 4 章和第 5 章，因为经典的图像处理算法基本上都封装在 imgproc 模块中；对于其他特殊需求，可以参考对应章节，如与特征点相关的知识可以参考第 10 章，与机器学习和深度学习相关的知识可以分别参考第 13 章和第 14 章。

本书不仅适合图像处理和计算机视觉领域的读者阅读，还适合高校师生、专业技术人员、图像处理爱好者、深度学习计算机视觉领域爱好者参考使用。

未经许可，不得以任何方式复制或抄袭本书之部分或全部内容。
版权所有，侵权必究。

图书在版编目（CIP）数据

学习 OpenCV 4：基于 Python 的算法实战 / 肖铃著. —北京：电子工业出版社，2022.3
ISBN 978-7-121-42439-7

Ⅰ. ①学… Ⅱ. ①肖… Ⅲ. ①图象处理软件－程序设计 Ⅳ. ①TP391.413

中国版本图书馆 CIP 数据核字（2021）第 241650 号

责任编辑：刘 伟　　　特约编辑：田学清
印　　刷：三河市良远印务有限公司
装　　订：三河市良远印务有限公司
出版发行：电子工业出版社
　　　　　北京市海淀区万寿路 173 信箱　　邮编：100036
开　　本：787×980　　1/16　　印张：28.75　　字数：703 千字
版　　次：2022 年 3 月第 1 版
印　　次：2022 年 3 月第 1 次印刷
定　　价：109.00 元

凡所购买电子工业出版社图书有缺损问题，请向购买书店调换。若书店售缺，请与本社发行部联系，联系及邮购电话：（010）88254888，88258888。

质量投诉请发邮件至 zlts@phei.com.cn，盗版侵权举报请发邮件至 dbqq@phei.com.cn。
本书咨询联系方式：010-51260888-819，faq@phei.com.cn。

前言

北宋大家张载的"横渠四句"非常有名,"为天地立心,为生民立命,为往圣继绝学,为万世开太平",历史上很多名人以之作为安身立命追求事业的勉励语。第一次读这句名言时,我非常激动,现在也时常想起,虽无法实现却也心向往之。

有幸与电子工业出版社再次合作,个人能力浅薄,不敢妄谈"为往圣继绝学",但也希望自己的书有一些实用价值,能在技术传播中贡献一份绵薄之力。在写作前,刘伟老师(本书编辑)和我反复推敲写作的目的是什么,这本书与别的图书有什么不同,能给读者带来什么帮助等命题。我们围绕上述命题提出了如下 4 个问题:为什么要学习图像处理?为什么要选用 OpenCV?为什么要用 OpenCV 4?这本书与别的图书有什么不同?

下面就通过对这 4 个问题的回答来介绍一下本书,让读者能够对本书的写作目的、写作思路,以及如何参考本书进行 OpenCV 的学习有一个初步的理解。

为什么要学习图像处理?

图像处理之所以重要,是因为人类对认知世界的需要。

我们都知道,当前智能手机的普及率非常高,当看到美景或有事情发生时,几乎所有人都会拿起手机拍照或录制视频,并将它们分享在各大平台。爱美之心,人皆有之,因此,相机软件也集成了越来越多的功能,如美颜、滤镜等,其中很多功能都是在照片采集后通过图像处理技术实现的。

各大手机厂商不遗余力地升级拍照或摄像功能,为了让用户能存储更多的照片和视频而不断提升手机的存储空间,这些都证明了用户对图像和视频的需求非常强烈,从侧面也反映了图像处理技术的广阔市场前景。

为什么要选用 OpenCV?

OpenCV 之所以大行其道,是因为其简单易用。

作为一个开源的计算机视觉库,OpenCV 由 C++语言编写,读者可以通过阅读源码对算法的实现细节进行学习,可读性好。另外,OpenCV 还实现了很多图像处理和计算机视觉中的通用算法,在 Linux、Windows、Android 和 macOS 等操作系统上均可运行,即一套代码能在多种系统中运行,通用性好。

OpenCV 支持 C++、Python、Java 等语言和 MATLAB 等多种程序接口,在最新版本中还提供了对 C#、Ch、Ruby、Go 语言的支持,不同语言中的接口名称和参数差异较小,用户在不同的开发环境中只需改写少量代码即可使用,一致性好。

OpenCV 在图像处理领域深耕多年,不仅对算法做了很多优化,还针对实用场景做了加速处理,如支持对 CUDA 等硬件的加速。另外,它还可以通过引入 Intel 的 IPP 高性能多媒体函数库获得更高的运行处理速度,高效性好。

OpenCV 大受欢迎还有一个重要的原因,就是当前深度学习和计算机视觉的飞速发展。在计算机视觉中进行模型训练和使用时，需要对输入图像做很多预处理和后处理，这些都需要用到 OpenCV 库。OpenCV 中很多算法的实现可以弥补深度学习模型推理耗时的缺陷，在某些特定场景下，OpenCV 中的传统算法可以获得与深度学习方法相差不大的结果，但在执行速度上远胜于深度学习方法，应用性好。

为什么要用 OpenCV 4？

不管什么软件，新版本的出现基本上都是因为对旧版本的代码做了很大力度的重构或增加了很多新特性，已经无法与旧版本兼容。

OpenCV 4 版本升级中废弃了一些旧的 API，对一些枚举类型的名称写法做了修改，对一些函数的参数做了调整，还有一个比较重要的变化就是增加了对 dnn 模块的支持。如果读者目前使用的是 OpenCV 3，那么也不必对这些变化有过多的焦虑，因为 OpenCV 的错误提示清晰易懂，错误定位与修改均比较快速、简单。

此外，建议读者在学习一门新技术时，最好选用较新的软件版本，因为很多软件在发展过程中会逐渐停止对旧版本的支持。

这本书与别的图书有什么不同？

OpenCV 对图像处理和计算机视觉算法做了系统封装，这些算法被划分为很多模块，其中最重要的就是 core 模块（第 3 章）和 imgproc 模块（第 4 章和第 5 章），而很多同类图书的内容就是围绕这两个模块展开的，但是本书内容没有局限于此。

本书以 OpenCV 模块来分章节作为写作思路，在每章的开始都有对模块的导读，导读内容依托 OpenCV 源码，讲述了本模块封装的算法函数，对其中的函数和数据结构做了清晰明了的注释，因此，导读内容既可以为读者学习算法提供帮助，又可以为读者开发查询提供参考。

本书在第 1 章讲述了 C++开发语言和 Python 开发语言，以及 Windows 和 Linux 系统中的 OpenCV 开发环境的搭建，只要读者按照案例中的操作步骤执行，就能够顺利地运行 OpenCV 库，不必为开发环境搭建中的各种问题焦头烂额。

core 模块和 imgproc 模块的重要性毋庸置疑，本书也针对这两个模块总结了很多案例，对这两个模块的算法做了透彻的讲解。但对于其他模块，如可视化模块 highgui、视频处理模块 videoio、视频分析模块 video、照片处理模块 photo、2D 特征模块 features2d、相机标定与三维重建模块 calib3d、传统目标检测模块 objdetect、机器学习模块 ml、深度学习模块 dnn，在其他图书中鲜有涉及，而本书对这些模块都做了详细导读和案例讲解。

这种系统化的讲解能够让读者对 OpenCV 有一个全面的认识，也能够让读者有机会了解 OpenCV 的强大功能，为读者在处理特殊需求时提供知识储备。

最后，感谢家人、朋友和出版社对本书写作的支持。希望本书能够给对 OpenCV 和图像处理有需求的读者以帮助，对于书中的疏漏之处，请不吝指教。如果有深度学习和计算机视觉学习需求的读者，敬请关注我的另一本图书《深度学习计算机视觉实战》，该书对深度学习和计算机视觉算法的基础、图像处理基础、计算机视觉案例、TensorFlow Lite，以及基于 TensorFlow Lite 在移动端和 PC 端的部署做了一站式的讲解，在此对您的支持表示衷心的感谢。

肖铃

2021 年 10 月

目录

第 1 章　OpenCV 快速入门 ········· 1
1.1　OpenCV 介绍 ··················· 1
1.1.1　OpenCV 概述 ············· 1
1.1.2　OpenCV 的代码结构 ······ 2
1.1.3　OpenCV 4 的新特性 ······ 3
1.2　OpenCV 开发环境搭建 ········ 5
1.2.1　案例 1：Windows 动态库开发环境搭建 ··········· 5
1.2.2　案例 2：Linux 动态库开发环境搭建 ···· 13
1.2.3　案例 3：Python 语言开发环境搭建 ···· 19
1.3　OpenCV 模块介绍 ············ 22
1.3.1　常用模块 ················ 22
1.3.2　扩展模块 ················ 23
1.4　OpenCV 源码编译 ············ 25
1.4.1　案例 4：OpenCV 编译 ···· 25
1.4.2　案例 5：OpenCV 裁剪编译 ···· 29
1.4.3　案例 6：扩展模块 opencv-contrib 编译 ··········· 30
1.5　进阶必备：OpenCV 入门参考 ···· 31
1.5.1　OpenCV 版本选择 ······· 31
1.5.2　如何学习 OpenCV ······· 31

第 2 章　图像读/写模块 imgcodecs ···· 33
2.1　模块导读 ····················· 33
2.2　图像读/写操作 ··············· 36
2.2.1　案例 7：图像读取 ········ 36
2.2.2　案例 8：图像保存 ········ 41
2.3　图像编/解码 ·················· 43
2.3.1　案例 9：图像编码应用 ···· 43
2.3.2　案例 10：图像解码应用 ···· 44
2.4　进阶必备：聊聊图像格式 ···· 45

第 3 章　核心库模块 core ············ 48
3.1　模块导读 ····················· 48
3.2　基本数据结构 ················· 60
3.2.1　案例 11：Mat 数据结构介绍及 C++ 调用 ················ 60
3.2.2　案例 12：Python 中的 Mat 对象操作 ···· 71
3.2.3　案例 13：Point 结构 ······ 73
3.2.4　案例 14：Rect 结构 ······· 76
3.2.5　案例 15：Size 结构 ······· 78
3.3　矩阵运算 ····················· 80
3.3.1　案例 16：四则运算 ······· 80
3.3.2　案例 17：位运算 ········· 82
3.3.3　案例 18：代数运算 ······· 83
3.3.4　案例 19：比较运算 ······· 86
3.3.5　案例 20：特征值与特征向量 ···· 89
3.3.6　案例 21：生成随机数矩阵 ···· 90
3.4　矩阵变换 ····················· 91
3.4.1　案例 22：矩阵转向量 ···· 91
3.4.2　案例 23：通道分离与通道合并 ···· 92
3.4.3　案例 24：图像旋转 ······· 93
3.4.4　案例 25：图像拼接 ······· 95
3.4.5　案例 26：图像边界拓展 ···· 96
3.4.6　案例 27：傅里叶变换 ···· 98
3.5　进阶必备：聊聊图像像素遍历与应用 ···· 99
3.5.1　案例 28：图像像素遍历 ···· 99
3.5.2　案例 29：提取拍照手写签名 ···· 101

第 4 章 图像处理模块 imgproc（一） …… 103
4.1 模块导读 …… 103
4.2 案例 30：颜色空间变换 …… 137
4.3 案例 31：图像尺寸变换 …… 138
4.4 基本绘制 …… 139
4.4.1 案例 32：绘制标记 …… 139
4.4.2 案例 33：绘制直线 …… 141
4.4.3 案例 34：绘制矩形 …… 142
4.4.4 案例 35：绘制圆 …… 143
4.4.5 案例 36：绘制椭圆 …… 144
4.4.6 案例 37：绘制文字 …… 145
4.5 形态学运算 …… 146
4.5.1 案例 38：腐蚀 …… 146
4.5.2 案例 39：膨胀 …… 148
4.5.3 案例 40：其他形态学运算 …… 149
4.6 图像滤波 …… 151
4.6.1 案例 41：方框滤波 …… 151
4.6.2 案例 42：均值滤波 …… 153
4.6.3 案例 43：高斯滤波 …… 153
4.6.4 案例 44：双边滤波 …… 154
4.6.5 案例 45：中值滤波 …… 155
4.7 边缘检测 …… 156
4.7.1 案例 46：Sobel 边缘检测 …… 157
4.7.2 案例 47：Scharr 边缘检测 …… 159
4.7.3 案例 48：Laplacian 边缘检测 …… 161
4.7.4 案例 49：Canny 边缘检测 …… 162
4.8 进阶必备：聊聊颜色模型 …… 164

第 5 章 图像处理模块 imgproc（二） …… 166
5.1 霍夫变换 …… 166
5.1.1 案例 50：霍夫线变换 …… 166
5.1.2 案例 51：霍夫圆变换 …… 169
5.2 案例 52：仿射变换 …… 171
5.3 案例 53：透视变换 …… 173
5.4 案例 54：重映射 …… 175
5.5 阈值化 …… 177
5.5.1 案例 55：基本阈值化 …… 177
5.5.2 案例 56：自适应阈值化 …… 179
5.6 图像金字塔 …… 180
5.6.1 案例 57：高斯金字塔 …… 180
5.6.2 案例 58：拉普拉斯金字塔 …… 182
5.7 直方图 …… 183
5.7.1 案例 59：直方图计算 …… 183
5.7.2 案例 60：直方图均衡化 …… 185
5.8 传统图像分割 …… 186
5.8.1 案例 61：分水岭算法 …… 186
5.8.2 案例 62：GrabCut 算法 …… 191
5.8.3 案例 63：漫水填充算法 …… 195
5.9 角点检测 …… 196
5.9.1 案例 64：Harris 角点检测 …… 196
5.9.2 案例 65：Shi-Tomasi 角点检测 …… 197
5.9.3 案例 66：亚像素角点检测 …… 199
5.10 图像轮廓 …… 201
5.10.1 案例 67：轮廓查找 …… 201
5.10.2 案例 68：轮廓绘制 …… 202
5.11 轮廓包裹 …… 204
5.11.1 案例 69：矩形边框 …… 204
5.11.2 案例 70：最小外接矩形 …… 205
5.11.3 案例 71：最小外接圆 …… 207
5.12 案例 72：多边形填充 …… 208
5.13 图像拟合 …… 210
5.13.1 案例 73：直线拟合 …… 210
5.13.2 案例 74：椭圆拟合 …… 211
5.13.3 案例 75：多边形拟合 …… 213
5.14 案例 76：凸包检测 …… 214
5.15 进阶必备：图像处理算法概述 …… 215

第 6 章　可视化模块 highgui ············ 217
6.1　模块导读 ············ 217
6.2　图像窗口 ············ 223
6.2.1　案例 77：创建与销毁窗口 ············ 223
6.2.2　案例 78：图像窗口操作 ············ 224
6.3　图像操作 ············ 225
6.3.1　案例 79：图像显示 ············ 225
6.3.2　案例 80：选取感兴趣区域 ············ 226
6.4　案例 81：键盘操作 ············ 227
6.5　案例 82：鼠标操作 ············ 228
6.6　案例 83：进度条操作 ············ 231
6.7　进阶必备：在 Qt 中使用 OpenCV ············ 233

第 7 章　视频处理模块 videoio ············ 240
7.1　模块导读 ············ 240
7.2　视频读取 ············ 254
7.2.1　案例 84：从文件读取视频 ············ 254
7.2.2　案例 85：从设备读取视频 ············ 256
7.3　视频保存 ············ 256
7.3.1　案例 86：从图片文件创建视频 ············ 257
7.3.2　案例 87：保存相机采集的视频 ············ 258
7.4　进阶必备：视频编/解码工具 FFMPEG ············ 259

第 8 章　视频分析模块 video ············ 261
8.1　运动分析 ············ 261
8.1.1　模块导读 ············ 261
8.1.2　案例 88：基于 MOG2 与 KNN 算法的运动分析 ············ 265
8.2　目标跟踪 ············ 267
8.2.1　模块导读 ············ 267
8.2.2　案例 89：基于 CamShift 算法的目标跟踪 ············ 273
8.2.3　案例 90：基于 meanShift 算法的目标跟踪 ············ 275
8.2.4　案例 91：稀疏光流法运动目标跟踪 ············ 277
8.2.5　案例 92：稠密光流法运动目标跟踪 ············ 279
8.3　进阶必备：深度学习光流算法 ············ 281

第 9 章　照片处理模块 photo ············ 284
9.1　模块导读 ············ 284
9.2　案例 93：基于 OpenCV 的无缝克隆 ············ 291
9.3　案例 94：基于 OpenCV 的图像对比度保留脱色 ············ 293
9.4　案例 95：基于 OpenCV 的图像修复 ············ 295
9.5　案例 96：基于 OpenCV 的 HDR 成像 ············ 298
9.6　图像非真实感渲染 ············ 301
9.6.1　案例 97：边缘保留滤波 ············ 301
9.6.2　案例 98：图像细节增强 ············ 303
9.6.3　案例 99：铅笔素描 ············ 303
9.6.4　案例 100：风格化图像 ············ 304
9.7　进阶必备：照片处理算法概述 ············ 305

第 10 章　2D 特征模块 features2d ············ 308
10.1　模块导读 ············ 308
10.2　特征点检测 ············ 326
10.2.1　案例 101：SIFT 特征点检测 ············ 326
10.2.2　案例 102：SURF 特征点检测 ············ 328
10.2.3　案例 103：BRISK 特征点检测 ············ 330
10.2.4　案例 104：ORB 特征点检测 ············ 331
10.2.5　案例 105：KAZE 特征点检测 ············ 332
10.2.6　案例 106：AKAZE 特征点检测 ············ 333
10.2.7　案例 107：AGAST 特征点检测 ············ 335
10.2.8　案例 108：FAST 特征点检测 ············ 336
10.3　特征点匹配 ············ 337
10.3.1　案例 109：Brute-Force 特征点匹配 ············ 337
10.3.2　案例 110：FLANN 特征点匹配 ············ 339
10.4　进阶必备：特征点检测算法概述 ············ 340

第 11 章 相机标定与三维重建模块 calib3d ··········342

- 11.1 模块导读 ··········342
- 11.2 单应性变换 ··········363
  - 11.2.1 案例 111：单应性变换矩阵 ··········363
  - 11.2.2 案例 112：单应性应用之图像插入 ··········365
- 11.3 相机标定 ··········367
  - 11.3.1 案例 113：棋盘角点检测并绘制 ··········367
  - 11.3.2 案例 114：消除图像失真 ··········369
- 11.4 进阶必备：聊聊镜头失真 ··········373

第 12 章 传统目标检测模块 objdetect ··········374

- 12.1 模块导读 ··········374
- 12.2 级联分类器的应用 ··········382
  - 12.2.1 案例 115：人脸检测 ··········382
  - 12.2.2 案例 116：人眼检测 ··········385
- 12.3 案例 117：HOG 描述符行人检测 ··········386
- 12.4 二维码应用 ··········387
  - 12.4.1 案例 118：二维码检测 ··········388
  - 12.4.2 案例 119：二维码解码 ··········389
- 12.5 进阶必备：聊聊条形码与二维码 ··········390

第 13 章 机器学习模块 ml ··········392

- 13.1 模块导读 ··········392
- 13.2 案例 120：基于 OpenCV 的 Logistic 回归 ··········409
- 13.3 案例 121：基于 OpenCV 的支持向量机 ··········412
- 13.4 案例 122：基于 OpenCV 的主成分分析 ··········415
- 13.5 进阶必备：机器学习算法概述 ··········417

第 14 章 深度学习模块 dnn ··········420

- 14.1 模块导读 ··········420
- 14.2 风格迁移 ··········433
  - 14.2.1 深度学习风格迁移 ··········433
  - 14.2.2 案例 123：OpenCV 实现风格迁移推理 ··········434
- 14.3 图像分类 ··········437
  - 14.3.1 深度学习图像分类 ··········437
  - 14.3.2 案例 124：基于 TensorFlow 训练 Fashion-MNIST 算法模型 ··········439
  - 14.3.3 案例 125：OpenCV 实现图像分类推理 ··········443
- 14.4 目标检测 ··········446
  - 14.4.1 深度学习目标检测 ··········446
  - 14.4.2 案例 126：OpenCV 实现目标检测推理 ··········446
- 14.5 图像超分 ··········448
  - 14.5.1 深度学习图像超分算法 ··········448
  - 14.5.2 案例 127：OpenCV 实现图像超分推理 ··········449
- 14.6 进阶必备：OpenCV 与计算机视觉 ··········450
  - 14.6.1 计算机视觉的发展 ··········451
  - 14.6.2 OpenCV 在计算机视觉中的应用 ··········451

第 1 章
OpenCV 快速入门

OpenCV（开源计算机视觉库）是一款个人和商用均免费的工具库，支持 Windows、Linux、macOS、iOS 和 Android 跨平台，提供 C/C++/Python/Java 函数。OpenCV 提供图像处理、视频采集与分析等基本功能，版本较新的库提供机器学习及深度学习的功能。

1.1 OpenCV 介绍

1.1.1 OpenCV 概述

OpenCV 是计算机视觉中的经典库，具有跨平台和多语言支持特性，功能非常强大。在 OpenCV 官网（见图 1.1），用户可以下载对应的库、访问 GitHub、阅读在线文档、学习 OpenCV 课程等。

图 1.1

OpenCV 在 2000 年发布第一个开源版本,到目前已经 20 多年了。2009 年,OpenCV 2.0 正式版本发布;2015 年,OpenCV 3.0 正式版本发布;2018 年,OpenCV 4.0 正式版本发布;当前最新版本为 OpenCV 4.5。

在 GitHub 上,可以追溯到的最早代码版本为 2.2,发布于 2010 年,其代码结构和最新版本的代码结构有些许差异,自 2.4.5 版本开始,形成了如今的代码结构,如图 1.2 所示。

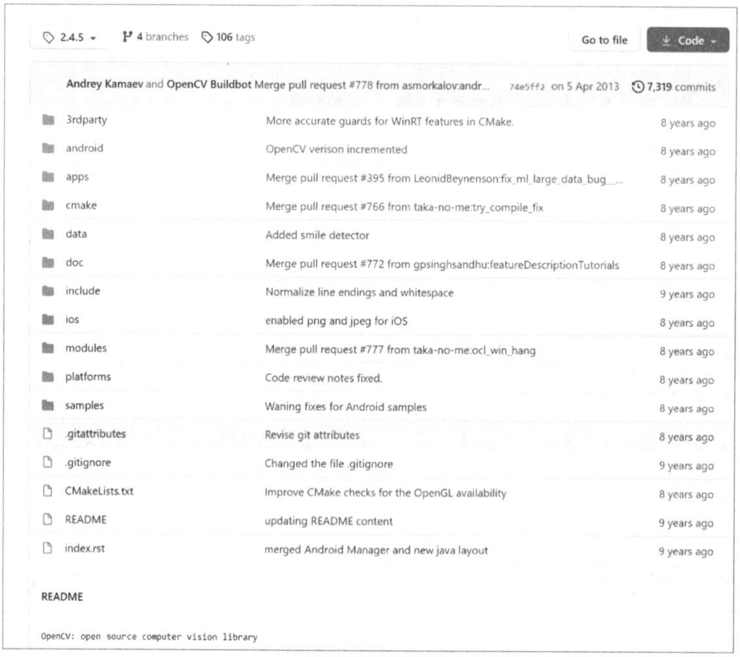

图 1.2

OpenCV 不同版本之间的差异来自引入最新算法、函数的完善和语言特性的丰富等方面。OpenCV 在发展过程中,逐步引入了算法加速、并行计算等先进技术,不断发展完善。目前,OpenCV 4 有广泛应用,OpenCV 3 也有较多应用,OpenCV 4 与 OpenCV 3 的函数兼容较好,但是 OpenCV 4 提供了更多特性,因此,读者可以参考升级。

1.1.2 OpenCV 的代码结构

图 1.2 是 OpenCV 的代码结构,包括 11 个文件夹和若干文件,各文件夹及文件的作用如下。

- 3rdparty。

3rdparty 文件夹存放 OpenCV 用到的第三方库的源码或下载脚本,如其中的 libjpeg 库为 jpeg 图片格式的解码库。

知识点：3rdparty 是常见的存放第三方依赖库的文件夹名称，有的项目中将其命名为 3rd，含义是相同的。

- apps。

apps 文件夹存放一些工具，如 traincascade 子目录为训练级联分类器的工具。

- cmake。

cmake 文件夹存放 cmake 编译生成项目工程时的依赖文件。

- data。

data 文件夹存放 OpenCV 样例用到的资源文件。

- doc。

doc 文件夹存放文档生成的脚本及资源文件。

- include。

include 文件夹包含 OpenCV 引入时包含的头文件，即：

```
#include "opencv/opencv.hpp"
```

- modules。

modules 文件夹存放 OpenCV 算法模块，是 OpenCV 代码的核心部分。

- platforms。

platforms 文件夹为 OpenCV 跨平台提供支持，文件夹中包含交叉编译工具链及实现跨平台编译所需的额外文件。

- samples。

samples 文件夹存放 OpenCV 官方提供的样例。

- CMakeLists.txt。

CMakeLists.txt 文件为 OpenCV cmake 编译脚本。

提示：CMake 为跨平台编译工具，CMake 写法表示软件名称，而小写的 cmake 为命令行中使用 CMake 软件编译项目的命令名称。

1.1.3 OpenCV 4 的新特性

在 OpenCV 4.0 正式版本发布之前，OpenCV 发布了另外两个版本：OpenCV 4.0.0-alpha 和 OpenCV 4.0.0-beta，如图 1.3 所示。

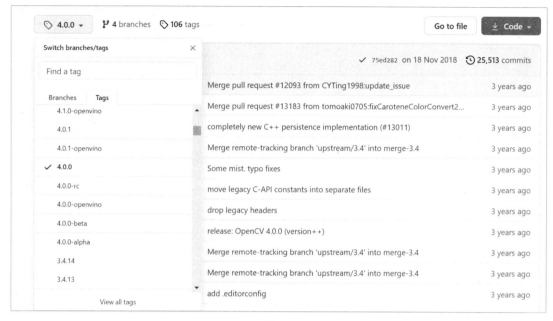

图 1.3

知识点 1：在代码版本管理中，alpha 版本为内测版本，此时处于开发测试阶段；beta 版本也属于内测版本，但是已经修复了 alpha 版本中的漏洞，可能存在未知问题，因此会在部分用户渠道发布，对于非开源软件，beta 版本使用较多。

知识点 2：在查看 GitHub 上的代码时，可以选择不同的 Branches 和 Tags。其中，Branches 为主分支；Tags 为对应不同节点的分支，在 Tags 中测试稳定后会合并到 master

在 OpenCV 3.x 的基础上，OpenCV 4.x 版本升级引入了如下新特性。

- 语言特性：正式引入 C++ 11 的库，用户可以方便地使用 C++ 11 的智能指针等特性。
- 新的库：引入 OpenVINO（Open Visual Inference and Neural Network Optimization，开源视觉推理与神经网络优化库），用于计算机视觉推理与深度学习推理开发。
- 算法库的扩展、重构与优化：如 DNN 模块引入 Vulkan backend，重构 opencv_stitching 模块，添加了新模块 G-API 等。
- 以前版本的问题修复：对于 OpenCV 的每一次版本更新，用户可以进入官网的库下载页面，选择"Release Notes"选项查看其更新细节，如图 1.4 所示。

第 1 章　OpenCV 快速入门

图 1.4

接着跳转进入 GitHub 中的 OpenCV 仓库，仓库中对本次版本的更新内容有详细说明，在"Wiki"选项中可以查看，如图 1.5 所示。

图 1.5

读者可以选择版本，查看其更新打印信息（ChangeLog）和更新情况。

1.2　OpenCV 开发环境搭建

1.2.1　案例 1：Windows 动态库开发环境搭建

在 Windows 上进行 C++开发，常用的 IDE（集成开发工具）为 Visual Studio，该软件由微软发布，最新版本为 Visual Studio 2019（Visual Studio 2022 正式版待发布）。本案例

基于 Visual Studio 2019 开发，Visual Studio 2019 的安装方法如下。

在安装前，读者需要去官网下载安装文件，对于个人开发者，可以选择下载社区版 Community 2019，如图 1.6 所示。

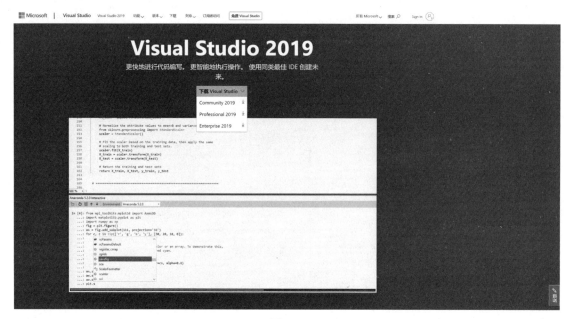

图 1.6

Community 2019 下载的文件名为 vs_Community.exe，双击该文件即可安装。在安装过程中，读者可以选择下载安装工作负载（不同开发环境）与单个组件，如图 1.7 所示。

Visual Studio 2019 支持 C++桌面开发、Python 开发、Node.js 开发等，读者可以选择自己需要的环境进行安装，本案例需要 C++桌面开发，勾选相应的复选框，安装完成后重启即可使用。

OpenCV 常用的两种开发语言是 C++和 Python，本节讲解在 Windows 下搭建 OpenCV C++语言开发环境。开发环境搭建方法有两种：安装官方发布的库文件和源码编译。

源码编译的方法参见 1.4 节，此处介绍使用官方发布的库文件进行安装的方法。

图 1.7

（1）进入官网，选择"Library"→"Releases"选项，进入 Releases 库文件包下载页面，根据相应的环境下载库文件，如图 1.8 所示。

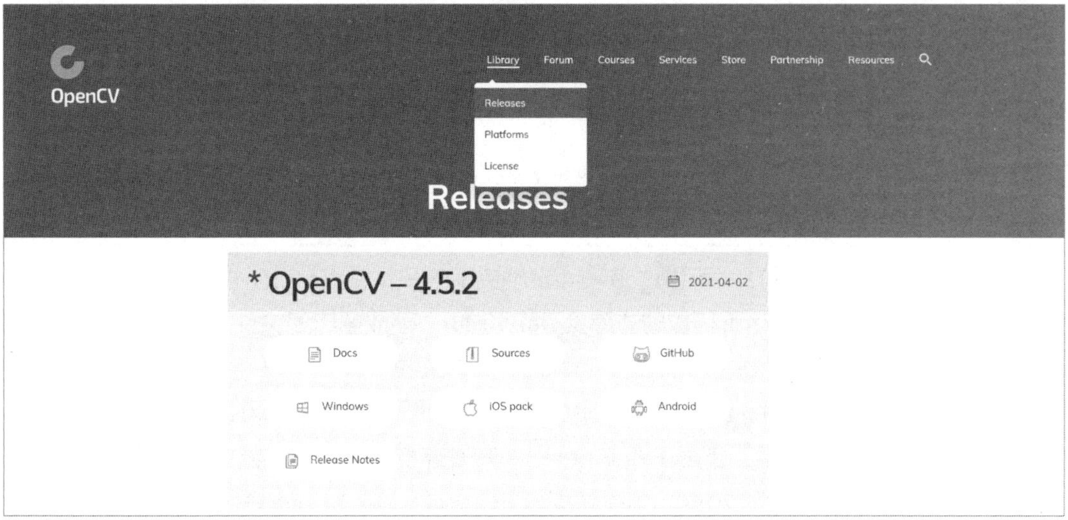

图 1.8

知识点：库文件分为 Release 和 Debug 两种形式，Release 为发行版本，其中不含调试信息，在生成文件包时编译器会做优化，文件包较小；Debug 为调试版本，含调试信息，用于开发人员开发调试，在正式发布产品中使用 Release 版本库文件。

本案例选择 OpenCV 4.5.2 版本，下载的文件名为 opencv-4.5.2-vc14_vc15.exe，双击该文件解压，如图 1.9 所示。

图 1.9

也可以右击该文件，在弹出的快捷菜单中选择"解压到当前文件夹"选项，得到文件夹 opencv，其中包含两个子文件夹。

- sources：用于存放当前版本的源码。
- build：用于存放编译的二进制库文件。进入 build 文件夹，此路径下有 6 个文件夹。
    - bin：存放第三方库依赖文件。
    - etc：存放资源文件。
    - include：存放头文件。
    - java：存放 Windows 系统 Java 库文件。
    - python：存放 Windows 系统 Python 语言安装文件。
    - x64：用于存放 Windows C++库文件，包含 vc14 和 vc15 两个版本。

知识点：vc 为 Visual Studio 的编译器，vc14 为 Visual Studio 2015 的编译器，vc15 为 Visual Studio 2017 的编译器，而 Visual Studio 2019 的编译器为 vc16。

对于 Visual Studio 2019，可以选用 x64 文件夹下的子文件夹 vc15 中对应的库文件，vc15 路径下包含两个文件夹：bin（存放动态库文件）和 lib（存放静态库文件）。OpenCV 工具库调用最重要的 3 个文件：头文件（存放于 include 文件夹中）、动态库（opencv_world452.dll 或 opencv_world452d.dll）及静态库（opencv_world452.lib 或 opencv_world452d.lib）。

（2）下载并解压完成后，需要配置 OpenCV 开发环境，并编写测试代码进行测试。

（3）打开 Visual Studio 2019，创建新项目。创建项目时，可以直接选择创建控制台应用，这样创建的项目中有主程序，并默认输出"Hello World"，如图 1.10 所示。

（4）配置新项目，包括配置项目名称和项目位置；解决方案名称可以使用默认的名称，与项目名称相同，如图 1.11 所示。

图 1.10

图 1.11

（5）在项目中，需要配置 OpenCV 库引用才能使用 OpenCV，需要配置 3 方面内容：包含目录、库目录和链接器依赖项，如图 1.12 所示。

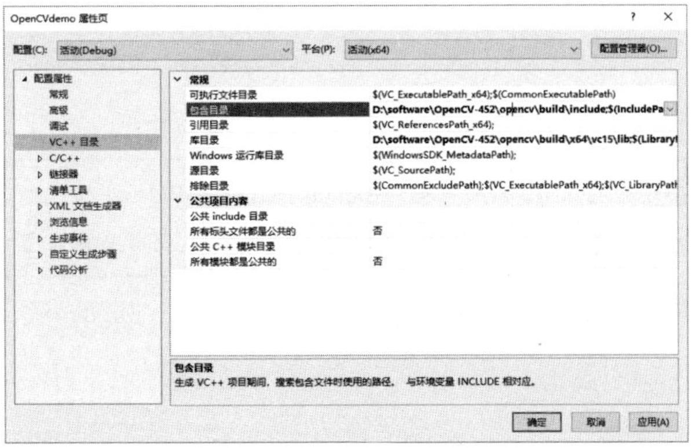

图 1.12

（6）在"链接器"→"输入"→"附加依赖项"中配置附加依赖项，即静态库文件名称，如图 1.13 所示（配置时注意选择 Debug 或 Release 对应的静态库文件）。

图 1.13

（7）依次单击"应用"和"确定"按钮完成配置，然后编写测试代码测试 OpenCV 库的使用。本案例的测试方法为读取一幅图像，并将图像显示出来：

```cpp
// OpenCVdemo.cpp ：此文件包含 main 函数。程序执行将在此处开始并结束

#include <iostream>
#include "opencv2/opencv.hpp"

int main()
{
    cv::Mat img = cv::imread("src.jpg", 1);        //图像读取
    cv::imshow("测试图像显示", img);                //图像显示
    cv::waitKey(0);                                 //等待读者操作
    cv::destroyWindow("测试图像显示");              //窗口对象销毁
}
```

注意：在开发过程中，需要根据是 Debug 模式还是 Release 模式来选择不同的 OpenCV 库。其中，Debug 模式配置 opencv_world452d.lib 库，而 Release 模式则配置 opencv_world452.lib 库。

（8）选择项目并右击，在弹出的快捷菜单中选择"生成"选项，编译项目。编译完成后，结果如图 1.14 所示。

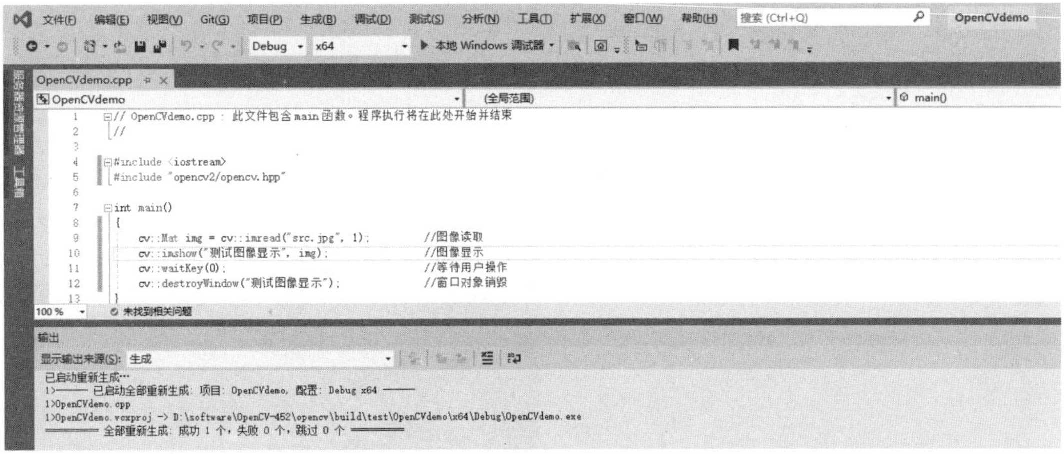

图 1.14

由于编译没有出错,所以此时可以单击图 1.14 中的"本地 Windows 调试器"按钮,或者按 F5 键运行项目。如果没有做额外的配置,则此时会弹出如图 1.15 所示的错误提示框。

图 1.15

这个错误称为运行时错误,其常见原因是动态库文件无法找到或不匹配。这种问题的解决办法有如下两种。

- 第一种方法是将动态库路径配置到环境变量中。可以选择"计算机"→"属性"→"高级系统设置"→"环境变量"选项,找到 Path 变量,将 OpenCV 文件包中的 opencv/build/x64/vc15/bin 路径配置到环境变量中,这样运行时就能找到动态库的路径了。
- 第二种方法就是将动态库复制到生成的可执行文件所在的路径下,如图 1.16 所示。

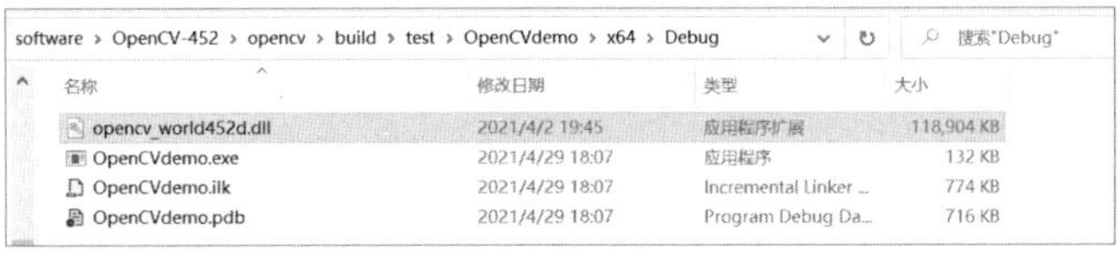

图 1.16

这两种方法都是为了让可执行文件 OpenCVdemo.exe 能在搜索路径中找到 OpenCV 的动态库文件,当前路径和环境变量都是 Windows 系统中可执行文件链接时的搜索路径。

再次执行"本地 Windows 调试器"命令，程序正常运行，如图 1.17 所示，表明在 Windows 系统中配置 OpenCV C++语言开发环境完成。

图 1.17

在使用 C++语言开发时，需要包含"opencv2/opencv.hpp"头文件，该头文件对所有模块的头文件做了引用：

```
#ifndef OPENCV_ALL_HPP
#define OPENCV_ALL_HPP

// 该头文件仅用于定义哪些模块参与了编译，定义了 HAVE_OPENCV_modulename 的值
#include "opencv2/opencv_modules.hpp"

// core 模块为必需的模块
#include "opencv2/core.hpp"

// 通过 HAVE_OPENCV_modulename 检查可选模块的引入
#ifdef HAVE_OPENCV_CALIB3D
#include "opencv2/calib3d.hpp"
#endif
#ifdef HAVE_OPENCV_FEATURES2D
#include "opencv2/features2d.hpp"
#endif
#ifdef HAVE_OPENCV_DNN
#include "opencv2/dnn.hpp"
#endif
#ifdef HAVE_OPENCV_FLANN
#include "opencv2/flann.hpp"
#endif
#ifdef HAVE_OPENCV_HIGHGUI
#include "opencv2/highgui.hpp"
#endif
#ifdef HAVE_OPENCV_IMGCODECS
```

```
#include "opencv2/imgcodecs.hpp"
#endif
#ifdef HAVE_OPENCV_IMGPROC
#include "opencv2/imgproc.hpp"
#endif
#ifdef HAVE_OPENCV_ML
#include "opencv2/ml.hpp"
#endif
#ifdef HAVE_OPENCV_OBJDETECT
#include "opencv2/objdetect.hpp"
#endif
#ifdef HAVE_OPENCV_PHOTO
#include "opencv2/photo.hpp"
#endif
#ifdef HAVE_OPENCV_STITCHING
#include "opencv2/stitching.hpp"
#endif
#ifdef HAVE_OPENCV_VIDEO
#include "opencv2/video.hpp"
#endif
#ifdef HAVE_OPENCV_VIDEOIO
#include "opencv2/videoio.hpp"
#endif

#endif
```

如果读者对 OpenCV 的结构比较熟悉，则在使用时可以只包含对应的某个模块，如图像滤波功能调用可以只包含"opencv2/imgproc.hpp"头文件，后续章节在讲解每个模块时，都会讲解包含该模块对应的头文件，通过导读内容可以查看当前模块导出的算法函数。

1.2.2 案例 2：Linux 动态库开发环境搭建

OpenCV Releases 中没有提供 Linux 下的安装包，因此需要读者在 Linux 系统下编译生成 Linux 的库文件。

本案例使用的 Linux 系统为 Ubuntu16.04，已经在虚拟机中进行了安装，系统中已经有 gcc、g++和 Make 等工具。另外，还需要安装 CMake 软件。

对于 CMake 软件，可以选择源码进行安装，也可以选择下载安装包进行安装，本案例直接下载安装包进行安装。

读者可以去 CMake 官网下载安装包，在"Download"中选择"Binary distributions"（二进制发布）选项，下载 Linux x86_64，如图 1.18 所示。

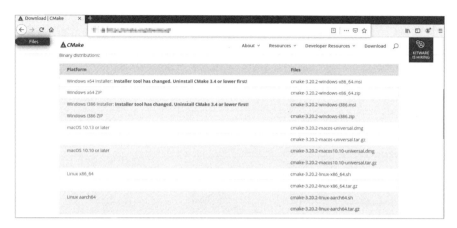

图 1.18

下载之后选择合适的安装位置并解压：
```
tar -xzf cmake-3.20.2-Linux-x86-64.tar.gz
```

然后将 cmake 添加到环境变量中，方法是打开 etc/profile，在最后添加 CMake 软件安装路径：
```
export PATH=$PATH:/home/lxiao217/software/cmake-3.20.2-linux-x86_64/bin
```

保存文件，重启机器，执行 cmake –version 命令，可以查看 CMake 软件的安装版本，并以此验证安装是否成功。

上述编译环境安装完成后，即可开始编译 OpenCV 源码，生成库文件，主要包括以下 4 步。

第 1 步，下载源码：
```
wget https://******.com/opencv/opencv/archive/4.5.2.zip
```

为了保持与 Windows 平台一致，本案例下载的是 4.5.2 版本的源码，下载之后放置到合适的位置，并解压。

第 2 步，源码编译。

在进行源码编译之前，需要先安装 OpenCV 编译需要的环境：
```
sudo apt-get install build-essential libgtk2.0-dev libgtk-3-dev libavcodec-dev libavformat-dev libjpeg-dev libswscale-dev libtiff5-dev
```

读者可以在源码目录下新建文件夹 build，如图 1.19 所示。

图 1.19

此时 build 文件夹与 CMake 编译脚本文件 CMakeLists.txt 文件处于同级目录,进入 build 文件夹,并右键选择打开终端(Open Terminal),使用 cmake 命令生成项目文件:

```
cmake ../ -D BUILD_opencv_world=ON -D OPENCV_GENERATE_PKGCONFIG=YES
```

编译选项说明如下。

- -D BUILD_opencv_world=ON:用于打开 opencv_world 编译开关,生成 opencv_world 的链接库。
- -D OPENCV_GENERATE_PKGCONFIG=YES:用于生成配置文件 opencv4.pc,OpenCV 4.x 默认不生成 opencv4.pc 文件。

cmake 执行完成后,会有"Configuring done"和"Generating done"的打印信息,如图 1.20 所示。

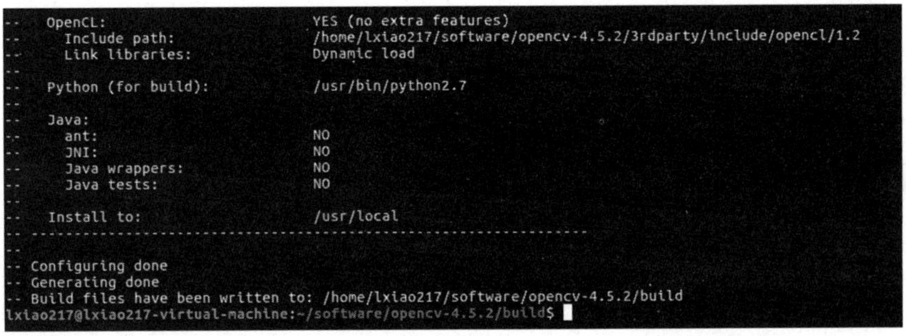

图 1.20

在打印信息中,会详细说明生成的项目文件的情况,如所需编译的 OpenCV 模块信息:

```
--   OpenCV modules:
--     To be built:                 calib3d core dnn features2d flann gapi highgui imgcodecs imgproc ml objdetect photo stitching ts video videoio world
--     Disabled:                    -
--     Disabled by dependency:      -
--     Unavailable:                 java python2 python3
--     Applications:                tests perf_tests apps
--     Documentation:               NO
--     Non-free algorithms:         NO
--
--   GUI:
--     GTK+:                        NO
--     VTK support:                 NO
--
--   Media I/O:
--     ZLib:                        zlib (ver 1.2.11)
--     JPEG:                        libjpeg-turbo (ver 2.0.6-62)
--     WEBP:                        build (ver encoder: 0x020f)
```

```
--      PNG:                         build (ver 1.6.37)
--      TIFF:                        build (ver 42 - 4.2.0)
--      JPEG 2000:                   build (ver 2.4.0)
--      OpenEXR:                     build (ver 2.3.0)
--      HDR:                         YES
--      SUNRASTER:                   YES
--      PXM:                         YES
--      PFM:                         YES
--
--    Video I/O:
--      DC1394:                      NO
--      FFMPEG:                      NO
--        avcodec:                   NO
--        avformat:                  NO
--        avutil:                    NO
--        swscale:                   NO
--        avresample:                NO
--      GStreamer:                   NO
--      v4l/v4l2:                    YES (linux/videodev2.h)
--
--    Parallel framework:            pthreads
--
--    Trace:                         YES (with Intel ITT)
--
--    Other third-party libraries:
--      VA:                          NO
--      Lapack:                      NO
--      Eigen:                       NO
--      Custom HAL:                  NO
--      Protobuf:                    build (3.5.1)
--
--    OpenCL:                        YES (no extra features)
--      Include path:                /home/lxiao217/software/opencv-4.5.2/3rdparty/include/opencl/1.2
--      Link libraries:              Dynamic load
--
--    Python (for build):            /usr/bin/python2.7
--
--    Java:
--      ant:                         NO
--      JNI:                         NO
--      Java wrappers:               NO
--      Java tests:                  NO
--
--    Install to:                    /usr/local
-- -----------------------------------------------------------------
```

在 build 目录下会生成项目工程文件 MakeFile，进入 build 目录，使用 make 命令进行项目编译，如图 1.21 所示。

图 1.21

make 编译成功之后，会在 build 的子文件夹 lib 中生成二进制库文件，如图 1.22 所示。

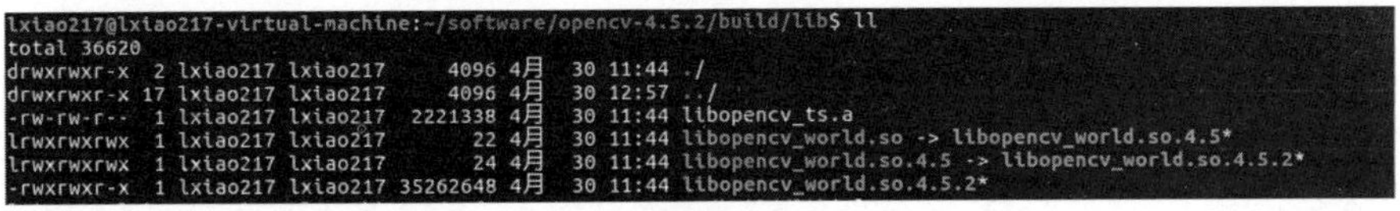

图 1.22

其中，libopencv_world.so 与 libopencv_world.so.4.5 两个文件为软链接，读者可以使用这里 libopencv_world.so 的库文件和头文件进行链接调用。

另外，读者还可以执行如下命令在本机安装 OpenCV 编译结果：

```
sudo make install
```

上述命令执行后即可在本机上安装 OpenCV，库文件和头文件被分别安装在 /usr/local/lib 和 /usr/local/include 路径下，如图 1.23 所示。

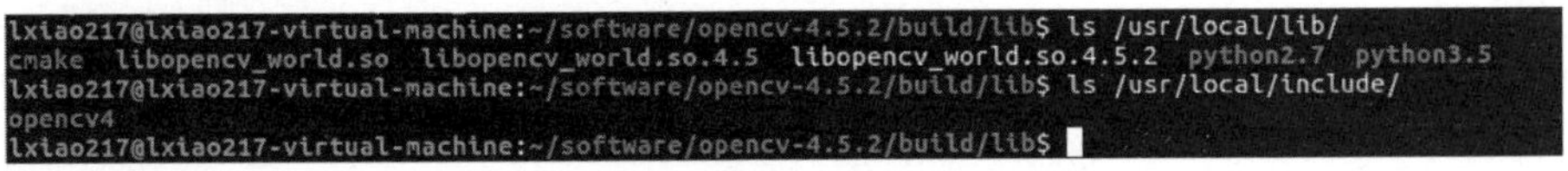

图 1.23

图 1.23 中的 include 文件夹下对应的头文件为 opencv4，opencv4 中才是常用的 OpenCV 的头文件的文件夹 opencv2，因此，读者需要手动将其复制出来放到外层，便于引用；否则需要在包含路径中多一层 opencv4 路径：

```
sudo cp -a /usr/local/include/opencv4/opencv2/ /usr/local/include/
```

第 3 步，环境配置。

环境配置包括以下两方面内容。

- 配置动态链接库加载路径。

打开/etc/ld.so.conf，在其中加上 OpenCV 动态链接库的路径：

```
include /usr/local/lib
```

- 环境变量配置。

在/etc/bash.bashrc 文件之后加上环境变量 PKG_CONFIG_PATH 的配置：

```
PKG_CONFIG_PATH=$PKG_CONFIG_PATH:/usr/local/lib/pkgconfig
export PKG_CONFIG_PATH
```

这样就真正完成了 OpenCV 的安装，并完成了环境配置。

第 4 步，代码测试。

在 Ubuntu 机器上安装 OpenCV 4 之后，读者可以编写测试程序以测试安装是否成功，本案例使用读取图像并显示来测试。

新建 C++文件，命名为 OpenCVTestDemo.cpp，并编写如下读取图像并显示的代码：

```cpp
#include <iostream>
#include "opencv2/opencv.hpp"

int main()
{
    cv::Mat img = cv::imread("src.jpg");        //图像读取
    cv::imshow("测试图像显示", img);              //图像显示
    cv::waitKey(0);                              //等待读者操作
    cv::destroyWindow("测试图像显示");            //窗口对象销毁
}
```

执行如下的代码编译命令：

```
g++ OpenCVTestDemo.cpp `pkg-config --cflags --libs opencv4`
```

执行后生成可执行文件，读者也可以编写 CMake 脚本，保存为 CMakeLists.txt：

```cmake
#指定需要的 CMake 的最低版本
cmake_minimum_required(VERSION 2.8)
#创建工程
project(OpenCVTest)
#指定 C++语言为 C++ 11
set(CMAKE_CXX_FLAGS "-std=c++11")

#查找 OpenCV 4.5.2 的安装路径
find_package(OpenCV 4.5.2 REQUIRED)
#引入 OpenCV 头文件路径
include_directories(${OpenCV_INCLUDE_DIRS})
#指定 OpenCVTestDemo.cpp 程序编译后生成可执行文件 OpenCVTestDemo
add_executable(OpenCVTestDemo OpenCVTestDemo.cpp)
```

```
#指定可执行文件 OpenCVTestDemo 链接 OpenCV 库
target_link_libraries(OpenCVTestDemo ${OpenCV_LIBS})
```

对于上述编译脚本，执行如下编译命令：

```
cmake      #生成项目文件
make       #编译项目
```

编译完成后，生成可执行文件 OpenCVTestDemo，执行该文件：

```
./OpenCVTestDemo
```

执行结果如图 1.24 所示。

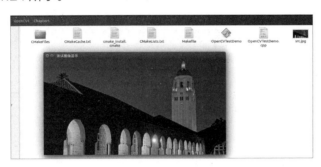

图 1.24

执行后，图像能够正常显示，表明 Linux OpenCV C++语言开发环境搭建完成。

1.2.3　案例 3：Python 语言开发环境搭建

在深度学习领域，Python 语言有着绝对的统治地位，因此很多的工具库都有 Python 语言安装包，而 OpenCV 是深度学习中常用的图像处理工具库，OpenCV Python 语言开发环境搭建在深度学习中更加适用。

OpenCV Python 语言直接使用命令行安装即可，在命令行之后加上期望安装的版本号，即可安装指定版本的 OpenCV：

```
pip install opencv-python==4.5
```

如果当前无此版本，则系统会报错，如图 1.25 所示。

```
D:\Code\OpenCV_Code\chapter9>pip install opencv-python==4.5
Collecting opencv-python==4.5
  ERROR: Could not find a version that satisfies the requirement opencv-python==4.5 (from versions: 3.4.8.29, 3.4.9.31,
3.4.9.33, 3.4.10.35, 3.4.10.37, 3.4.11.39, 3.4.11.41, 3.4.11.43, 3.4.11.45, 3.4.13.47, 4.1.2.30, 4.2.0.32, 4.2.0.34, 4.3
.0.36, 4.3.0.38, 4.4.0.40, 4.4.0.42, 4.4.0.44, 4.4.0.46, 4.5.1.48)
ERROR: No matching distribution found for opencv-python==4.5
WARNING: You are using pip version 19.2.3, however version 21.0.1 is available.
You should consider upgrading via the 'python -m pip install --upgrade pip' command.
```

图 1.25

在图 1.25 中，错误打印信息给出了可以安装的版本号。从中还可以发现，Python 语言安装包的版本号和 C++语言安装包的版本号不同，C++安装使用的是 4.5.2 版本，而 Python 语言中则没有该版本，为了对应，本案例选取最近的版本 4.5.1.48 进行安装。

```
pip install opencv-python==4.5.1.48
```

命令执行后，首先下载 OpenCV Python 的安装包，在网络不好时会存在下载超时失败的情况，如图 1.26 所示。

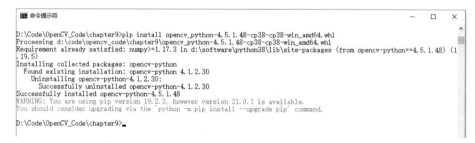

图 1.26

在此种情况下，读者可以从下载打印信息中打开链接进行下载：

```
Downloading
https://files.******.org/packages/00/84/1c26cfa8d202c8c42fe9db27ea0925382b2ed8f16af5d7e5d93a
62c780d8/opencv_python-4.5.1.48-cp38-cp38-win_amd64.whl
```

下载完成后，执行命令安装：

```
pip install opencv_python-4.5.1.48-cp38-cp38-win_amd64.whl
```

最终的结果如图 1.27 所示。

图 1.27

同理，读者可以选择安装 opencv-contrib-python 库（opencv-contrib 库的用途在以后章节中会有介绍，读者可以选择安装）：

```
pip install opencv-contrib-python==4.5.1.48
```

安装时最好选择和 OpenCV 版本相同的 opencv-contrib 库进行安装，安装成功的结果如图 1.28 所示。

图 1.28

安装完成后，可以编写测试程序测试安装是否成功，本案例测试程序为图像读取并显示：

```
#OpenCV 库引入
import cv2

print(cv2.version.opencv_version)          #OpenCV 版本打印
img = cv2.imread("src.jpg")                #图像读取
cv2.imshow("OpenCV Python Test", img)      #图像显示
cv2.waitKey(0)                             #等待读者操作
cv2.destroyWindow("OpenCV Python Test")    #窗口对象销毁
```

测试结果如图 1.29 所示。

图 1.29

图像显示正常，版本号打印输出为 4.5.1.48，说明安装成功。

1.3 OpenCV 模块介绍

OpenCV 为开源软件，读者可以阅读其文档和源码来了解其代码结构，整个算法库的框架如图 1.30 所示。

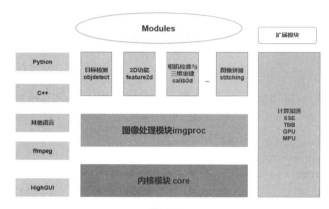

图 1.30

在 OpenCV 的源码中，在 modules 路径下存放着 OpenCV 的所有模块，最重要的模块就是内核模块 core 与图像处理模块 imgproc。本书后续章节将分模块介绍，结合案例讲解 OpenCV 的使用，本节介绍常用模块及扩展模块的功能。

1.3.1 常用模块

OpenCV 4 中包含的模块有 21 个，每个模块都被保存在一个单独的子目录中，如图 1.31 所示。

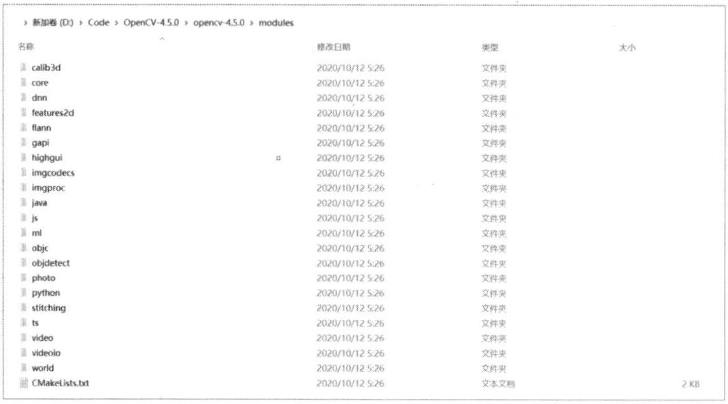

图 1.31

主要模块说明如下。

- calib3d：由相机校准（calibration）和三维重建（3d）两部分组成，主要用于相机标定与三维视觉等。
- core：OpenCV 的内核模块，定义了基础数据结构与基础计算。
- dnn：主要用于深度学习推理部署，不支持模型训练。
- features2d：主要用于特征点处理，如特征点检测与匹配等。
- flann：FLANN 为快速最近邻算法（Fast Library for Approximate Nearest Neighbors）的缩写，该模块包含快速近似最近邻搜索和聚类等功能。
- gapi：对图像处理算法做了加速处理，不属于 OpenCV 的功能模块。
- highgui：用于创建图像化界面操作，如创建和操作图像显示窗口、鼠标与键盘事件处理、进度条等图像化交互操作。
- imgcodecs：负责图像文件的读/写，如图像的读取与保存。
- imgproc：是 OpenCV 图像处理最重要的模块，主要功能有图像滤波、图像几何变换、直方图操作等。
- ml：机器学习模块，包含常见的机器学习算法，如支持向量机和随机森林等。
- objdetect：主要用于图像目标检测，如 Haar 特征检测等。
- photo：主要负责照片处理，如照片修复和去噪等。
- stitching：负责图像拼接，功能包括图像特征点寻找与匹配等图像拼接技术。
- video：用于视频分析，如运动估计、背景分离等。
- videoio：负责视频的读/写，如主要视频文件的读取和写入。

1.3.2 扩展模块

OpenCV 在视觉算法中的功能非常强大，其中一个原因就是该算法库一直在与时俱进地更新算法，对于具有专利的算法（如 SURF），以及一些还没有稳定的算法，OpenCV 会将其置于扩展模块中，这些扩展模块包含在 opencv-contrib 代码库中；而稳定的算法会被移到 OpenCV 主仓库代码中，因此，读者需要谨慎地使用 opencv-contrib，因为不同版本的函数可能存在差异。

在 OpenCV 3.x 版本之后，opencv-contrib 就不再包含于 OpenCV 源码中了，opencv-contrib 的源码可以在 GitHub 上下载，然后参与 OpenCV 源码编译，编译方法将在案例 5 中讲解。

opencv-contrib 的模块及其功能说明如表 1.1 所示。

表 1.1

模 块 名 称	功　　能
alphamat	Alpha Matting 信息流算法
aruco	增强现实标记算法
barcode	条形码检测与解码方法
bgsegm	增强背景-前景分割算法
bioinspired	仿生学视觉模型和衍生工具
ccalib	用于三维重建的自定义校准模式
cudaarithm	CUDA 矩阵运算
cudabgsegm	CUDA 背景分割
cudacodec	CUDA 视频编解码
cudafeatures2d	CUDA 特征检测与描述
cudafilters	CUDA 图像滤波
cudaimgproc	CUDA 图像处理
cudalegacy	CUDA 传统支持
cudaobjdetect	CUDA 目标检测
cudaoptflow	CUDA 光流算法
cudastereo	CUDA 立体匹配
cudawarping	CUDA 图像扭曲
cudev	CUDA 设备层
cvv	计算机视觉程序交互式可视化调试的 GUI
datasets	用于处理不同数据集的框架
dnn_objdetect	基于 DNN 的目标检测
dnn_superres	基于 DNN 的超分
dpm	基于可变形零件的模型
face	人脸分析
freetype	使用 freetype/harfbuzz 绘制 UTF-8 字符串
fuzzy	基于模糊数学的图像处理
hdf	分层数据格式 I/O 例程
hfs	基于层次特征选择的图像分割方法
img_hash	提供不同的图像哈希算法的实现
intensity_transform	提供用于调整图像对比度的强度变换算法的实现
julia	OpenCV Julia 绑定
line_descriptor	用于从图像中提取线条的二进制描述符
mcc	Macbeth 图表模块
optflow	光流算法
ovis	OGRE 三维可视化器
phase_unwrapping	相位展开 API

续表

模 块 名 称	功　　能
plot	Mat 数据绘制函数
quality	图像质量分析 API
rapid	基于轮廓的三维目标跟踪
reg	图像配准
rgbd	RGB 深度处理
saliency	显著性 API
sfm	运动结构分析
shape	形状距离与匹配
stereo	立体匹配算法
structured_light	结构光 API
superres	超分模块
surface_matching	曲面匹配
text	场景文字检测与识别
tracking	追踪 API
videostab	视频稳定
viz	三维可视化器
wechat_qrcode	微信二维码检测器，用于检测和解析二维码
xfeatures2d	features2d 扩展模块
ximgproc	imgproc 扩展模块
xobjdetect	objdetect 扩展模块
xphoto	photo 扩展模块

1.4　OpenCV 源码编译

通过前面章节的讲解，读者已经对 OpenCV 源码，以及其扩展模块 opencv-contrib 有了一个初步的了解，在 OpenCV 源码中，提供了 CMake 编译脚本 CMakeLists.txt 文件，读者可以通过源码编译的方式安装 OpenCV。

在进行 Linux OpenCV C++语言开发环境搭建时，已经讲解了在 Linux 系统下编译 OpenCV，并配置使用环境的方法，本节不再赘述 Linux 系统下的编译，只讲解 Windows 系统下的编译。

1.4.1　案例 4：OpenCV 编译

在 Linux 系统下的 OpenCV 编译已经在 1.2.2 节做了介绍，因此，本节案例讲解 Windows

系统下 OpenCV 源码的编译。

在编译之前，需要去 CMake 官网下载 CMake 安装包并安装，下载完成之后，双击它可执行文件进行安装。

安装完成之后，可以通过 CMake 生成项目文件；然后通过 Visual Studio 2019 编译项目，生成二进制库文件。

CMake 生成项目文件有两种方法，第一种是使用 CMake 图像化界面操作，第二种是通过命令行执行（编译命令请参考 1.2.2 节案例 2 中的源码编译部分）。

下面是使用 CMake 图像化界面生成项目文件的步骤。

第 1 步，打开 CMake（cmake-gui），在其中配置源码路径和二进制文件生成路径，如图 1.32 所示。

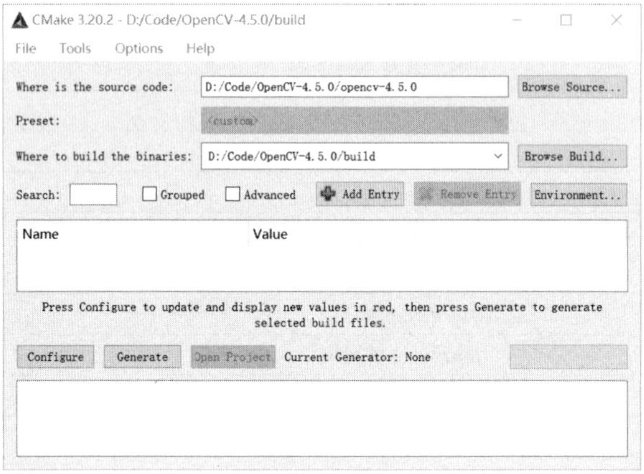

图 1.32

第 2 步，配置项目。

单击"Configure"按钮配置项目，此时会弹出如图 1.33 所示的对话框，读者需要配置项目的生成器、平台及工具集等。

配置完成且出现"Configuring done"打印信息之后，会因为有的配置不支持而发出警告（配置界面为红色），如果不是环境错误问题，则可以忽略，重新单击"Configure"按钮，红色警告会消除。

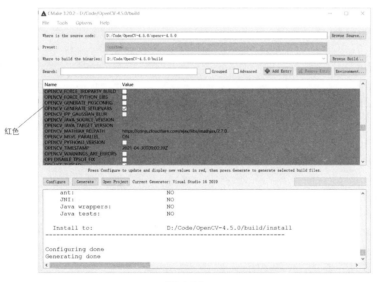

图 1.33

第 3 步，生成项目文件。

单击"Generate"按钮，生成项目文件，生成结果如图 1.34 所示。

图 1.34

提示：用户需要在配置中勾选"Build_opencv_world"编译选项才会生成 opencv_world 库文件。

第 4 步，项目编译。

单击图 1.32 中的"Open Project"按钮，使用 Visual Studio 2019 打开项目文件，读者

需要分别编译 Debug 和 Release 版本的 OpenCV。编译完成后，在编译结果保存文件夹 build 中有 lib 和 bin 两个子文件夹。其中，lib 文件夹存放编译的静态库文件，bin 文件夹存放编译的动态库文件，结果如图 1.35～图 1.37 所示。

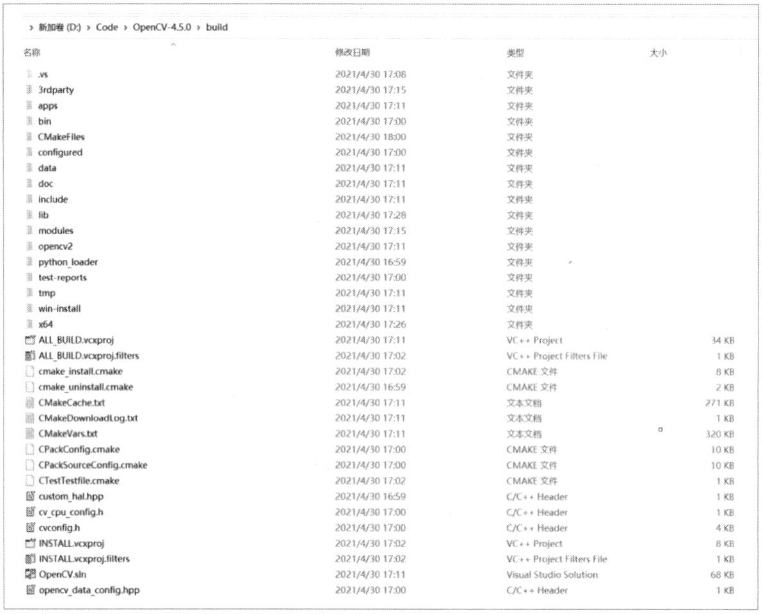

图 1.35

图 1.36

图 1.37

图 1.35～图 1.37 展示了编译结果的保存层级结构，图 1.37 中的 opencv_world450.dll 为 OpenCV 的 Release 动态库，而在 Debug 目录下保存的则是 opencv_world450d.dll，读者可以将此编译结果按照案例 1 中的配置方法配置使用。

1.4.2 案例 5：OpenCV 裁剪编译

OpenCV 包含很多模块，功能齐全，但是对于很多读者来说，这些模块并不一定都用得上，而且编译了这些模块之后，二进制库会比较大。对于很多软件部署来说，这会造成资源的浪费，因此，本节介绍 OpenCV 的裁剪编译，裁剪不必要的模块使之不参与编译。

在 CMake 配置（Configure）完成之后，读者可以勾选要编译的模块，如图 1.38 所示。

图 1.38

OpenCV 的所有模块都有对应的编译选项，如 core 模块对应的编译选项是 BUILD_opencv_core，图 1.38 是 OpenCV 的默认编译配置。对于某些读者来说，可能只需使用 OpenCV 读/写图像，做一些基本的变换操作，此时，只需编译 core、imgproc 和 imgcodecs 这 3 个模块即可。如果读者只需编译 C++的二进制库文件，不需要编译其他语言的包，则可以去掉其他语言选项，如 BUILD_JAVA 语言选项。

本案例裁剪后只保留 core、imgproc 和 imgcodecs 这 3 个模块的编译选项，读者记得勾选"BUILD_opencv_world"复选框，编译的 Release 结果如图 1.39 所示。

图 1.39

在图 1.39 中，opencv_world450.dll 的大小为 36997KB，对比图 1.37 中的 opencv_world450.dll（58516KB），可以看出，裁剪后占用空间变小，方便部署使用。

读者也可以通过编译命令控制模块是否参与编译，如不编译 dnn 模块，可以将其开关关掉：
cmake ../opencv-4.5.0 -DBUILD_opencv_dnn=OFF -DBUILD_opencv_world=ON

同理，其他模块可以通过 -DBUILD_opencv_*=OFF 关闭编译选项，通过 -DBUILD_opencv_*=ON 打开编译选项。

1.4.3　案例 6：扩展模块 opencv-contrib 编译

OpenCV 为了保证二进制的兼容性，为读者提供良好的性能和稳定性，对于新开发且没有经过充分测试的 API，不会作为 OpenCV 官方发行版的一部分，这些新的 API 被置于扩展模块 opencv-contrib 中。

当模块开发测试成熟并具备发行资格之后，会被移动到正式库 OpenCV 中，开发团队也会为该模块提供可靠性支持，而 opencv-contrib 中的功能模块可能会在不同的版本中有差异，稳定性不能保证，读者需要谨慎选择。本案例讲解在进行 OpenCV 源码编译时，选择加入扩展模块 opencv-contrib 的源码编译。

如案例 4 所述，读者在使用 cmak-gui 打开 OpenCV 源码，配置编译输出文件后，单击 "Configure" 按钮，可以配置编译选项。编译扩展模块 opencv-contrib 需要通过选项 "OPENCV_EXTRA_MODULES_PATH" 设置扩展模块源码所在路径，如图 1.40 所示。

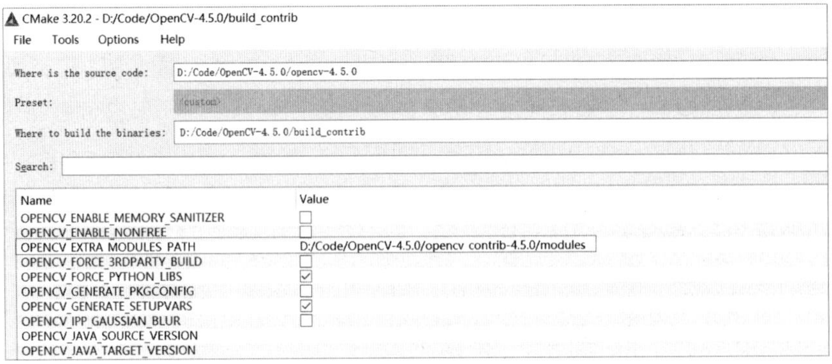

图 1.40

配置完成后，重新单击"Configure"和"Generate"按钮，即可生成项目工程。出现"Generating done"打印信息后，可以单击"Open Project"按钮，打开工程进行编译。

可以在命令行中配置扩展模块的路径进行编译：

```
cmake ../opencv-4.5.0
-DOPENCV_EXTRA_MODULES_PATH=D:\Code\OpenCV-4.5.0\opencv_contrib-4.5.0\modules
```

编译完成后，可以进行环境配置并使用，其过程可以参考案例 1，此处不再赘述。

1.5 进阶必备：OpenCV 入门参考

1.5.1 OpenCV 版本选择

很多读者在学习 OpenCV 时都比较关心 OpenCV 版本的选择问题，本节提供版本选择的两条建议。

- 根据机器硬件选择 x86 或 x64。

在进行 OpenCV 开发环境搭建时，需要考虑编程语言、操作系统类型与处理器的位数，本章讲解了 Python 和 C++两种主流语言的开发环境搭建。编程语言确定后，可以根据另外两个参数选择合适的 OpenCV 安装包来安装。

- 根据使用意图选择 OpenCV 3.x 或 OpenCV 4.x。

目前，使用比较多的版本为 OpenCV 3.x 和 OpenCV 4.x，官方已经不再对 OpenCV 2.x 提供技术支持，因此不建议使用。OpenCV 在更新过程中，API 接口保持了较好的一致性，因此，在常用模块上，无论是 OpenCV 3.x 还是 OpenCV 4.x，其函数名称等不存在差异，即读者不管是选择 3.x 还是选择 4.x，本书的案例都能得到很好的运行。

如果你是刚开始学习 OpenCV，则建议使用新版本，目前，OpenCV 的最新版本为 OpenCV 4.5，在 OpenCV 4.x 版本中，对深度学习模块 dnn 做了很多改进与优化。

如果是在已有项目中引入了 OpenCV，则需要按照项目中现有的版本学习；如果在项目中新引入 OpenCV，则需要考虑项目的开发环境等。例如，在使用 Visual Studio 2010 做项目开发时，如果引入 OpenCV 4.x，则可能存在编译出错的问题，而使用 Visual Studio 2019 编译 OpenCV 则不存在问题。

1.5.2 如何学习 OpenCV

OpenCV 是图像处理和计算机视觉中的重要工具库，对常用的图像处理算法做了很好的封装，这对于读者来说更加方便，因此本节给出了递进式学习的建议，读者可以根据实际需求选择合适的学习层次。

- 第一层次：开发环境搭建，案例学习。

读者在学习 OpenCV 时，可以选择自己熟悉的语言，搭建开发环境，动手将本书的案例实现出来，这样有助于掌握 OpenCV 中 API 的使用，有案例辅助，可以增强学习的信心。

- 第二层次：明确需求，带着问题学习。

OpenCV 是一个辅助工具，用于帮助读者实现实际需求，因此，读者可以在借鉴案例的基础上使用 OpenCV 解决自己面临的问题。

- 第三层次：算法学习，源码分析。

对于有意向深入研究图像处理或计算机视觉的读者，可以在掌握 OpenCV 的使用并能解决实际问题的基础上研究算法背后的实现原理，阅读 OpenCV 开源代码以得到进一步的提升。

第 2 章
图像读/写模块 imgcodecs

图像读/写是指从磁盘读取图像或将图像保存到磁盘,是图像处理中最基本的操作,该操作定义在 imgcodecs 模块中。

2.1 模块导读

imgcodecs 模块需要通过包含头文件"opencv2/imgcodecs.hpp"引入,通过阅读该头文件中的内容,读者可以初步了解该模块的功能。如果读者想深入阅读算法实现源码,则可以通过该头文件进行跳转。

imgcodecs.hpp 文件的介绍如下:

```
#ifndef OPENCV_IMGCODECS_HPP
#define OPENCV_IMGCODECS_HPP

#include "opencv2/core.hpp"


//////////////////////////////// image codec ////////////////////////////////
namespace cv
{

//图像读取方式
enum ImreadModes {
    IMREAD_UNCHANGED                    = -1,
```

```cpp
        IMREAD_GRAYSCALE              = 0,
        IMREAD_COLOR                  = 1,
        IMREAD_ANYDEPTH               = 2,
        IMREAD_ANYCOLOR               = 4,
        IMREAD_LOAD_GDAL              = 8,
        IMREAD_REDUCED_GRAYSCALE_2    = 16,
        IMREAD_REDUCED_COLOR_2        = 17,
        IMREAD_REDUCED_GRAYSCALE_4    = 32,
        IMREAD_REDUCED_COLOR_4        = 33,
        IMREAD_REDUCED_GRAYSCALE_8    = 64,
        IMREAD_REDUCED_COLOR_8        = 65,
        IMREAD_IGNORE_ORIENTATION     = 128
    };

//图像保存方式
enum ImwriteFlags {
        IMWRITE_JPEG_QUALITY             = 1,
        IMWRITE_JPEG_PROGRESSIVE         = 2,
        IMWRITE_JPEG_OPTIMIZE            = 3,
        IMWRITE_JPEG_RST_INTERVAL        = 4,
        IMWRITE_JPEG_LUMA_QUALITY        = 5,
        IMWRITE_JPEG_CHROMA_QUALITY      = 6,
        IMWRITE_PNG_COMPRESSION          = 16,
        IMWRITE_PNG_STRATEGY             = 17,
        IMWRITE_PNG_BILEVEL              = 18,
        IMWRITE_PXM_BINARY               = 32,
        IMWRITE_EXR_TYPE                 = (3 << 4) + 0,
        IMWRITE_EXR_COMPRESSION          = (3 << 4) + 1,
        IMWRITE_WEBP_QUALITY             = 64,
        IMWRITE_PAM_TUPLETYPE            = 128,
        IMWRITE_TIFF_RESUNIT             = 256,
        IMWRITE_TIFF_XDPI                = 257,
        IMWRITE_TIFF_YDPI                = 258,
        IMWRITE_TIFF_COMPRESSION         = 259,
        IMWRITE_JPEG2000_COMPRESSION_X1000 = 272
    };

//EXR 格式保存方式
enum ImwriteEXRTypeFlags {
        /*IMWRITE_EXR_TYPE_UNIT = 0, //目前不支持 */
        IMWRITE_EXR_TYPE_HALF  = 1,
        IMWRITE_EXR_TYPE_FLOAT = 2
    };

//EXR 压缩保存方式
enum ImwriteEXRCompressionFlags {
        IMWRITE_EXR_COMPRESSION_NO        = 0,
```

```cpp
        IMWRITE_EXR_COMPRESSION_RLE      = 1,
        IMWRITE_EXR_COMPRESSION_ZIPS     = 2,
        IMWRITE_EXR_COMPRESSION_ZIP      = 3,
        IMWRITE_EXR_COMPRESSION_PIZ      = 4,
        IMWRITE_EXR_COMPRESSION_PXR24    = 5,
        IMWRITE_EXR_COMPRESSION_B44      = 6,
        IMWRITE_EXR_COMPRESSION_B44A     = 7,
        IMWRITE_EXR_COMPRESSION_DWAA     = 8,
        IMWRITE_EXR_COMPRESSION_DWAB     = 9,
    };

//imwrite 用于优化压缩算法的 PNG 特定标志
enum ImwritePNGFlags {
        IMWRITE_PNG_STRATEGY_DEFAULT      = 0,
        IMWRITE_PNG_STRATEGY_FILTERED     = 1,
        IMWRITE_PNG_STRATEGY_HUFFMAN_ONLY = 2,
        IMWRITE_PNG_STRATEGY_RLE          = 3,
        IMWRITE_PNG_STRATEGY_FIXED        = 4
    };

//imwrite PAM 特定的 tupletype 标志，用于定义 PAM 文件的"TUPETYPE"字段
enum ImwritePAMFlags {
        IMWRITE_PAM_FORMAT_NULL             = 0,
        IMWRITE_PAM_FORMAT_BLACKANDWHITE    = 1,
        IMWRITE_PAM_FORMAT_GRAYSCALE        = 2,
        IMWRITE_PAM_FORMAT_GRAYSCALE_ALPHA  = 3,
        IMWRITE_PAM_FORMAT_RGB              = 4,
        IMWRITE_PAM_FORMAT_RGB_ALPHA        = 5,
    };

/** 图像读取，读取方式由 cv::ImreadModes 定义*/
CV_EXPORTS_W Mat imread( const String& filename, int flags = IMREAD_COLOR );

/** 从文件中读取多幅图像*/
CV_EXPORTS_W bool imreadmulti(const String& filename, CV_OUT std::vector<Mat>& mats, int flags
= IMREAD_ANYCOLOR);

/** 图像文件保存*/
CV_EXPORTS_W bool imwrite( const String& filename, InputArray img,
              const std::vector<int>& params = std::vector<int>());

/** 将多幅图像保存到文件中，函数未导出*/
CV_WRAP static inline
bool imwritemulti(const String& filename, InputArrayOfArrays img,
                  const std::vector<int>& params = std::vector<int>())
{
```

```
        return imwrite(filename, img, params);
}

/** 从数据流中读取图像*/
CV_EXPORTS_W Mat imdecode( InputArray buf, int flags );
CV_EXPORTS Mat imdecode( InputArray buf, int flags, Mat* dst);

/** 将图像文件编码为字节流*/
CV_EXPORTS_W bool imencode( const String& ext, InputArray img,
                            CV_OUT std::vector<uchar>& buf,
                            const std::vector<int>& params = std::vector<int>());

/** 如果文件能够被 OpenCV 解码，则返回 true*/
CV_EXPORTS_W bool haveImageReader( const String& filename );

/** 如果图像能够编码为指定文件名，则返回 true*/
CV_EXPORTS_W bool haveImageWriter( const String& filename );

}

#endif //OPENCV_IMGCODECS_HPP
```

在该头文件中，定义了图像读取与保存的方式，这些方式由枚举值表示；定义了从文件读取图像或将数据流解码为图像，以及将图像文件保存为文件或编码为数据流的函数。

2.2 图像读/写操作

2.2.1 案例7：图像读取

单幅图像读取操作由 imread 函数完成，该函数从图像文件中载入一幅图像并返回，若读取失败，则返回空矩阵（Mat::data==NULL）。

目前支持的图片格式如表 2.1 所示。

表 2.1

格　式	文件后缀名	格　式	文件后缀名
Windows Bitmaps	.bmp，.dib	PFM	.pfm
JPEG 文件	.jpeg，.jpg，.jpe	Sun 光栅文件	.sr，.ras
JPEG 2000	.jp2	TIFF	.tiff，.tif
PNG	.png	OpenEXR 图像文件	.exr
WebP	.webp	Radiance HDR 文件	.hdr，.pic
便携式图像格式	.pbm，.pgm，.ppm，.pxm，.pnm	—	—

imread 函数的 C++语言函数定义如下：
```
CV_EXPORTS_W Mat imread( const String& filename, int flags = IMREAD_COLOR );
```
imread 函数的 Python 语言函数定义如下：
```
retval = imread(filename, flags=None)
```
imread 函数的参数说明如下。
- filename：待读取的图像文件路径。
- flags：图像读取模式，模式定义在 cv::ImreadModels 中，默认值为 IMREAD_COLOR。
- retval：读取的图像矩阵，返回值类型为 Mat（Mat 数据结构将在第 3 章讲解）。

imread 函数通过文件内容而不是文件扩展名来确定图像的类型，对于彩色图像，图像将按照 BGR 的通道顺序解码。

> 知识点：OpenCV 中所有的接口均定义在命令空间 cv 中，因此，C++语言开发需要使用命名空间 cv，如 cv::imread。

图像读取模式 ImreadModels 的定义如下：
```
enum ImreadModes {
    IMREAD_UNCHANGED            = -1, //按原样返回加载的图像（会带上 alpha 通道）。忽略 EXIF 方向
    IMREAD_GRAYSCALE            = 0,  //将图像转为单通道灰度图像
    IMREAD_COLOR                = 1,  //将图像转为 BGR 三通道彩色图像
    IMREAD_ANYDEPTH             = 2,  //如果图像深度为 16bit/32bit，则会返回该深度图像；否则返回
8bit 图像
    IMREAD_ANYCOLOR             = 4,  //按照任意颜色图像格式读取
    IMREAD_LOAD_GDAL            = 8,  //使用 gdal 驱动程序加载图像
    IMREAD_REDUCED_GRAYSCALE_2  = 16, //将图像转为单通道灰度图像且图像尺寸变为原始图像尺寸的 1/2
    IMREAD_REDUCED_COLOR_2      = 17, //将图像转为 BGR 三通道彩色图像且图像尺寸变为原始图像尺寸的 1/2
    IMREAD_REDUCED_GRAYSCALE_4  = 32, //将图像转为单通道灰度图且图像尺寸变为原始图像尺寸的 1/4
    IMREAD_REDUCED_COLOR_4      = 33, //将图像转为 BGR 三通道彩色图像且图像尺寸变为原始图像尺寸的 1/4
    IMREAD_REDUCED_GRAYSCALE_8  = 64, //将图像转为单通道灰度图且图像尺寸变为原始图像尺寸的 1/8
    IMREAD_REDUCED_COLOR_8      = 65, //将图像转为 BGR 三通道彩色图像且图像尺寸变为原始图像尺寸的 1/8
    IMREAD_IGNORE_ORIENTATION   = 128 //忽略 EXIF 中的方向标识，不旋转图像
};
```

ImreadModes 为枚举类型，因此，读者在传入该参数时，可以使用 cv::IMREAD_GRAYSCALE 枚举值形式，也可以直接传入枚举值的数值 1，这两种形式等价。

提示：为了便于讲解，在案例中会先用到后续章节所讲的 OpenCV 函数，分别如下。
- imshow：图像显示。
- waitKey：等待读者按键操作。
- destroyWindow：窗口销毁。

读者如果想了解这些函数的功能细节，则可以跳转到第 6 章。

本案例实现图像读取的 Python 代码如下：

```
#OpenCV 库引入
import cv2

img = cv2.imread("src.jpg", 1)              #图像读取
cv2.imshow("Image Read", img)               #图像显示
cv2.waitKey(0)                              #等待读者操作
cv2.destroyWindow("Image Read")             #窗口对象销毁
```

如果在目录中没有图像文件 src.jpg，则图像读取会失败，返回 None，在进行代码调试时，可以查看变量的值，如图 2.1 所示。

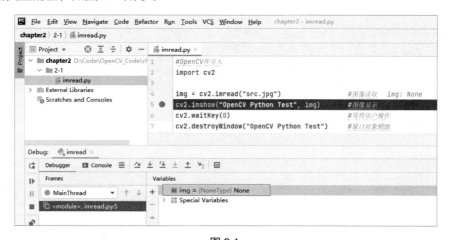

图 2.1

此时，如果继续运行程序，则会报错，如图 2.2 所示。

图 2.2

图 2.2 中的报错原因为断言失败。因为在显示图像之前，程序会断言读取的图像矩阵的宽高尺寸均大于 0（size.width>0 && size.height>0），而读取空图像返回值为 None，显然不满足断言条件，所以出现断言失败报错，程序崩溃。

因此，在读取一幅图像并将读取的图像矩阵用于后续操作之前，需要先判断读取的内容是否为空，若为空则不进行相关操作；否则可能会出现意想不到的错误。

```
#OpenCV 库引入
import cv2

img = cv2.imread("src.jpg")            #图像读取
if img is None:
    print("Image Read Error!")
else:
    cv2.imshow("Image Read", img)      #图像显示
    cv2.waitKey(0)                      #等待读者操作
    cv2.destroyWindow("Image Read")    #窗口对象销毁
```

这样，程序执行就不会出错了，只会打印图像读取错误信息。

知识点：在进行图像读取操作之前，判断图像矩阵是否为空或图像的尺寸是否大于 0 是一个好的编码习惯，在 C、C++ 语言中，对指针等使用前的判空也是如此。

图像读取失败的原因一般有 3 种：文件不存在、没有读取权限、文件格式错误。本案例的原因为图像文件不存在，因此，在路径下添加图像文件之后，读取结果如图 2.3 所示。

图像读取还有另外一个函数 imreadmulti，作用是从一个文件中读取多幅图像。

imreadmulti 函数的 C++语言函数定义如下：
```
CV_EXPORTS_W bool imreadmulti(const String& filename,
CV_OUT std::vector<Mat>& mats, int flags =
IMREAD_ANYCOLOR);
```

imreadmulti 函数的 Python 语言函数定义如下：
```
retval, mats = imreadmulti (filename, mats=None,
flags=None)
```

图 2.3

imreadmulti 函数的参数说明如下。
- filename：待读取的图像文件路径。
- mats：读取的图像文件集合（返回值）。
- flags：图像读取模式，模式定义在 cv::ImreadModels 中。

- retval：函数调用是否成功的状态值，bool 类型。

从一个文件中读取多幅图像案例的代码如下：

```python
import cv2

res = cv2.imreadmulti("src_tiff.tiff")           #图像读取
if not res:
    print("Image Read Error!")
else:
    index = 1
    for img in res[1]:
        cv2.imshow("Image Read" + str(index), img)   #图像显示
        cv2.waitKey(0)                                #等待读者操作
        index += 1
    cv2.destroyAllWindows()                           #销毁所有窗口对象
```

案例中首先读取图像文件，然后将读取到的多幅图像显示出来。在第 7 行打断点调试，可以查看 imreadmulti 的返回值，如图 2.4 所示。

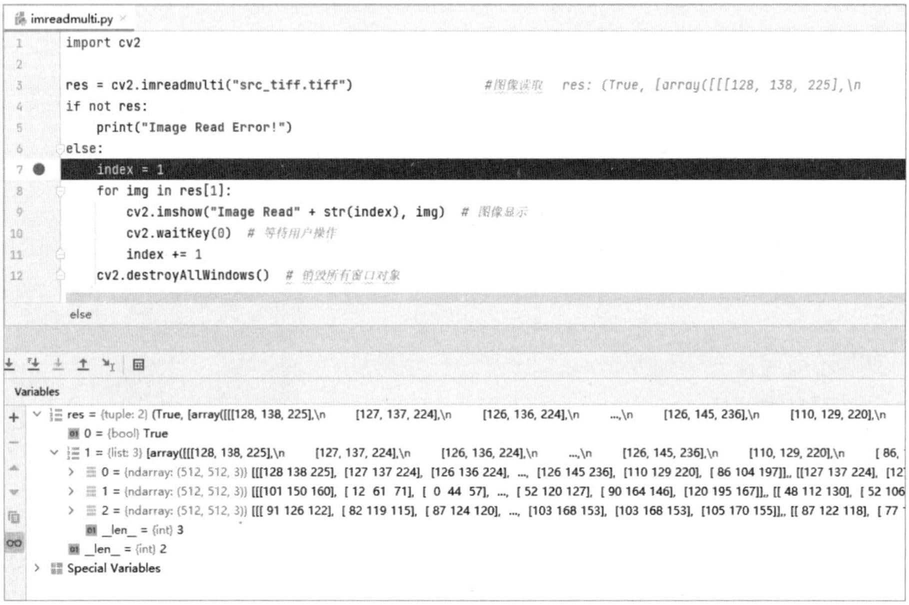

图 2.4

函数 imreadmulti 返回的结果 res 的类型为 tuple，包含两个值，第一个值为 bool 类型，该值表示函数调用的状态，即图像文件读取是否成功；若成功则第二个参数为 list 类型，保存的就是读取的图像结果。

读取的图像结果显示如图 2.5 所示（读取的图像为 imwritemulti 函数保存多幅图像的结果）。

图 2.5

2.2.2 案例 8：图像保存

图像处理之后，经常需要将处理结果保存到本地文件中，此时需要使用 OpenCV 中的图像保存函数 imwrite。

imwrite 函数的 C++语言函数定义如下：

```
CV_EXPORTS_W bool imwrite( const String& filename, InputArray img,
            const std::vector<int>& params = std::vector<int>());
```

imwrite 函数的 Python 语言函数定义如下：

```
retval = imwrite(filename, img, params=None)
```

imwrite 函数的参数说明如下。

- filename：保存的文件名称。
- img：Mat 或 vector<Mat>类型的图像。
- params：格式化编码为成对的特定参数，该参数可选，定义在 cv::ImwriteFlags 中。
- retval：图像保存是否成功的标志，bool 类型，保存成功返回 true，保存失败返回 false（返回值）。

图像保存标志 cv::ImwriteFlags 的定义如下：

```
enum ImwriteFlags {
    IMWRITE_JPEG_QUALITY          = 1,  //对于 JPEG 图像，该值设置图像质量，值从 0 到 100，越高越好
                                        //默认值为 95
    IMWRITE_JPEG_PROGRESSIVE      = 2,  //允许 JPEG 特征，值为 0 或 1，默认为 False
    IMWRITE_JPEG_OPTIMIZE         = 3,  //允许 JPEG 特征，值为 0 或 1，默认为 False
    IMWRITE_JPEG_RST_INTERVAL     = 4,  //JPEG 重启间隔，值从 0 到 65535，默认值为 0
    IMWRITE_JPEG_LUMA_QUALITY     = 5,  //分离 luma 质量等级，值从 0 到 100，默认值为 0
    IMWRITE_JPEG_CHROMA_QUALITY   = 6,  //分离 chroma 质量等级，值从 0 到 100，默认值为 0
    IMWRITE_PNG_COMPRESSION       = 16, //压缩等级，值从 0 到 9，默认值为 1
    IMWRITE_PNG_STRATEGY          = 17, //cv::ImwritePNGFlags 之一，默认值为 IMWRITE_PNG_STRATEGY_RLE
```

```
    IMWRITE_PNG_BILEVEL              = 18, //PNG 二进制等级，值为 0 或 1，默认值为 0
    IMWRITE_PXM_BINARY               = 32, //对于 PPM、PGM 或 PBM，该值为格式标志，值为 0 或 1，默认值为 1
    IMWRITE_EXR_TYPE                 = (3 << 4) + 0, /* 即 48 */ //重载 EXR 存储类型
    IMWRITE_WEBP_QUALITY             = 64, //对于 WEBP，该值为质量 1 到 100
    IMWRITE_PAM_TUPLETYPE            = 128, //对于 PAM，将 TUPLETYPE 字段设置为格式定义的相应字符串值
    IMWRITE_TIFF_RESUNIT             = 256, //对于 TIFF，用于指定要设置的 DPI
    IMWRITE_TIFF_XDPI                = 257, //对于 TIFF，用于指定 X 方向的 DPI
    IMWRITE_TIFF_YDPI                = 258, //对于 TIFF，用于指定 Y 方向的 DPI
    IMWRITE_TIFF_COMPRESSION         = 259, //对于 TIFF，用于指定图像压缩方案
    IMWRITE_JPEG2000_COMPRESSION_X1000= 272 //对于 JPEG2000，用于指定目标压缩率，值从 0 到 1000
                                            //默认值为 1000
};
```

本案例调用 imread 函数读取图像，读取模式为灰度图像并将尺寸缩减为原始图像尺寸的 1/2；然后调用 imwrite 函数进行图像保存。Python 语言案例代码如下：

```
import cv2

img = cv2.imread("src.jpg", cv2.IMREAD_REDUCED_GRAYSCALE_2)      #图像读取
if img is None:
    print("Image Read Error!")
else:
    stat = cv2.imwrite("dst_2.jpg", img)                          #图像保存
    if(stat):
        print("Image Write Success!")
```

执行完成后，会在当前路径下保存文件名为 dst_2.jpg 的图像文件，并输出打印信息"Image Write Success!"，表明文件保存成功。

多幅图像保存为一个图像文件的函数 imwritemulti 没有 C++ 函数导出（有内联函数，见 2.1 节），因为多幅图像写入文件功能被 imwrite 函数重载了。imwritemulti 函数的 Python 语言函数定义如下：

```
retval = imwritemulti(filename, img, params=None)
```

imwritemulti 函数的参数说明如下。
- filename：保存的图像文件名称。
- img：图像数据。
- params：格式化编码为成对的特定参数，该参数可选，定义在 cv::ImwriteFlags 中。
- retval：图像保存是否成功的标志，bool 类型，保存成功返回 true，保存失败返回 false（返回值）。

读取多幅图像，并调用 imwritemulti 函数保存多幅图像到一个图像文件的案例代码如下：

```
import cv2
```

```
img1 = cv2.imread("src.jpg")                    #读取第一幅图像
img2 = cv2.imread("src1.jpg")                   #读取第二幅图像
img3 = cv2.imread("src2.jpg")                   #读取第三幅图像
img = [img1, img2, img3]
stat = cv2.imwritemulti("dst_tiff.tiff", img)   #图像保存
if (stat):
    print("Image Write Success!")
```

保存后在路径下生成 dst_tiff.tiff 文件，并输出保存成功的打印信息。

2.3 图像编/解码

在网络中传输文件时，传输的是数据流，因此，如果需要在网络中传输一幅图像，就需要对图像进行编码；对于网络接收到的图像数据流，需要使用图像解码还原为图像。OpenCV 中提供了用于图像编码与解码的函数 imencode 与 imdecode。

2.3.1 案例 9：图像编码应用

在 OpenCV 中，用于图像编码的函数是 imencode。imencode 函数的 C++语言函数定义如下：

```
CV_EXPORTS_W bool imencode( const String& ext, InputArray img,
                    CV_OUT std::vector<uchar>& buf,
                    const std::vector<int>& params = std::vector<int>());
```

imencode 函数的 Python 语言函数定义如下：

```
retval, buf = imencode(ext, img, params=None)
```

imencode 函数的参数说明如下。

- ext：文件扩展名，以决定输出格式。
- img：待编码图像数据。
- params：格式化编码为成对的特定参数，该参数可选，由 cv::ImwriteFlags 定义。
- buf：输出的数据流（返回值）。
- retval：编码是否成功的标志，类型为 bool，编码成功返回 true，编码失败返回 false（返回值）。

本案例将读取的图像进行编码，然后保存到 txt 文件中，如果读者在网络中传输图像，则可以直接将编码后的数据流用于网络数据传输。本案例的代码实现如下：

```
import cv2
import numpy as np
```

```
img = cv2.imread("src.jpg")                    #图像读取
img_encode = cv2.imencode('.jpg', img)[1]      #图像编码
data_encode = np.array(img_encode)
bytes_encode = data_encode.tobytes()           #数据类型转换
print(type(bytes_encode))                      #类型打印为：<class 'bytes'>

#将数据保存到txt文件中
with open("imencode.txt", "wb") as f:
    f.write(bytes_encode)
```

在调试状态下，可以看到各变量的数值，如图2.6所示，编码后的二进制数据流保存在变量 bytes_encode 中。

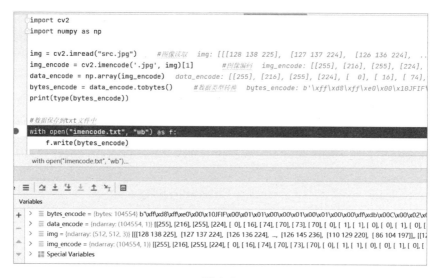

图2.6

执行完成后，会在路径下生成 imencode.txt 文件，其中保存着编码后的图像数据。

2.3.2 案例10：图像解码应用

对于接收到的图像数据流，需要使用图像解码功能解码为图像数据。

在 OpenCV 中，用于图像解码的函数为 imdecode。imdecode 函数的 C++语言函数定义如下：

```
CV_EXPORTS_W Mat imdecode( InputArray buf, int flags )
// 重载函数
CV_EXPORTS Mat imdecode( InputArray buf, int flags, Mat* dst)
```

imdecode 函数的 Python 语言函数定义如下：

```
retval = imdecode(buf, flags)
```

imdecode 函数的参数说明如下。
- buf：bytes 类型的输入数据流。
- flags：解码标志，由 cv::ImreadModes 定义。
- retval：解码的图像结果（返回值）。

本案例从 txt 中读取已保存的图像数据流，然后将数据流解码为图像并显示。本案例的代码如下：

```python
import numpy as np
import cv2

#从文件中读取图像数据
with open("imencode.txt", "rb") as f:
    data_encode = f.read()
img_array = np.frombuffer(data_encode, np.uint8)
#图像解码
img_decode = cv2.imdecode(img_array, cv2.IMREAD_COLOR)
#解码后的图像显示
cv2.imshow("img_decode", img_decode)
cv2.waitKey(0)
cv2.destroyAllWindows()
```

解码后的图像显示如图 2.7 所示。

图 2.7

2.4 进阶必备：聊聊图像格式

图像格式即图像文件存放在存储卡等介质上的格式，常用的图像格式有 BMP、JPEG、TIFF、RAW 等，受到存储容量的限制，图像文件通常都会经过压缩再存储，表 2.1 列举了 OpenCV 中图像读/写支持的格式，本节对一些常用格式做简单介绍。

（1）BMP 格式。

Windows BMP 格式图像文件又称位图，后缀为.bmp 或.dib，是 Windows 系统中最常见的图像格式，也是 Windows 环境中图像数据处理的一种标准格式，因此，Windows 环境中的图像处理软件都支持 BMP 格式。

BMP 格式采用位映射存储格式，除了图像深度可以设置（图像深度可设置为 1bit、4bit、8bit 及 24bit），不采用其他的压缩方式，因此，存储 BMP 文件所需的空间会很大。对于 BMP 文件，在进行数据存储时，采用从左到右、从下到上的图像扫描方式。

（2）JPEG 格式。

JPEG（Joint Photographic Experts Group，联合图像专家小组）是面向连续色调静止图像的一种压缩标准，该标准由国际标准化组织（ISO）制定。JPEG 格式图像文件采用 JPEG 标准，是目前最常用的图像文件格式，后缀名为.jpg 或.jpeg。

JPEG 格式是一种先进的压缩格式，可以去除图像中的冗余数据，该格式压缩比通常为 10:1 到 40:1，用 JPEG 格式存储的文件大小是其他类型文件大小的 1/20～1/10，能够将图像压缩在很小的存储空间内。JPEG 格式属于有损压缩格式，压缩比越大，图像的品质就越低，因此，如果要求高品质图像，则压缩比不宜设置得过大。

JPEG 格式可分为标准 JPEG、渐进式 JPEG 及 JPEG2000 3 种格式。

标准 JPEG 格式在网页中加载时，需要图像文件全部加载完毕才能展示图像。

渐进式 JPEG 在网页中加载时，会先呈现图像的粗略外观，然后逐渐呈现图像细节，因而称为渐进式 JPEG。渐进式 JPEG 格式图像文件比标准 JPEG 格式图像文件小，因此，网页端图像的展示建议使用这种格式。

JPEG2000 是新一代图像压缩方法，压缩品质更高，JPEG2000 格式文件后缀为.jp2。在无线传输图像时，经常会遇到信号不稳造成的马赛克现象，或者图像位置错乱问题，JPEG2000 可以改善这种情况下的图像传输品质。JPEG2000 的压缩率比标准 JPEG 的压缩率高约 30%，支持有损压缩和无损压缩两种方式，支持渐进式传输，也支持设定感兴趣区域（指定图片上感兴趣区域的压缩质量），还可以选择图像中的某一部分先行解压。

（3）PNG 格式。

PNG（Portable Network Graphics，便携式网络图形）格式是一种采用无损压缩数据算法的位图格式，后缀为.png，该格式当初的设计意图是替代有专利的 GIF 和 TIFF 文件格式，是目前比较常用的一种图像格式。

（4）WebP 格式。

WebP 由 Google 发布，文件后缀为.webp，提供了有损压缩与无损压缩两种方式。该格式派生于 VP8，支持的最大像素数量是 16383×16383。相较于 JPEG 格式，WebP 格式采用有损压缩，在保持与 JPEG 格式相同的图片质量的情况下，文件大小会比 JPEG 格式文件小，WebP 无损压缩的图像比 PNG 图像少了 45%的存储大小，因此能够有效地减少图像在网络上的传输时间。

（5）TIFF 格式。

TIFF（Tag Image File Format，标签图像文件）格式是一种灵活的位图格式，文件后缀为.tiff 或.tif。TIFF 格式采用 3 级体系结构，内部结构分为 3 部分：文件头信息区、标识信息区和图像数据区。文件头信息区存储 TIFF 文件解析必需的信息，标识信息区包含了有关于图像的所有信息，图像数据区存储图像信息。

TIFF 格式应用广泛，拥有多种压缩方案，可以描述多种类型的图像，不依赖硬件，具有可移植性。

OpenCV 在进行图像编/解码时，在 Windows 和 MacOSX 操作系统上，默认情况下使用 OpenCV 自带的编/解码器（libjpeg、libpng、libtiff 和 libjasper），因此，在这两种系统上，OpenCV 总是可以读取 JPEG、PNG 和 TIFF 格式的图像文件。在 macOS 上，还可以选择使用 macOS 本机的图像读取器。

在 Linux、BSD 或其他类 UNIX 的操作系统上，OpenCV 寻找与操作系统镜像一起提供的编/解码器，在安装相关软件包时，不要忘记安装开发文件，如 Debian 和 Ubuntu 中的 libjpeg dev。

第 3 章
核心库模块 core

core 模块中定义了 OpenCV 的基础数据结构与基础运算,是整个库的核心模块,因此,在引入 OpenCV 时,必须包含 core 模块。

3.1 模块导读

core 模块对应的头文件为"opencv2/core.hpp",该头文件中定义了一些枚举类型标志或类型,用于函数中的参数,如排序标志 SortFlags;定义了基础运算函数,如 add、subtract、multiply 和 divide 的四则运算等。另外,该头文件中还定义了异常类 Exception,该类继承于 std::exception,用于处理 OpenCV 中的异常。

"opencv2/core.hpp"头文件的定义与说明如下:

```
#ifndef OPENCV_CORE_HPP
#define OPENCV_CORE_HPP

#ifndef __cplusplus
#  error core.hpp header must be compiled as C++
#endif

#include "opencv2/core/cvdef.h"
#include "opencv2/core/base.hpp"
#include "opencv2/core/cvstd.hpp"
#include "opencv2/core/traits.hpp"
#include "opencv2/core/matx.hpp"
#include "opencv2/core/types.hpp"
```

```cpp
#include "opencv2/core/mat.hpp"
#include "opencv2/core/persistence.hpp"
```

OpenCV 中的所有内容都定义在命名空间 cv 中：
```cpp
namespace cv {
```
下面代码定义了异常类，这个类封装了关于程序中发生的错误的所有或几乎所有必要的信息。异常通常是通过 CV_Error 和 CV_Error_ 宏来隐式构造与抛出的。
```cpp
class CV_EXPORTS Exception : public std::exception
{
public:

    //构造函数
    Exception();
    Exception(int _code, const String& _err, const String& _func, const String& _file, int _line);
    virtual ~Exception() throw();

    /*! 抛出异常，以字符串形式返回错误描述信息和上下文信息*/
    virtual const char *what() const throw() CV_OVERRIDE;
    void formatMessage();

    String msg;        //格式化错误信息

    int code;          //异常码，由 CVStatus 定义
    String err;        //错误描述
    String func;       //函数名
    String file;       //发生错误的文件名
    int line;          //源码中发生错误的行号
};
```
下面代码定义了一些枚举类型的标志，这些标志在后续的算法函数中用于参数传递：
```cpp
//排序标志
enum SortFlags { SORT_EVERY_ROW    = 0, //每行独立排序
                 SORT_EVERY_COLUMN = 1, //矩阵每列均独立排序
                 SORT_ASCENDING    = 0, //每行升序排序
                 SORT_DESCENDING   = 16 //每行降序排序
               };

//协变量计算方式标志
enum CovarFlags {
    COVAR_SCRAMBLED= 0,
    COVAR_NORMAL   = 1,
    COVAR_USE_AVG  = 2,
    COVAR_SCALE    = 4,
    COVAR_ROWS     = 8,
```

```cpp
    COVAR_COLS      = 16
};

// k-Means 算法标志
enum KmeansFlags {
    KMEANS_RANDOM_CENTERS           = 0,
    KMEANS_PP_CENTERS               = 2,
    KMEANS_USE_INITIAL_LABELS       = 1
};

enum ReduceTypes { REDUCE_SUM = 0,      //输出和
                   REDUCE_AVG = 1,      //输出平均值
                   REDUCE_MAX = 2,      //输出最大值
                   REDUCE_MIN = 3       //输出最小值
                 };
```

core 模块中定义了很多基础运算，如矩阵交换、加减乘除等运算（可以参见案例 16）：

```cpp
/** 矩阵交换*/
CV_EXPORTS void swap(Mat& a, Mat& b);
CV_EXPORTS void swap( UMat& a, UMat& b );

/** 计算外推像素的源位置*/
CV_EXPORTS_W int borderInterpolate(int p, int len, int borderType);

/** 在图像周围生成边框*/
CV_EXPORTS_W void copyMakeBorder(InputArray src, OutputArray dst,
                        int top, int bottom, int left, int right,
                        int borderType, const Scalar& value = Scalar() );

/** 两个矩阵逐元素相加*/
CV_EXPORTS_W void add(InputArray src1, InputArray src2, OutputArray dst,
                    InputArray mask = noArray(), int dtype = -1);

/** 两个矩阵逐元素相减*/
CV_EXPORTS_W void subtract(InputArray src1, InputArray src2, OutputArray dst,
                    InputArray mask = noArray(), int dtype = -1);


/** 两个矩阵逐元素相乘*/
CV_EXPORTS_W void multiply(InputArray src1, InputArray src2,
                    OutputArray dst, double scale = 1, int dtype = -1);

/** 两个矩阵逐元素相除*/
CV_EXPORTS_W void divide(InputArray src1, InputArray src2, OutputArray dst,
                    double scale = 1, int dtype = -1);
CV_EXPORTS_W void divide(double scale, InputArray src2,
```

```cpp
                    OutputArray dst, int dtype = -1);

/** 计算一个数组乘以系数 alpha 后和另一个数组的和*/
CV_EXPORTS_W void scaleAdd(InputArray src1, double alpha, InputArray src2,
                           OutputArray dst);

/** 计算两个矩阵的加权和，计算方式为 dst = src1*alpha + src2*beta + gamma;*/
CV_EXPORTS_W void addWeighted(InputArray src1, double alpha, InputArray src2,
                              double beta, double gamma, OutputArray dst,
                              int dtype = -1);

/** 缩放、计算绝对值，并将结果转换为 8 位*/
CV_EXPORTS_W void convertScaleAbs(InputArray src, OutputArray dst,
                                  double alpha = 1, double beta = 0);

/** 将数组转换为半精度浮点数*/
CV_EXPORTS_W void convertFp16(InputArray src, OutputArray dst);

/** 执行数组的查找表转换*/
CV_EXPORTS_W void LUT(InputArray src, InputArray lut, OutputArray dst);

/** 计算矩阵元素的和*/
CV_EXPORTS_AS(sumElems) Scalar sum(InputArray src);

/** 统计非零元素的数量*/
CV_EXPORTS_W int countNonZero( InputArray src );

/** 对给定的二值矩阵计算非零像素的位置*/
CV_EXPORTS_W void findNonZero( InputArray src, OutputArray idx );

/** 矩阵求均值*/
CV_EXPORTS_W Scalar mean(InputArray src, InputArray mask = noArray());

/** 计算数组元素的平均值和标准偏差*/
CV_EXPORTS_W void meanStdDev(InputArray src, OutputArray mean,
                             OutputArray stddev,
                             InputArray mask=noArray());

/** 计算数组的绝对范数*/
CV_EXPORTS_W double norm(InputArray src1, int normType = NORM_L2,
                         InputArray mask = noArray());

/** 计算绝对差范数或相对差范数*/
CV_EXPORTS_W double norm(InputArray src1, InputArray src2,
                         int normType = NORM_L2, InputArray mask = noArray());
CV_EXPORTS double norm( const SparseMat& src, int normType );
```

```
/** 计算峰值信噪比（PSNR）图像质量度量*/
CV_EXPORTS_W double PSNR(InputArray src1, InputArray src2, double R=255.);

/** 朴素最近邻查找器*/
CV_EXPORTS_W void batchDistance(InputArray src1, InputArray src2,
                    OutputArray dist, int dtype, OutputArray nidx,
                    int normType = NORM_L2, int K = 0,
                    InputArray mask = noArray(), int update = 0,
                    bool crosscheck = false);

/** 归一化数组的范数或值范围*/
CV_EXPORTS_W void normalize( InputArray src, InputOutputArray dst,
                    double alpha = 1, double beta = 0,
                    int norm_type = NORM_L2, int dtype = -1,
                    InputArray mask = noArray());
CV_EXPORTS void normalize( const SparseMat& src, SparseMat& dst,
                    double alpha, int normType );
```

如下为与矩阵中最大值/最小值处理相关的函数：

```
/** 寻找矩阵的全局最小值或最大值及其位置*/
CV_EXPORTS_W void minMaxLoc(InputArray src, CV_OUT double* minVal,
                    CV_OUT double* maxVal = 0, CV_OUT Point* minLoc = 0,
                    CV_OUT Point* maxLoc = 0,
                    InputArray mask = noArray());
CV_EXPORTS void minMaxLoc(const SparseMat& a, double* minVal,
                    double* maxVal, int* minIdx = 0, int* maxIdx = 0);

/** 寻找矩阵的全局最小值或最大值及其索引*/
CV_EXPORTS void minMaxIdx(InputArray src, double* minVal, double* maxVal = 0,
                    int* minIdx = 0, int* maxIdx = 0,
                    InputArray mask = noArray());
```

如下为与矩阵降维、图像通道处理相关的函数（可以参见案例23）：

```
/** 矩阵降维为vector*/
CV_EXPORTS_W void reduce(InputArray src, OutputArray dst, int dim, int rtype,
                    int dtype = -1);

/** 将多个单通道矩阵合并为一个多通道矩阵*/
CV_EXPORTS void merge(const Mat* mv, size_t count, OutputArray dst);
CV_EXPORTS_W void merge(InputArrayOfArrays mv, OutputArray dst);

/** 将一个多通道矩阵拆分为多个单通道矩阵*/
CV_EXPORTS void split(const Mat& src, Mat* mvbegin);
CV_EXPORTS_W void split(InputArray m, OutputArrayOfArrays mv);
```

/** 将输入矩阵中的指定通道拷贝到输出矩阵的指定通道上，该函数提供了一种高级机制来打乱图像通道*/

```cpp
CV_EXPORTS void mixChannels(const Mat* src, size_t nsrcs,
                            Mat* dst, size_t ndsts,
                            const int* fromTo, size_t npairs);
CV_EXPORTS void mixChannels(InputArrayOfArrays src,
                            InputOutputArrayOfArrays dst,
                            const int* fromTo, size_t npairs);
CV_EXPORTS_W void mixChannels(InputArrayOfArrays src,
                              InputOutputArrayOfArrays dst,
                              const std::vector<int>& fromTo);

/** 提取源图像中的特定通道*/
CV_EXPORTS_W void extractChannel(InputArray src, OutputArray dst, int coi);

/** 向输出中插入输入的单通道*/
CV_EXPORTS_W void insertChannel(InputArray src, InputOutputArray dst,
                                int coi);
```

如下为与矩阵旋转或翻转相关的操作（见案例24）：

```cpp
/** 沿水平、垂直或两个轴翻转矩阵*/
CV_EXPORTS_W void flip(InputArray src, OutputArray dst, int flipCode);

//旋转方式
enum RotateFlags {
    ROTATE_90_CLOCKWISE = 0,           //顺时针90°旋转
    ROTATE_180 = 1,                    //顺时针180°旋转
    ROTATE_90_COUNTERCLOCKWISE = 2,    //顺时针270°旋转
};
/** 以90°的倍数旋转矩阵，旋转方式由RotateFlags定义*/
CV_EXPORTS_W void rotate(InputArray src, OutputArray dst, int rotateCode);

/** 用输入重复填充输出*/
CV_EXPORTS_W void repeat(InputArray src, int ny, int nx, OutputArray dst);
CV_EXPORTS Mat repeat(const Mat& src, int ny, int nx);
```

如下为与矩阵拼接相关的操作（可参见案例25）：

```cpp
/** 两个矩阵水平拼接*/
CV_EXPORTS void hconcat(const Mat* src, size_t nsrc, OutputArray dst);
CV_EXPORTS void hconcat(InputArray src1, InputArray src2, OutputArray dst);
CV_EXPORTS_W void hconcat(InputArrayOfArrays src, OutputArray dst);

/** 两个矩阵垂直拼接*/
CV_EXPORTS void vconcat(const Mat* src, size_t nsrc, OutputArray dst);
CV_EXPORTS void vconcat(InputArray src1, InputArray src2, OutputArray dst);
CV_EXPORTS_W void vconcat(InputArrayOfArrays src, OutputArray dst);
```

如下为矩阵的按位操作（可以参见案例17）：

```cpp
/** 两个矩阵逐元素按位与*/
```

```cpp
CV_EXPORTS_W void bitwise_and(InputArray src1, InputArray src2,
                              OutputArray dst, InputArray mask = noArray());

/** 两个矩阵逐元素按位或*/
CV_EXPORTS_W void bitwise_or(InputArray src1, InputArray src2,
                             OutputArray dst, InputArray mask = noArray());

/** 两个矩阵逐元素按位异或*/
CV_EXPORTS_W void bitwise_xor(InputArray src1, InputArray src2,
                              OutputArray dst, InputArray mask = noArray());

/** 对输入矩阵逐元素求非*/
CV_EXPORTS_W void bitwise_not(InputArray src, OutputArray dst,
                              InputArray mask = noArray());
```

如下是计算两个矩阵之间逐元素绝对差值的代码:

```cpp
CV_EXPORTS_W void absdiff(InputArray src1, InputArray src2, OutputArray dst);
```

如下为矩阵拷贝、矩阵比较及求最大/最小值的函数（见案例 19）:

```cpp
/** 用于矩阵拷贝*/
void CV_EXPORTS_W copyTo(InputArray src, OutputArray dst, InputArray mask);
/** 检查数组元素是否位于 lowerb 和 upperb 两个数组的元素之间*/
CV_EXPORTS_W void inRange(InputArray src, InputArray lowerb,
                          InputArray upperb, OutputArray dst);

/** 逐元素比较两个矩阵*/
CV_EXPORTS_W void compare(InputArray src1, InputArray src2, OutputArray dst,
                          int cmpop);

/** 求两个矩阵的最小值*/
CV_EXPORTS_W void min(InputArray src1, InputArray src2, OutputArray dst);
CV_EXPORTS void min(const Mat& src1, const Mat& src2, Mat& dst);
CV_EXPORTS void min(const UMat& src1, const UMat& src2, UMat& dst);

/** 求两个矩阵的最大值*/
CV_EXPORTS_W void max(InputArray src1, InputArray src2, OutputArray dst);
CV_EXPORTS void max(const Mat& src1, const Mat& src2, Mat& dst);
CV_EXPORTS void max(const UMat& src1, const UMat& src2, UMat& dst);
```

如下为与矩阵代数运算相关的函数（见案例 18）:

```cpp
/** 计算每个元素的平方根*/
CV_EXPORTS_W void sqrt(InputArray src, OutputArray dst);

/** 计算每个元素的幂运算*/
CV_EXPORTS_W void pow(InputArray src, double power, OutputArray dst);
```

```cpp
/** 计算每个数组元素的指数*/
CV_EXPORTS_W void exp(InputArray src, OutputArray dst);

/** 计算每个数组元素的自然对数*/
CV_EXPORTS_W void log(InputArray src, OutputArray dst);

/** 根据向量的大小和角度计算二维向量的 x 和 y 坐标*/
CV_EXPORTS_W void polarToCart(InputArray magnitude, InputArray angle,
                              OutputArray x, OutputArray y,
                              bool angleInDegrees = false);

/** 计算二维向量的幅值和角度*/
CV_EXPORTS_W void cartToPolar(InputArray x, InputArray y,
                              OutputArray magnitude, OutputArray angle,
                              bool angleInDegrees = false);

/** 计算二维向量的旋转角度*/
CV_EXPORTS_W void phase(InputArray x, InputArray y, OutputArray angle,
                        bool angleInDegrees = false);

/** 计算二维向量的幅值*/
CV_EXPORTS_W void magnitude(InputArray x, InputArray y,
                            OutputArray magnitude);

/**检查输入数组的每个元素是否存在无效值*/
CV_EXPORTS_W bool checkRange(InputArray a, bool quiet = true,
                             CV_OUT Point* pos = 0,
                             double minVal = -DBL_MAX,
                             double maxVal = DBL_MAX);

/** 将无效值 NaNs 用指定数值替换*/
CV_EXPORTS_W void patchNaNs(InputOutputArray a, double val = 0);

/** 执行广义矩阵乘法*/
CV_EXPORTS_W void gemm(InputArray src1, InputArray src2, double alpha,
                       InputArray src3, double beta,
                       OutputArray dst, int flags = 0);

/** 计算矩阵与其转置的乘积*/
CV_EXPORTS_W void mulTransposed( InputArray src, OutputArray dst, bool aTa,
                                 InputArray delta = noArray(),
                                 double scale = 1, int dtype = -1 );

/** 矩阵变换*/
CV_EXPORTS_W void transpose(InputArray src, OutputArray dst);
```

```cpp
/** 矩阵按元素变换*/
CV_EXPORTS_W void transform(InputArray src, OutputArray dst, InputArray m );

/** 透视变换*/
CV_EXPORTS_W void perspectiveTransform(InputArray src, OutputArray dst, InputArray m );

/** 将方阵的下半部分或上半部分复制到另一半*/
CV_EXPORTS_W void completeSymm(InputOutputArray m, bool lowerToUpper = false);

/** 初始化缩放的单位矩阵*/
CV_EXPORTS_W void setIdentity(InputOutputArray mtx, const Scalar& s = Scalar(1));

/** 返回平方浮点矩阵的行列式*/
CV_EXPORTS_W double determinant(InputArray mtx);

/** 矩阵求迹*/
CV_EXPORTS_W Scalar trace(InputArray mtx);

/** 求矩阵的逆或伪逆*/
CV_EXPORTS_W double invert(InputArray src, OutputArray dst, int flags = DECOMP_LU);

/** 解决一个或多个线性系统或最小二乘问题*/
CV_EXPORTS_W bool solve(InputArray src1, InputArray src2,
                    OutputArray dst, int flags = DECOMP_LU);

/** 矩阵按行或列排序*/
CV_EXPORTS_W void sort(InputArray src, OutputArray dst, int flags);

/** 矩阵按行或列排序并保存排序像素值原始的位置信息*/
CV_EXPORTS_W void sortIdx(InputArray src, OutputArray dst, int flags);

/** 求三次方程的实根*/
CV_EXPORTS_W int solveCubic(InputArray coeffs, OutputArray roots);

/** 求多项式方程的实根或复数根*/
CV_EXPORTS_W double solvePoly(InputArray coeffs, OutputArray roots,
                          int maxIters = 300);

/** 计算对称矩阵的特征值和特征向量*/
CV_EXPORTS_W bool eigen(InputArray src, OutputArray eigenvalues,
                    OutputArray eigenvectors = noArray());

/** 计算非对称矩阵的特征值和特征向量（仅限实特征值）*/
CV_EXPORTS_W void eigenNonSymmetric(InputArray src, OutputArray eigenvalues,
                                OutputArray eigenvectors);
```

```
/** 计算一组向量的协方差矩阵*/
CV_EXPORTS void calcCovarMatrix( const Mat* samples, int nsamples,
                                 Mat& covar, Mat& mean,
CV_EXPORTS_W void calcCovarMatrix( InputArray samples, OutputArray covar,
                                   InputOutputArray mean, int flags,
                                   int ctype = CV_64F);
```

如下为与主成分分析相关的函数：

```
/**封装 PCA::operator() */
CV_EXPORTS_W void PCACompute(InputArray data, InputOutputArray mean,
                             OutputArray eigenvectors, int maxComponents = 0);

/**封装 PCA::operator()并添加特征值输出参数*/
CV_EXPORTS_AS(PCACompute2) void PCACompute(InputArray data,
                                           InputOutputArray mean,
                                           OutputArray eigenvectors,
                                           OutputArray eigenvalues,
                                           int maxComponents = 0);

/**封装 PCA::operator() */
CV_EXPORTS_W void PCACompute(InputArray data, InputOutputArray mean,
                             OutputArray eigenvectors,
                             double retainedVariance);

/**封装 PCA::operator() 并添加特征值输出参数 */
CV_EXPORTS_AS(PCACompute2) void PCACompute(InputArray data,
                                           InputOutputArray mean,
                                           OutputArray eigenvectors,
                                           OutputArray eigenvalues,
                                           double retainedVariance);

/**封装 PCA::project */
CV_EXPORTS_W void PCAProject(InputArray data, InputArray mean,
                             InputArray eigenvectors, OutputArray result);

/**封装 PCA::backProject */
CV_EXPORTS_W void PCABackProject(InputArray data, InputArray mean,
                                 InputArray eigenvectors, OutputArray result);
```

如下为与矩阵奇异值分解相关的函数：

```
/**封装 SVD::compute */
CV_EXPORTS_W void SVDecomp( InputArray src, OutputArray w, OutputArray u,
                            OutputArray vt, int flags = 0 );

/**封装 SVD::backSubst */
CV_EXPORTS_W void SVBackSubst( InputArray w, InputArray u, InputArray vt,
```

```
                        InputArray rhs, OutputArray dst );
```

如下函数用来计算两个向量之间的马氏距离：
```
CV_EXPORTS_W double Mahalanobis(InputArray v1, InputArray v2,
                        InputArray icovar);
```

如下为用于矩阵傅里叶变换、余弦变换等变换的函数（见案例 27）：
```
/** 计算一维或二维矩阵的离散傅里叶变换*/
CV_EXPORTS_W void dft(InputArray src, OutputArray dst, int flags = 0,
                int nonzeroRows = 0);

/** 计算一维或二维矩阵的离散傅里叶逆变换*/
CV_EXPORTS_W void idft(InputArray src, OutputArray dst, int flags = 0,
                int nonzeroRows = 0);

/** 计算一维或二维数组的离散余弦变换*/
CV_EXPORTS_W void dct(InputArray src, OutputArray dst, int flags = 0);

/**计算一维或二维矩阵的离散余弦逆变换*/
CV_EXPORTS_W void idct(InputArray src, OutputArray dst, int flags = 0);

/**执行两个傅里叶谱的对应元素乘法*/
CV_EXPORTS_W void mulSpectrums(InputArray a, InputArray b, OutputArray c,
                        int flags, bool conjB = false);

/** 基于给定向量大小返回最佳 DFT 大小*/
CV_EXPORTS_W int getOptimalDFTSize(int vecsize);
```

如下为与随机数矩阵相关的运算函数（见案例 21）：
```
/** 返回默认随机数生成器*/
CV_EXPORTS RNG& theRNG();

/** 设置默认随机数生成器的状态*/
CV_EXPORTS_W void setRNGSeed(int seed);

/**生成均匀分布随机数数组*/
CV_EXPORTS_W void randu(InputOutputArray dst, InputArray low,
                InputArray high);

/**生成正态分布随机数数组*/
CV_EXPORTS_W void randn(InputOutputArray dst, InputArray mean,
                InputArray stddev);

/**随机打乱数组元素*/
CV_EXPORTS_W void randShuffle(InputOutputArray dst, double iterFactor = 1.,
                RNG* rng = 0);
```

如下为一些类的声明：
```
/** 主成分分析*/
class CV_EXPORTS PCA

/**线性判别分析*/
class CV_EXPORTS LDA

/** 奇异值分解*/
class CV_EXPORTS SVD

/**随机数生成器*/
class CV_EXPORTS RNG


/** @brief Mersenne Twister 随机数生成器*/
class CV_EXPORTS RNG_MT19937
```

函数 kmeans 实现了一个 k-Means 算法，该算法可以找到聚类的中心，并对聚类周围的输入样本进行分组：

```
CV_EXPORTS_W double kmeans( InputArray data, int K, InputOutputArray bestLabels,
                    TermCriteria criteria, int attempts,
                    int flags, OutputArray centers = noArray() );
```

如下为一些类的声明：

```
//Mat 格式化输出
class CV_EXPORTS Formatted

class CV_EXPORTS Algorithm;
//OpenCV 算法基类
class CV_EXPORTS_W Algorithm

} //namespace cv

#include "opencv2/core/operations.hpp"
#include "opencv2/core/cvstd.inl.hpp"
#include "opencv2/core/utility.hpp"
#include "opencv2/core/optim.hpp"
#include "opencv2/core/ovx.hpp"

#endif /*OPENCV_CORE_HPP*/
```

3.2 基本数据结构

本节介绍 OpenCV 中的基本数据结构，在编写 C++ 代码时，需要为定义的变量指定类型，了解基本数据结构有助于读者在开发时选择合适的类型来存储数据。

3.2.1 案例 11：Mat 数据结构介绍及 C++ 调用

Mat 数据结构是 OpenCV 中最重要的数据结构，是 OpenCV 中图像最常用的存储格式。Mat 数据结构定义在头文件 "opencv2/core/mat.hpp" 中。

如下为 Mat 类的定义：

```
class CV_EXPORTS Mat
{
public:
```

此处展示 Mat 的构造函数，多数情况下使用默认构造函数，构造的矩阵可以被重新分配空间。图像矩阵可以通过 Mat 构造函数构造，也可以通过 Mat 类的成员函数 create 构建。

```
    Mat();
    //通过行、列、类型 3 个参数构造
    Mat(int rows, int cols, int type);
    //通过 Size 类型参数传递矩阵大小、类型两个参数构造
    Mat(Size size, int type);
    //通过行、列、类型、初始化值 4 个参数构造
    Mat(int rows, int cols, int type, const Scalar& s);
    //通过 Size 类型参数传递矩阵大小、类型、初始化值 3 个参数构造
    Mat(Size size, int type, const Scalar& s);
    //通过维度、数组传递的 sizes、类型 3 个参数构造
    Mat(int ndims, const int* sizes, int type);
    //通过 vector 传递的 sizes、类型两个参数构造
    Mat(const std::vector<int>& sizes, int type);
    //通过维度、数组传递的 sizes、类型、初始化值 4 个参数构造
    Mat(int ndims, const int* sizes, int type, const Scalar& s);
    //通过 vector 传递的 sizes、类型、初始化值 3 个参数构造
    Mat(const std::vector<int>& sizes, int type, const Scalar& s);
    //通过已有 Mat 构造
    Mat(const Mat& m);
    //通过行、列、类型、读者数据、每行所占字节数 5 个参数构造
    Mat(int rows, int cols, int type, void* data, size_t step=AUTO_STEP);
    //通过 Size 类型参数传递矩阵大小、类型、读者数据、每行所占字节数 4 个参数构造
    Mat(Size size, int type, void* data, size_t step=AUTO_STEP);
    //通过维度、数组传递的 sizes、类型、读者数据、每行所占字节数 5 个参数构造
```

```cpp
Mat(int ndims, const int* sizes, int type, void* data, const size_t* steps=0);
//通过 vector 传递的 sizes、类型、读者数据、每行所占字节数 4 个参数构造
Mat(const std::vector<int>& sizes, int type, void* data, const size_t* steps=0);
//通过已有 Mat、Mat 中选取的行范围、Mat 中选取的列范围构造
Mat(const Mat& m, const Range& rowRange, const Range& colRange=Range::all());
//通过已有 Mat、感兴趣区域 roi 构造
Mat(const Mat& m, const Rect& roi);
//通过已有 Mat、Range 数组类型选取的范围构造
Mat(const Mat& m, const Range* ranges);
//通过已有 Mat、vector<Range>类型选取的范围构造
Mat(const Mat& m, const std::vector<Range>& ranges);
```

如下为 Mat 构造的模板函数：

```cpp
template<typename _Tp> explicit Mat(const std::vector<_Tp>& vec, bool copyData=false);

template<typename _Tp, typename = typename std::enable_if<std::is_arithmetic<_Tp>::value>::type>
explicit Mat(const std::initializer_list<_Tp> list);

template<typename _Tp> explicit Mat(const std::initializer_list<int> sizes, const std::initializer_list<_Tp> list);

template<typename _Tp, size_t _Nm> explicit Mat(const std::array<_Tp, _Nm>& arr, bool copyData=false);

template<typename _Tp, int n> explicit Mat(const Vec<_Tp, n>& vec, bool copyData=true);

template<typename _Tp, int m, int n> explicit Mat(const Matx<_Tp, m, n>& mtx, bool copyData=true);

template<typename _Tp> explicit Mat(const Point_<_Tp>& pt, bool copyData=true);

template<typename _Tp> explicit Mat(const Point3_<_Tp>& pt, bool copyData=true);

template<typename _Tp> explicit Mat(const MatCommaInitializer_<_Tp>& commaInitializer);

//! 从 GpuMat 下载数据
explicit Mat(const cuda::GpuMat& m);
```

如下为 Mat 类的析构函数：

```cpp
~Mat();
```

如下为 Mat 类重载的赋值运算符：

```cpp
Mat& operator = (const Mat& m);
Mat& operator = (const MatExpr& expr);
```

如下为 Mat 对象元素操作的函数：

```cpp
//! 从 Mat 中取得 UMat
UMat getUMat(AccessFlag accessFlags, UMatUsageFlags usageFlags = USAGE_DEFAULT) const;

/** 为指定的矩阵行创建矩阵头*/
Mat row(int y) const;

/** 为指定的矩阵列创建矩阵头*/
Mat col(int x) const;

/** 为指定的行范围创建矩阵头*/
Mat rowRange(int startrow, int endrow) const;
Mat rowRange(const Range& r) const;

/** 为指定的列范围创建矩阵头*/
Mat colRange(int startcol, int endcol) const;
Mat colRange(const Range& r) const;

/** 从矩阵中提取对角线*/
Mat diag(int d=0) const;

/** 创建对角矩阵*/
static Mat diag(const Mat& d);

/** 创建数组和基础数据的完整副本*/
Mat clone() const CV_NODISCARD;

/** 拷贝一个矩阵到另一个中*/
void copyTo( OutputArray m ) const;
void copyTo( OutputArray m, InputArray mask ) const;

/** 将数组转换为具有可选缩放比例的其他数据类型*/
void convertTo( OutputArray m, int rtype, double alpha=1, double beta=0 ) const;

/** 矩阵赋值，convertTo 的函数形式*/
void assignTo( Mat& m, int type=-1 ) const;

/** 赋值运算符*/
Mat& operator = (const Scalar& s);

/** 根据 mask 将矩阵中的元素设置为指定值*/
Mat& setTo(InputArray value, InputArray mask=noArray());

/** 维度变换*/
Mat reshape(int cn, int rows=0) const;
```

```cpp
Mat reshape(int cn, int newndims, const int* newsz) const;
Mat reshape(int cn, const std::vector<int>& newshape) const;
```

如下为 Mat 对象变换的一些操作：

```cpp
/** 转置矩阵*/
MatExpr t() const;

/** 求矩阵的逆*/
MatExpr inv(int method=DECOMP_LU) const;

/** 对两个矩阵执行按元素的乘法或除法*/
MatExpr mul(InputArray m, double scale=1) const;

/** 计算两个三元素向量的叉积*/
Mat cross(InputArray m) const;

/** 计算两个 vectors 的点积*/
double dot(InputArray m) const;
```

生成特定类型的矩阵，如全 0 矩阵、全 1 矩阵等：

```cpp
/** 返回指定大小与类型的全 0 矩阵*/
static MatExpr zeros(int rows, int cols, int type);
static MatExpr zeros(Size size, int type);
static MatExpr zeros(int ndims, const int* sz, int type);

/** 返回指定大小与类型的全 1 矩阵*/
static MatExpr ones(int rows, int cols, int type);
static MatExpr ones(Size size, int type);
static MatExpr ones(int ndims, const int* sz, int type);

/** 返回指定大小与类型的单位矩阵*/
static MatExpr eye(int rows, int cols, int type);
static MatExpr eye(Size size, int type);
```

如下为 Mat 对象创建函数：

```cpp
/** 重新分配空间，这是 Mat 的一个关键方法*/
void create(int rows, int cols, int type);
void create(Size size, int type);
void create(int ndims, const int* sizes, int type);
void create(const std::vector<int>& sizes, int type);
```

如下为 Mat 对象的一些操作函数：

```cpp
/** 增加引用计数*/
void addref();

/** 减少引用计数，必要时（引用计数为 0）释放内存空间*/
void release();
```

```cpp
//! 内部函数,释放空间
void deallocate();
//! 内部函数
void copySize(const Mat& m);

/** 为一定数量的行保留空间*/
void reserve(size_t sz);

/** 为一定数量的字节保留空间*/
void reserveBuffer(size_t sz);

/** 调整矩阵大小*/
void resize(size_t sz);
void resize(size_t sz, const Scalar& s);

//! 内部函数
void push_back_(const void* elem);

/** 在矩阵底部增加元素*/
template<typename _Tp> void push_back(const _Tp& elem);
template<typename _Tp> void push_back(const Mat_<_Tp>& elem);
template<typename _Tp> void push_back(const std::vector<_Tp>& elem);
void push_back(const Mat& m);

/** 从矩阵底部删除元素*/
void pop_back(size_t nelems=1);

/** 在父矩阵中查找矩阵头*/
void locateROI( Size& wholeSize, Point& ofs ) const;

/** 在父矩阵中调整子矩阵的大小和位置*/
Mat& adjustROI( int dtop, int dbottom, int dleft, int dright );

/** 提取一个矩形子矩阵*/
Mat operator()( Range rowRange, Range colRange ) const;
Mat operator()( const Rect& roi ) const;
Mat operator()( const Range* ranges ) const;
Mat operator()(const std::vector<Range>& ranges) const;
template<typename _Tp> operator std::vector<_Tp>() const;
template<typename _Tp, int n> operator Vec<_Tp, n>() const;
template<typename _Tp, int m, int n> operator Matx<_Tp, m, n>() const;
template<typename _Tp, std::size_t _Nm> operator std::array<_Tp, _Nm>() const;
```

如下为返回 Mat 对象属性(如矩阵的类型等)的函数:

```cpp
/** 查看矩阵是否连续*/
```

```cpp
bool isContinuous() const;

//! 判断一个矩阵是否是另一个矩阵的子矩阵
bool isSubmatrix() const;

/** 返回以字节为单位的矩阵元素大小*/
size_t elemSize() const;

/** 返回每个矩阵元素通道的大小（单位为字节）*/
size_t elemSize1() const;

/** 返回矩阵元素类型*/
int type() const;

/** 返回矩阵深度*/
int depth() const;

/** 返回矩阵通道数*/
int channels() const;

/** 返回归一化步数*/
size_t step1(int i=0) const;

/** 判断矩阵是否为空*/
bool empty() const;

/** 返回矩阵总的元素数目*/
size_t total() const;
size_t total(int startDim, int endDim=INT_MAX) const;

/** 检查 Mat 是否为 Vector，用于检查传入的数据格式是否正确*/
int checkVector(int elemChannels, int depth=-1, bool requireContinuous=true) const;
```

如下函数返回矩阵指定行的指针：

```cpp
uchar* ptr(int i0=0);
const uchar* ptr(int i0=0) const;
uchar* ptr(int row, int col);
const uchar* ptr(int row, int col) const;
uchar* ptr(int i0, int i1, int i2);
const uchar* ptr(int i0, int i1, int i2) const;
uchar* ptr(const int* idx);
const uchar* ptr(const int* idx) const;
template<int n> uchar* ptr(const Vec<int, n>& idx);
template<int n> const uchar* ptr(const Vec<int, n>& idx) const;
template<typename _Tp> _Tp* ptr(int i0=0);
template<typename _Tp> const _Tp* ptr(int i0=0) const;
```

```
template<typename _Tp> _Tp* ptr(int row, int col);
template<typename _Tp> const _Tp* ptr(int row, int col) const;
template<typename _Tp> _Tp* ptr(int i0, int i1, int i2);
template<typename _Tp> const _Tp* ptr(int i0, int i1, int i2) const;
template<typename _Tp> _Tp* ptr(const int* idx);
template<typename _Tp> const _Tp* ptr(const int* idx) const;
template<typename _Tp, int n> _Tp* ptr(const Vec<int, n>& idx);
template<typename _Tp, int n> const _Tp* ptr(const Vec<int, n>& idx) const;
```

如下函数返回对指定数组元素的引用：

```
template<typename _Tp> _Tp& at(int i0=0);
template<typename _Tp> const _Tp& at(int i0=0) const;
template<typename _Tp> _Tp& at(int row, int col);
template<typename _Tp> const _Tp& at(int row, int col) const;
template<typename _Tp> _Tp& at(int i0, int i1, int i2);
template<typename _Tp> const _Tp& at(int i0, int i1, int i2) const;
template<typename _Tp> _Tp& at(const int* idx);
template<typename _Tp> const _Tp& at(const int* idx) const;
template<typename _Tp, int n> _Tp& at(const Vec<int, n>& idx);
template<typename _Tp, int n> const _Tp& at(const Vec<int, n>& idx) const;
template<typename _Tp> _Tp& at(Point pt);
template<typename _Tp> const _Tp& at(Point pt) const;
```

如下函数完成与 Mat 矩阵迭代器相关的操作：

```
/** 返回矩阵迭代器的首元素迭代器*/
template<typename _Tp> MatIterator_<_Tp> begin();
template<typename _Tp> MatConstIterator_<_Tp> begin() const;

 /** 返回矩阵迭代器的尾元素后一个位置的迭代器*/
template<typename _Tp> MatIterator_<_Tp> end();
template<typename _Tp> MatConstIterator_<_Tp> end() const;

/** 类似于 C++ 11 中的 foreach 函数的功能*/
template<typename _Tp, typename Functor> void forEach(const Functor& operation);
template<typename _Tp, typename Functor> void forEach(const Functor& operation) const;

Mat(Mat&& m);
Mat& operator = (Mat&& m);

enum { MAGIC_VAL  = 0x42FF0000, AUTO_STEP = 0, CONTINUOUS_FLAG = CV_MAT_CONT_FLAG,
SUBMATRIX_FLAG = CV_SUBMAT_FLAG };
enum { MAGIC_MASK = 0xFFFF0000, TYPE_MASK = 0x00000FFF, DEPTH_MASK = 7 };
```

如下为 Mat 对象的成员函数：

```
/*! 类成员变量*/
int flags;
```

```
    int dims;              //维度
    int rows, cols;        //行和列
    uchar* data;           //元素数据
    const uchar* datastart;
    const uchar* dataend;
    const uchar* datalimit;
```

如下为标准内存分配器:

```
    MatAllocator* allocator;
    static MatAllocator* getStdAllocator();
    static MatAllocator* getDefaultAllocator();
    static void setDefaultAllocator(MatAllocator* allocator);

    //! 内部函数
    void updateContinuityFlag();

    //! 与 UMat 交互
    UMatData* u;

    MatSize size;
    MatStep step;

protected:
    template<typename _Tp, typename Functor> void forEach_impl(const Functor& operation);
};
```

Mat 类中定义了多种 Mat 的构造析构方式，还有非常多的图像矩阵操作，如图像矩阵元素的增加与删除，提取子矩阵，获取图像矩阵元素类型，获取图像矩阵深度、通道数、宽、高、元素数量，判断图像矩阵是否为空等。

本案例使用 C++语言开发，对 Mat 对象的构造及一些常见操作做介绍，可以让读者对图像存储类型 Mat 有一个清晰的了解，3.2.2 节案例将讲解 Python 语言中 Mat 对象的操作。

使用 Mat 构造函数创建图像矩阵的代码如下：

```cpp
#include <iostream>
#include "opencv2/opencv.hpp"

using namespace std;
using namespace cv;

int main()
{
    Mat m1 = Mat();                              //默认构造函数
    if (m1.empty())
        cout << "Empty Matrix!" << endl;
```

```
    Mat m2 = Mat(3, 3, CV_8U);                      //构造 3×3（单位为像素）的图像
    cout << "m2 type: " << m2.type() << endl;       //输出图像矩阵元素类型
    cout << "m2 size: " << m2.size() << endl;       //输出图像矩阵尺寸

    //Mat 的大小由 Size 类型定义，初始值由 Scalar 类型定义
    Mat m3 = Mat(Size(512, 512), CV_8U, Scalar(128));
    if (!m3.empty())
    {
        imshow("m3", m3);
        waitKey(0);
    }
}
```

本案例中选取了常用的 3 种 Mat 构造方式来创建图像矩阵 m1、m2、m3。其中，m1 调用默认构造函数，创建的图像矩阵为空；m2 创建 3×3 的矩阵，并指定类型为 CV_8U；m3 创建 512×512 的矩阵，指定类型为 CV_8U，并初始化初值为 128。

执行后的打印信息如图 3.1 所示。

图 3.1

CV_8U 为创建的图像矩阵的类型，该类型由宏定义，几个常用的类型定义如下：

```
#define CV_8U   0
#define CV_8S   1
#define CV_16U  2
#define CV_16S  3
#define CV_32S  4
#define CV_32F  5
#define CV_64F  6
#define CV_16F  7
```

因此，m2 type 的输出结果为数值 0。

图像矩阵 m3 的显示结果如图 3.2 所示，由于其像素初值为 128，所以为灰色。

图 3.2

Mat 的构造函数有众多的重载函数,读者可以选取合适的方式构造。读取图像后,可以调用成员函数或成员变量获取图像的信息。

如下案例代码用于读取本地图像文件,获取图像中的元素总量,图像的宽(cols,列数)、高(rows,行数)、通道数、深度:

```
//读取图像
Mat m4 = imread("src.jpg", IMREAD_COLOR);
cout << "m4 total element: " << m4.total() << endl;    //输出总的元素数量
cout << "m4 rows: " << m4.rows << endl;                //输出行数
cout << "m4 cols: " << m4.cols << endl;                //输出列数
cout << "m4 channels: " << m4.channels() << endl;      //输出通道数
cout << "m4 depth: " << m4.depth() << endl;            //输出深度
```

执行后的输出结果如图 3.3 所示。

```
m4 total element: 262144
m4 rows: 512
m4 cols: 512
m4 channels: 3
m4 depth: 0
```

图 3.3

如图 3.3 所示,总的元素数量为宽和高的乘积,图像的读取方式为 IMREAD_COLOR,该方式读取的图像为 3 通道的 BGR 彩色图像,因此 m4 的通道数为 3。

对于读取的图像 m4,调用 clone() 方法,拷贝一份给 m5;也可以调用 copyTo() 方法,将图像数据拷贝给新的 Mat 对象:

```
Mat m5 = m4.clone();                    //图像 clone
Mat m6;
m5.copyTo(m6);
if (!m6.empty())
{
    imshow("m6", m6);
    waitKey(0);
}
```

在上面的案例代码中,将 m4 的数据拷贝(clone)一份给 m5,然后调用 copyTo() 方法,将 m5 的数据拷贝给 m6,m6 的结果显示如图 3.4 所示。

Mat 类的成员函数 resize 可以改变图像的行数,代码如下:

```
//改变矩阵行数
m6.resize(m6.rows / 2);
imshow("m6-resize", m6);
waitKey(0);
```

将 m6 图像的行数变为原始图像行数的 1/2,变换后的结果如图 3.5 所示。

图 3.4

图 3.5

实际上，很多读者的需求是调整图像的宽高尺寸，此时应该调用图像处理模块 imgproc 中的 resize 函数（第 4 章会详细介绍该函数的使用方法），代码如下：

```
Mat m7;
resize(m5, m7, Size(m5.rows / 2, m5.cols / 2));
imshow("resize", m7);
waitKey(0);
```

在以上代码中，首先创建 Mat 对象 m7；然后调用 resize 函数，将 m5 图像矩阵调整尺寸后存入 m7 中，调整后的 m7 图像的尺寸为 m5 尺寸的 1/2。resize 函数执行后的结果如图 3.6 所示。

有时需要对图像矩阵中的元素进行操作，Mat 类对象可以执行图像元素的访问操作：

```
//元素遍历
Mat m8 = (Mat_<int>(3, 3) << 121, 212, 231, 108, 210, 130, 100, 110, 120);
for (int i = 0; i < m8.rows; i++)            //遍历行
{
    for (int j = 0; j < m8.cols; j++)        //遍历列
        cout << m8.at<uint>(i, j) << " ";
    cout << endl;
}

cout << "-----------" << endl;
//获取行指针方式遍历
for (int k = 0; k < m8.rows; k++)
{
    uint* prow = m8.ptr<uint>(k);
    //使用指针进行元素访问
```

```
            cout << prow[0] << " " << prow[1] << " " << prow[2] << endl;
    }
```

在上面的代码中,创建 Mat 对象 m8 并初始化,第一种遍历方式为按行列遍历 m8 并输出元素值;第二种遍历方式是获取每一行的指针,然后使用指针访问元素。

执行后的打印结果如图 3.7 所示。

图 3.6

图 3.7

3.2.2 案例 12:Python 中的 Mat 对象操作

在 Python 语言中,Mat 类型的对象构造操作可以通过 numpy 来实现(创建图像矩阵)。如下代码用来创建全 0 矩阵 m1,然后将所有值初始化为 128,并将结果显示:

```
import cv2
import numpy as np

#创建全 0 矩阵
m1 = np.zeros([512, 512], dtype=np.uint8)
m1[:] = 128                  #初始化
cv2.imshow("m1", m1)         #图像显示
cv2.waitKey(0)
cv2.destroyAllWindows()
```

m1 的显示结果如图 3.8 所示。

也可以对其中的部分元素值进行修改:

```
m1[128:384, 128:384] = 0
cv2.imshow("m1", m1)
cv2.waitKey(0)
```

在上面的代码中，将行、列为 128 到 384 区域的像素值赋值为 0（黑色），修改像素值后的图像显示结果如图 3.9 所示。

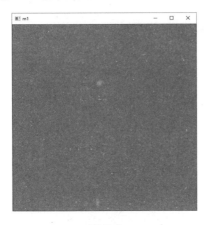
图 3.8

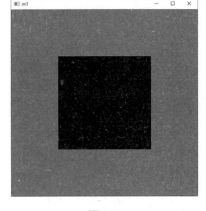
图 3.9

读取本地图像文件、获取图像的宽高、调整图像的大小的操作案例代码如下：

```
#图像读取
m2 = cv2.imread("src.jpg", cv2.IMREAD_COLOR)
#显示类型
print("m2 type: ", type(m2))

#获取图像尺寸与通道
shape = m2.shape
rows = shape[0]
cols = shape[1]
channels = shape[2]
print("m2 rows: ", rows)
print("m2 cols: ", cols)
print("m2 channels: ", channels)

#调整图像尺寸
mask = np.ones(shape, dtype=np.uint8)              #创建掩模
m3 = cv2.copyTo(m2, mask)                          #图像拷贝
m3 = cv2.resize(m3, (int(rows/2), int(cols/2)))    #图像尺寸调整
cv2.imshow("m3", m3)                               #图像显示
cv2.waitKey(0)
cv2.destroyAllWindows()
```

本案例使用的输入图像如图 3.10 所示，该图像为后面很多案例的源图像。

图像大小调整后的结果显示如图 3.11 所示。

图 3.10

图 3.11

3.2.3 案例 13：Point 结构

在 OpenCV 中，点的定义由 Point 类实现。Point 类的定义如下：

```
typedef Point_<int> Point2i;
typedef Point_<int64> Point2l;
typedef Point_<float> Point2f;
typedef Point_<double> Point2d;
typedef Point2i Point;
```

可以看出，Point 为 Point2i 类型的别名，Point2i 为 Point_<int>的别名，Point_为模板类，其定义如下：

```
template<typename _Tp> class Point_
{
public:
    typedef _Tp value_type;

    //! 构造函数
    Point_();
    Point_(_Tp _x, _Tp _y);
    Point_(const Point_& pt);
    Point_(Point_&& pt) CV_NOEXCEPT;
    Point_(const Size_<_Tp>& sz);
    Point_(const Vec<_Tp, 2>& v);

    //! 赋值运算符
    Point_& operator = (const Point_& pt);
    Point_& operator = (Point_&& pt) CV_NOEXCEPT;
    //! 转换为其他数据类型
    template<typename _Tp2> operator Point_<_Tp2>() const;

    //! 转换为旧风格 C 结构
```

```cpp
    operator Vec<_Tp, 2>() const;

    //! 点乘运算
    _Tp dot(const Point_& pt) const;
    //! 双精度点乘
    double ddot(const Point_& pt) const;
    //! 叉乘
    double cross(const Point_& pt) const;
    //! 检查点是否在指定的矩形中
    bool inside(const Rect_<_Tp>& r) const;
    _Tp x; //!< 点的 x 坐标
    _Tp y; //!< 点的 y 坐标
};
```

目前提供的坐标点的数据类型为 int（Point2i）、int64（Point2l）、float（Point2f）、double（Point2d）几种类型。

另外，还有 3 个值的坐标，如 Point3i、Point3f、Point3d，其定义如下：

```cpp
typedef Point3_<int> Point3i;
typedef Point3_<float> Point3f;
typedef Point3_<double> Point3d;
```

3 个值的坐标不仅包括 x 和 y 坐标，还包括第三个值 z 坐标。

Point 类型的应用案例代码如下：

```cpp
#include <iostream>
#include "opencv2/opencv.hpp"

using namespace std;
using namespace cv;

int main()
{
    // 点构造，调用默认构造函数
    Point2i pt1;
    cout << "pt1: " << pt1 << endl;

    //点构造并初始化为[1.2, 2.3, 3.4]
    Point3f pt2(1.2f, 2.3f, 3.4f);
    cout << "pt2: " << pt2.x << ',' << pt2.y << ',' << pt2.z << endl;

    //点乘
    Point3f pt3(2.0, 3.0, 4.0);
    float dot_product = pt2.dot(pt3);
    cout << "dot product: " << dot_product << endl;

    //叉乘
```

```
    Point3f pt4(1, 0, 0);                    //x方向单位向量
    Point3f pt5(0, 1, 0);                    //y方向单位向量
    Point3f pt_cross = pt4.cross(pt5);       //计算结果为z方向单位向量
    cout << "cross product: " << pt_cross << endl;

    //判断点是否在矩阵内，该操作仅适用于二维点
    Rect2f rect(0, 0, 4, 5);                 //矩形定义
    Point2f pt6(3.0, 2.5);
    bool is_inside = pt6.inside(rect);
    if (is_inside)
    {
        cout << "Point pt4 is inside of Rectangle rect!" << endl;
    }
    return 0;
}
```

案例中调用默认构造函数创建点对象 pt1，pt2 由对象构造并初始化而产生，案例中还展示了点乘与叉乘计算，以及判断点是否处于矩形内的操作。案例的计算结果打印信息如图 3.12 所示。

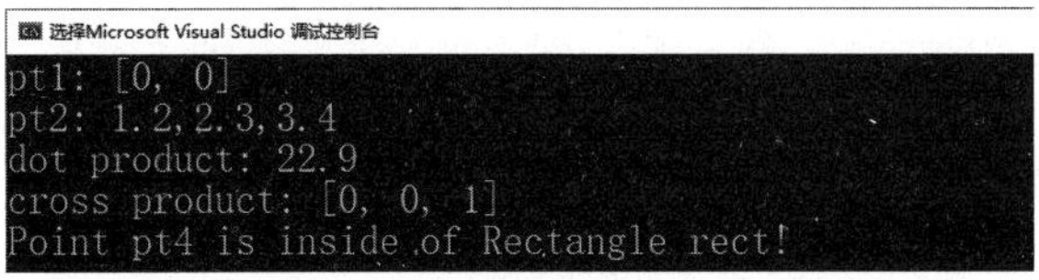

图 3.12

在 Python 语言中，Point 类型可以由 tuple 数据类型表示，如下案例代码展示了点的表示，以及利用 numpy 进行点乘和叉乘的计算：

```python
import numpy as np

#表示点
pt1 = (10, 20)
print("pt1 type: ", type(pt1))

#点乘
pt2 = (1, 2)
res = np.dot(pt1, pt2)
print("dot result: ", res)

#叉乘
pt3 = (1, 0, 0)
pt4 = (0, 1, 0)
pt5 = np.cross(pt3, pt4)
print("cross result: ", pt5)
```

执行上述代码，计算结果如图 3.13 所示。

```
D:\software\python38\python.exe D:/Code/OpenCV_Code/chapter3/3-1/Point.py
pt1 type:   <class 'tuple'>
dot result:  50
cross result:   [0 0 1]

Process finished with exit code 0
```

图 3.13

3.2.4　案例 14：Rect 结构

OpenCV 中定义了矩形的表示类 Rect，和 Point 类似，Rect 的定义如下：

```
typedef Rect_<int> Rect2i;
typedef Rect_<float> Rect2f;
typedef Rect_<double> Rect2d;
typedef Rect2i Rect;
```

如上定义了不同数据类型的矩形，有 int 类型的 Rect2i、float 类型的 Rect2f、double 类型的 Rect2d。

模板类 Rect_的定义如下：

```
template<typename _Tp> class Rect_
{
public:
    typedef _Tp value_type;

    //! 默认构造函数
    Rect_();
    Rect_(_Tp _x, _Tp _y, _Tp _width, _Tp _height);
    Rect_(const Rect_& r);
    Rect_(Rect_&& r) CV_NOEXCEPT;
    Rect_(const Point_<_Tp>& org, const Size_<_Tp>& sz);
    Rect_(const Point_<_Tp>& pt1, const Point_<_Tp>& pt2);

    Rect_& operator = ( const Rect_& r );
    Rect_& operator = ( Rect_&& r ) CV_NOEXCEPT;
    //! 定义左上角的点
    Point_<_Tp> tl() const;
    //! 定义右下角的点
    Point_<_Tp> br() const;

    //! 矩形尺寸（宽、高）
    Size_<_Tp> size() const;
    //! 矩形面积
    _Tp area() const;
```

```cpp
    //! 矩形是否为空
    bool empty() const;

    //! 数据类型转换
    template<typename _Tp2> operator Rect_<_Tp2>() const;

    //! 检查矩形是否包含点
    bool contains(const Point_<_Tp>& pt) const;

    _Tp x;      //!< 左上角点的 x 坐标
    _Tp y;      //!< 左上角点的 y 坐标
    _Tp width;  //!< 矩形宽度
    _Tp height; //!< 矩形高度
};
```

案例中展示了 Rect 对象的构造，Rect 顶点坐标、宽高、面积等信息的获取操作，以及矩形是否包含点的判断操作，案例代码如下：

```cpp
#include <iostream>
#include "opencv2/opencv.hpp"

using namespace std;
using namespace cv;

int main()
{
    // 矩形构造，调用默认构造函数
    Rect2i r1;
    if (r1.empty())         //判断矩形是否为空
    {
        cout << "Rect r1 is empty!" << endl;
    }

    //矩形信息输出
    Rect2i r2(10, 10, 20, 20);
    cout << "r2 top-left: " << r2.tl() << endl;              //左上角点
    cout << "r2 bottom-right: " << r2.br() << endl;          //右下角点
    cout << "r2 width: " << r2.width << endl;                //矩形宽度
    cout << "r2 height: " << r2.height << endl;              //矩形高度
    cout << "r2 area: " << r2.area() << endl;                //矩形面积

    //判断矩形 r2 是否包含点 pt1
    Point2i pt1(15, 15);
    if (r2.contains(pt1))
    {
        cout << "Rect r2 contains Point pt1!" << endl;
```

```
    }
    return 0;
}
```

执行结果如图 3.14 所示。

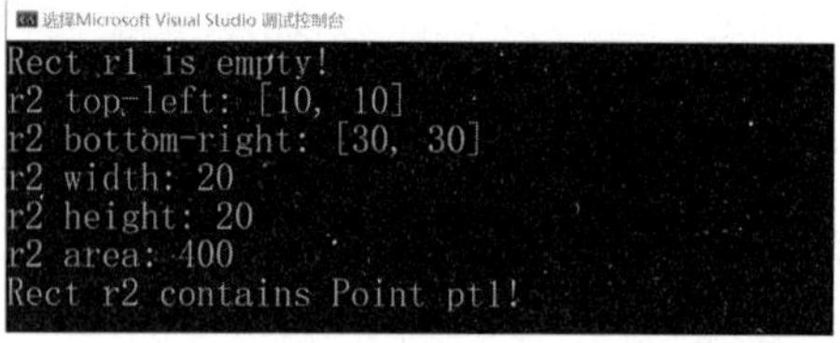

```
Rect r1 is empty!
r2 top-left: [10, 10]
r2 bottom-right: [30, 30]
r2 width: 20
r2 height: 20
r2 area: 400
Rect r2 contains Point pt1!
```

图 3.14

如图 3.14 所示，r1 构造后没有进行初始化，因此为空；r2 构造时传入了左上角的坐标(10, 10)，矩形宽度 20，矩形高度 20；点 pt1 在矩形 r2 内部，因此 r2.contains(pt1)返回为 true，打印操作的判断条件为真。

OpenCV Python 语言中的 Rect 是通过 tuple 类型的数据表示，即左上角点坐标(x,y)及矩形宽高 4 个值表示的 Rect 为(x, y, w, h)的形式，在需要传入 Rect 数据类型的地方，可以按照 tuple 类型的数据传入。

3.2.5 案例 15：Size 结构

OpenCV 的很多函数中都需要传入 Size 类型的参数，如 3.2.1 节中的 Mat 类的一个构造函数为 Mat(Size size, int type)，需要通过 Size 类型对象传入构造矩阵的尺寸。Size 类型对象的作用为指定图像或矩形的尺寸，包含两个成员变量 width 和 height。Size 类的定义如下：

```
template<typename _Tp> class Size_
{
public:
    typedef _Tp value_type;

    //! 默认构造函数
    Size_();
    Size_(_Tp _width, _Tp _height);
    Size_(const Size_& sz);
    Size_(Size_&& sz) CV_NOEXCEPT;
    Size_(const Point_<_Tp>& pt);

    Size_& operator = (const Size_& sz);
    Size_& operator = (Size_&& sz) CV_NOEXCEPT;
    //! 计算面积（width*height）
    _Tp area() const;
    //! 宽高比 (width/height)
```

```cpp
    double aspectRatio() const;
    //! 若为空则返回 true
    bool empty() const;

    //! 数据类型的转换
    template<typename _Tp2> operator Size_<_Tp2>() const;

    _Tp width; //!< 宽
    _Tp height; //!< 高
};

typedef Size_<int> Size2i;
typedef Size_<int64> Size2l;
typedef Size_<float> Size2f;
typedef Size_<double> Size2d;
typedef Size2i Size;
```

Size 类型的 C++代码调用案例如下：

```cpp
#include <iostream>
#include "opencv2/opencv.hpp"

using namespace std;
using namespace cv;

int main()
{
    // 调用默认构造函数
    Size size1;
    if (size1.empty())                                              //判断 Size 对象是否为空
    {
        cout << "size1 is empty!" << endl;
    }

    //Size 对象信息输出
    Size2i size2(10, 20);
    cout << "size2 width: " << size2.width << endl;                 //输出宽度
    cout << "size2 height: " << size2.height << endl;               //输出高度
    cout << "size2 area: " << size2.area() << endl;                 //输出面积
    cout << "size2 aspect ratio: " << size2.aspectRatio() << endl;  //输出宽高比
    return 0;
}
```

调用后的输出结果如图 3.15 所示。

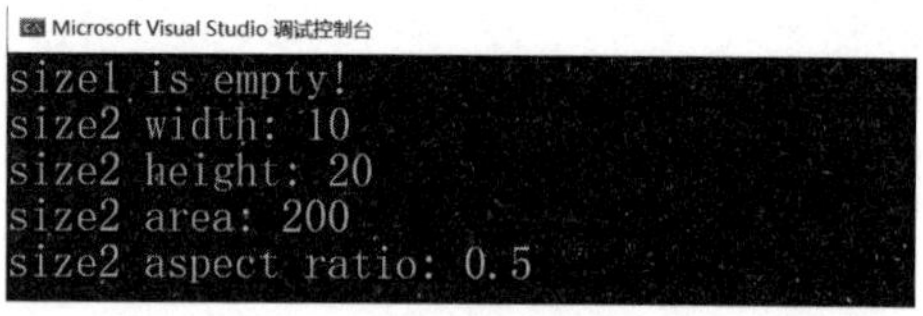

```
┌─────────────────────────────────┐
│ Microsoft Visual Studio 调试控制台 │
│ size1 is empty!                 │
│ size2 width: 10                 │
│ size2 height: 20                │
│ size2 area: 200                 │
│ size2 aspect ratio: 0.5         │
└─────────────────────────────────┘
```

图 3.15

如图 3.15 所示，案例中调用了 Size 对象的成员函数创建对象 size1，判断 size1 是否为空；然后创建了第二个对象 size2，输出了 size2 的宽度、高度、面积、宽高比。

OpenCV Python 语言中的 Size 对象是通过 tuple 类型的数据表示，即(w, h)的形式传入宽高数据的，在需要传入 Size 数据类型的地方，可以按照 tuple 类型的数据传入。

3.3 矩阵运算

core 模块中提供了 OpenCV 图像矩阵的基本运算，这些基本运算是复杂算法的基础，OpenCV 对这些基本运算都做了函数封装，如四则运算、位运算等。

3.3.1 案例 16：四则运算

OpenCV 中封装的四则运算函数分别为加法（add）、减法（subtract）、乘法（multiply）、除法（divide）。这 4 个函数的 Python 语言函数定义如下：

```
dst = add(src1, src2, dst=None, mask=None, dtype=None)
dst = subtract(src1, src2, dst=None, mask=None, dtype=None)
```

参数说明如下。

- src1：输入矩阵 1。
- src2：输入矩阵 2。
- dst：输出结果（返回值）
- mask：掩模，用于指定参与计算的矩阵。
- dtype：输出序列深度。

```
dst = multiply(src1, src2, dst=None, scale=None, dtype=None)
dst = divide(src1, src2, dst=None, scale=None, dtype=None)
```

参数说明如下。

- src1：输入矩阵 1。
- src2：输入矩阵 2。
- dst：输出结果（返回值）
- scale：比例因子。

- dtype：输出序列深度。

四则运算的案例代码如下：

```
import cv2
import numpy as np

#创建矩阵 m1 和 m2
m1 = np.array([[1.,2.,3.], [4.,5.,6.], [7.,8.,9.]])
m2 = np.array([[11.,12.,13.], [14.,15.,16.], [17.,18.,19.]])

#矩阵加法
m_add = cv2.add(m1, m2)
print("m1,m2 add result:\n ", m_add)
#矩阵减法
m_sub = cv2.subtract(m1, m2)
print("m1,m2 subtract result:\n ", m_sub)
#矩阵乘法
m_mul = cv2.multiply(m1, m2)
print("m1,m2 multiply result: \n", m_mul)
#矩阵除法
m_div = cv2.divide(m1, m2)
print("m1,m2 divide result: \n", m_div)
```

执行上述代码，结果如图 3.16 所示。

```
m1,m2 add result:
  [[12. 14. 16.]
 [18. 20. 22.]
 [24. 26. 28.]]
m1,m2 subtract result:
  [[-10. -10. -10.]
 [-10. -10. -10.]
 [-10. -10. -10.]]
m1,m2 multiply result:
[[ 11.  24.  39.]
 [ 56.  75.  96.]
 [119. 144. 171.]]
m1,m2 divide result:
[[0.09090909 0.16666667 0.23076923]
 [0.28571429 0.33333333 0.375      ]
 [0.41176471 0.44444444 0.47368421]]
```

图 3.16

对于上述四则运算，读者也可以直接通过运算符（+、-、*、/）来实现，代码如下：

```
#矩阵加法
m_add = m1 + m2
print("m1,m2 add result:\n ", m_add)
#矩阵减法
m_sub = m1 - m2
print("m1,m2 subtract result:\n ", m_sub)
#矩阵乘法
```

```
m_mul = m1 * m2
print("m1,m2 multiply result: \n", m_mul)
#矩阵除法
m_div = m1 / m2
print("m1,m2 divide result: \n", m_div)
```

直接使用运算符的计算结果与图 3.16 所示的结果相同。

3.3.2 案例 17：位运算

OpenCV 中定义了用于位运算的函数：bitwise_and（按位与运算），bitwise_or（按位或运算），bitwise_xor（异或运算），bitwise_not（非运算）。它们的 Python 语言函数定义如下：

```
dst = bitwise_and(src1, src2, dst=None, mask=None)
dst = bitwise_or(src1, src2, dst=None, mask=None)
dst = bitwise_xor(src1, src2, dst=None, mask=None)
```

参数说明如下。

- src1：输入矩阵 1。
- src2：输入矩阵 2。
- dst：输出结果（返回值）。
- mask：掩模，用于指定参与计算的矩阵。

```
dst = bitwise_not(src, dst=None, mask=None)
```

参数说明如下。

- src：输入矩阵。
- dst：输出结果（返回值）
- mask：掩模，用于指定参与计算的矩阵。

位运算的案例代码如下：

```
import cv2
import numpy as np

m1 = np.array([[1,2], [3,1]])
m2 = np.array([[5,3], [4,6]])

#按位与
m_and = cv2.bitwise_and(m1, m2)
print("m1,m2 bitwise_and result:\n ", m_and)
#按位或
m_or = cv2.bitwise_or(m1, m2)
print("m1,m2 bitwise_or result:\n ", m_or)
#异或运算
m_xor = cv2.bitwise_xor(m1, m2)
print("m1,m2 bitwise_xor result: \n", m_xor)
```

```
#非运算
m_not = cv2.bitwise_not(m1)
print("m1 bitwise_not result: \n", m_not)
```

上述案例代码的执行结果如图 3.17 所示。

```
m1,m2 bitwise_and result:
 [[1 2]
  [0 0]]
m1,m2 bitwise_or result:
 [[5 3]
  [7 7]]
m1,m2 bitwise_xor result:
 [[4 1]
  [7 7]]
m1 bitwise_not result:
 [[-2 -3]
  [-4 -2]]
```

图 3.17

在位运算中，各种运算的运算方法此处不做深入讲解（如按位与运算，当对应位均为 1 时，计算结果为 1，其他情况的结果均为 0），m1 和 m2 中的数字按照二进制的方式编码如下：

m1:[[001, 010], [011, 001]]
m2:[[101, 011], [100, 110]]

以按位与为例，若 m1 和 m2 对应位均为 1，则结果为 1；否则为 0。因此计算的 m_and 的二进制结果为[[001, 010], [000, 000]]，对应的十进制的结果为[[1, 2], [0, 0]]，与图 3.17 中的计算结果相同。同理，读者可以自行计算验证其他位运算的结果。

3.3.3 案例 18：代数运算

OpenCV 提供了计算矩阵均值的函数 mean，在调用 mean 函数时，每个通道的结果独立计算并返回。mean 函数的定义如下：

```
retval = mean(src, mask=None)
```

参数说明如下。
- src：输入矩阵。
- mask：掩模。
- retval：计算结果（返回值）。

mean 函数的使用案例如下：

```
import cv2
import numpy as np

m1 = np.array([[1,2],[3,4]])
mean_result = cv2.mean(m1)
```

```
print("m1 mean is: ", mean_result)
```
求均值案例的计算结果为:
```
m1 mean is:  (2.5, 0.0, 0.0, 0.0)
```
normalize 函数用于对矩阵做归一化,其定义如下:
```
dst = normalize(src, dst, alpha=None, beta=None, norm_type=None, dtype=None, mask=None)
```
参数说明如下。

- src:输入矩阵。
- dst:输出结果(返回值)。
- alpha:第一个参数,表示归一化的下界。
- beta:第二个参数,表示归一化的上界。
- norm_type:归一化类型,由 NormTypes 定义。
- dtype:输出类型。
- mask:掩模。

其中,NormTypes 的定义如下:
```
enum NormTypes {
            NORM_INF            = 1,
            NORM_L1             = 2,
            NORM_L2             = 4,
            NORM_L2SQR          = 5,
            NORM_HAMMING2       = 7,
            NORM_TYPE_MASK      = 7,
            NORM_RELATIVE       = 8,
            NORM_MINMAX         = 32
         };
```
normalize 函数中可以使用的归一化类型参数如下:
```
NORM_INF = 1        #归一化数组的 L∞范数
NORM_L1 = 2         #归一化数组的 L1 范数
NORM_L2 = 4         #归一化数组的 L2 范数
NORM_MINMAX = 32    #线性归一化
```
normalize 函数的使用案例如下:
```
import cv2
import numpy as np

m1 = np.array([[2.5, 7.5], [12.5, 10.0]])
dst = np.array([])
norm_result = cv2.normalize(m1, dst, norm_type=cv2.NORM_MINMAX)
print("m1 normalize result is: \n", norm_result)
```
归一化案例的输出结果为:
```
m1 normalize result is:
```

```
[[0.   0.5 ]
 [1.   0.75]]
```

知识点：归一化是机器学习和深度学习中很重要的一个技巧，可以减小数据之间的差异对模型训练结果的影响，还可以加速模型的收敛。

OpenCV中提供了用于计算平方根（sqrt）、幂运算（pow）、指数（exp）和对数运算（log）的函数，其定义如下：

`dst = sqrt(src, dst=None)`

参数说明如下。
- src：输入矩阵。
- dst：输出矩阵（返回值）。

`dst = pow(src, power, dst=None)`

参数说明如下。
- src：输入矩阵。
- power：幂次数。
- dst：输出矩阵（返回值）。

`dst = exp(src, dst=None)`

参数说明如下。
- src：输入矩阵。
- dst：输出矩阵（返回值）。

`dst = log(src, dst=None)`

参数说明如下。
- src：输入矩阵。
- dst：输出矩阵（返回值）。

案例代码如下：

```python
import cv2
import numpy as np

m1 = np.array([[1., 2.], [3., 4.]])
#平方根运算
sqrt_result = cv2.sqrt(m1)
print("m1 sqrt result is: \n", sqrt_result)
#幂运算
pow_result = cv2.pow(m1, 2)
print("m1 pow result is: \n", pow_result)
#指数运算
exp_result = cv2.exp(m1)
```

```
print("m1 exp result is: \n", exp_result)
#对数运算
log_result = cv2.log(m1)
print("m1 log result is: \n", log_result)
```

计算结果如下：

```
m1 sqrt result is:
 [[1.         1.41421356]
 [1.73205081 2.        ]]
m1 pow result is:
 [[ 1.  4.]
 [ 9. 16.]]
m1 exp result is:
 [[ 2.71828183  7.3890561 ]
 [20.08553692 54.59815003]]
m1 log result is:
 [[0.         0.69314718]
 [1.09861229 1.38629436]]
```

3.3.4 案例 19：比较运算

OpenCV 提供了矩阵比较运算函数，如两个矩阵的比较(compare)，求最大值(max)、最小值（min），排序（sort），其定义如下：

```
dst = compare(src1, src2, cmpop, dst=None)
```

参数说明如下。

- src1：输入矩阵 1。
- src2：输入矩阵 2。
- cmpop：比较方式。
- dst：输出结果（返回值）。

```
dst = max(src1, src2, dst=None)
dst = min(src1, src2, dst=None)
```

参数说明如下。

- src1：输入矩阵 1。
- src2：输入矩阵 2。
- dst：输出矩阵（返回值）。

```
minVal, maxVal, minLoc, maxLoc = minMaxLoc(src, mask=None)
```

参数说明如下。

- src：输入矩阵。
- mask：掩模。
- minVal：输出的最小值（返回值）。

- maxVal：输出的最大值（返回值）。
- minLoc：输出的最小值位置（返回值）。
- maxLoc：输出的最大值位置（返回值）。

```
dst = sort(src, flags, dst=None)
```

参数说明如下。
- src：输入矩阵。
- flags：排序方式，由 SortFlags 定义（见 3.1 节）。
- dst：输出排序结果。

比较两个矩阵的案例代码如下：

```
import cv2
import numpy as np

m1 = np.array([[6,2],[3,9]])
m2 = np.array([[5,2],[4,9]])

#矩阵比较
compare_result = cv2.compare(m1, m2, cv2.CMP_EQ)
print("m1,m2 compare result is:\n", compare_result)
```

比较运算的结果如下：

```
m1,m2 compare result is:
[[  0 255]
 [  0 255]]
```

compare 函数执行两个矩阵的逐元素比较操作，根据 CmpTypes 定义的比较方式，如果比较结果为 true，则输出矩阵中该位置的值为 255；否则为 0。compare 函数中的前两个参数为输入矩阵 src1、src2；第三个参数为 cmpop，该值由 CmpTypes 定义，表示比较方式，其取值如下：

```
CMP_EQ = 0          #src1 与 src2 相等
CMP_GE = 2          #src1 大于 src2
CMP_GT = 1          #src1 大于或等于 src2
CMP_LE = 4          #src1 小于 src2
CMP_LT = 3          #src1 小于或等于 src2
CMP_NE = 5          #src1 不等于 src2
```

求两个矩阵中的最大值和最小值的案例代码如下：

```
import cv2
import numpy as np

m1 = np.array([[6,2],[3,9]])
m2 = np.array([[5,2],[4,9]])
```

```
#求两个矩阵中的最大值
max_result = cv2.max(m1, m2)
print("max result in m1,m2 is:\n", max_result)

#求两个矩阵中的最小值
min_result = cv2.min(m1, m2)
print("min result in m1,m2 is:\n", min_result)
```

结果输出如下：

```
max result in m1,m2 is:
 [[6 2]
 [4 9]]
min result in m1,m2 is:
 [[5 2]
 [3 9]]
```

OpenCV 提供了获取矩阵中最小值和最大值及其位置的函数 minMaxLoc，使用的案例代码如下：

```
import cv2
import numpy as np

m1 = np.array([[6,2],[3,9]])
m2 = np.array([[5,2],[4,9]])

#求矩阵中的最小值和最大值及其位置
minmax_loc = cv2.minMaxLoc(m1)
print("min max location in m1 is:\n", minmax_loc)
```

输出结果如下：

```
min max location in m1 is:
 (2.0, 9.0, (1, 0), (1, 1))
```

minMaxLoc 函数调用后的返回值为 tuple 类型，其中第一个值为最小值，第二个值为最大值，第三个值为最小值的坐标，第四个值为最大值的坐标。

排序案例代码如下：

```
#升序排序
sort_result = cv2.sort(m1, cv2.SORT_ASCENDING)
print("sort ascending result:\n ", sort_result)
```

升序排序结果如下：

```
sort ascending result:
 [[2 6]
 [3 9]]
```

3.3.5 案例 20：特征值与特征向量

特征值和特征向量在机器学习中很常见，OpenCV 提供了函数 eigen，用于计算对称矩阵的特征值和特征向量，其定义如下：

```
retval, eigenvalues, eigenvectors = eigen(src, eigenvalues=None, eigenvectors=None)
```

参数说明如下。
- src：输入矩阵。
- eigenvalues：特征值（返回值）。
- eigenvectors：特征向量（返回值）。
- retval：计算状态（返回值）。

使用 eigen 计算特征值和特征向量的案例代码如下：

```
import cv2
import numpy as np

m1 = np.array([[1.,2.],[3.,4.]])
#求特征值和特征向量
eigen_result = cv2.eigen(m1)
print("m1 eigen result is:\n", eigen_result)
```

输出结果为：

```
m1 eigen result is:
 (True, array([[5.],
       [0.]]), array([[ 0.4472136 ,  0.89442719],
       [ 0.89442719, -0.4472136 ]]))
```

返回的 tuple 类型数据中的第一个值 True 表示可以计算特征值与特征向量，第二个值为计算得到的特征值，第三个值为计算得到的特征向量。

另外，OpenCV 中还有计算非对称矩阵的特征值和特征向量的函数 eigenNonSymmetric，其定义如下：

```
eigenvalues, eigenvectors = eigenNonSymmetric(src, eigenvalues=None, eigenvectors=None)
```

参数说明如下。
- src：输入矩阵。
- eigenvalues：特征值（返回值）。
- eigenvectors：特征向量（返回值）。

案例代码如下：

```
import cv2
import numpy as np
```

```
m1 = np.array([[1.,2.],[3.,4.]])
#计算非对称矩阵的特征值和特征向量
eigen_result1 = cv2.eigenNonSymmetric(m1)
print("m1 eigen result is:\n", eigen_result1)
```

计算结果如下：

```
m1 eigen result is:
 (array([[ 5.37228132],
       [-0.37228132]]), array([[-0.42222915, -0.92305231],
       [-0.82456484,  0.56576746]]))
```

3.3.6 案例 21：生成随机数矩阵

OpenCV 提供了生成随机数矩阵的函数 randn 和 randu。其中，randn 生成的矩阵服从正态分布，randu 生成的矩阵服从均匀分布，其定义如下：

```
dst = randn(dst, mean, stddev)
```

参数说明如下。

- dst：输出随机数矩阵（返回值）。
- mean：均值（期望）。
- stddev：生成的随机数的标准差。

```
dst = randu(dst, low, high)
```

参数说明如下。

- dst：输出随机数矩阵（返回值）。
- low：随机数生成下界。
- high：随机数生成上界。

randn 函数使用的案例代码如下：

```
import cv2
import numpy as np

#创建全 0 矩阵
dst = np.zeros((3,3), np.int8)
#生成服从正态分布的矩阵
cv2.randn(dst, 3, 1)
print("array generated by randn:\n", dst)
```

生成的随机矩阵如下：

```
array generated by randn:
 [[3 3 2]
 [3 4 3]
 [2 3 4]]
```

案例中调用函数 randn 传入 3 个参数：dst（存放生成的矩阵）、mean（生成矩阵的期望）、

stddev（生成矩阵的标准差）。

调用函数 randu 的案例代码如下：

```python
import cv2
import numpy as np

#创建全 0 矩阵
dst = np.zeros((3,3), np.int8)
#生成服从均匀分布的矩阵
cv2.randu(dst, 1, 10)
print("array generated by randu:\n", dst)
```

生成的矩阵如下：

```
array generated by randu:
 [[8 6 5]
 [3 5 8]
 [6 3 8]]
```

案例中调用函数 randu 传入 3 个参数：dst（存放生成的矩阵）、low（生成矩阵的值的下界）、high（生成矩阵的值的上界）。

3.4 矩阵变换

core 模块中提供了图像矩阵的基础变换操作，如矩阵转向量、通道分离与通道合并、傅里叶变换等，还定义了一些基础操作，如图像旋转、图像拼接、透视变换等。

3.4.1 案例 22：矩阵转向量

OpenCV 中提供了用于矩阵转向量的函数 reduce，该函数将矩阵的行/列视为一组一维向量，并对向量执行指定的操作（操作由 ReduceTypes 定义），直到获得单个行/列，从而将矩阵缩减为一个向量，该函数可用于计算光栅图像的水平和垂直投影。

reduce 函数的 Python 语言函数定义如下：

```python
dst = reduce(src, dim, rtype, dst=None, dtype=None)
```

该函数的参数说明如下。

- src：输入图像。
- dim：降维的维度索引，0 代表矩阵降维为单行，1 代表矩阵降维为单列。
- rtype：ReduceTypes 定义的降维操作。
- dst：输出 vector 类型（返回值）。
- dtype：输出 vector 类型。

其中，ReduceTypes 的定义如下：

```
REDUCE_AVG = 1          #输出为输入矩阵所有行/列的平均值
REDUCE_MAX = 2          #输出为输入矩阵所有行/列的最大值
REDUCE_MIN = 3          #输出为输入矩阵所有行/列的最小值
REDUCE_SUM = 0          #输出为输入矩阵所有行/列的和
```

reduce 函数的使用案例代码如下：

```
import cv2

src = cv2.imread("src.jpg")
print("src shape: ", src.shape)

#矩阵转向量操作
dst = cv2.reduce(src, 0, cv2.REDUCE_MAX)
print("dst shape: ", dst.shape)
```

执行后的结果如下：

```
src shape:  (512, 512, 3)
dst shape:  (1, 512, 3)
```

案例中 reduce 函数降维的维度为 0，即会将输入中的每一列取最大值，组成一个单行的 vector，每个通道单独计算，因此最终输出的维度为 1×512×3。

3.4.2 案例 23：通道分离与通道合并

对于一副 RGB 格式的图像，包括 R、G、B 3 个通道，OpenCV 中提供了将一幅图像的 3 个通道分离的函数 split，也提供了将 3 个单通道图像合并为一幅 3 通道图像的函数 merge。

split 函数的定义如下：

```
mv = split(m, mv=None)
```

参数说明如下。

- m：输入的多通道矩阵。
- mv：分离后的矩阵集合（返回值）。

split 使用案例如下：

```
import cv2

src = cv2.imread("src.jpg")
#颜色空间分离
b,g,r = cv2.split(src)
#显示每个通道的图像
cv2.imshow("b", b)
cv2.imshow("g", g)
cv2.imshow("r", r)
```

```
cv2.waitKey(0)
cv2.destroyAllWindows()
```

本案例使用的输入图像如图 3.10 所示，显示的每个通道的图像如图 3.18 所示。

图 3.18

merge 函数的定义如下：

```
dst = merge(mv, dst=None)
```

参数说明如下。

- mv：矩阵集合。
- dst：合并后的多通道矩阵结果（返回值）。

通道合并的案例如下：

```
import cv2

src = cv2.imread("src.jpg")
#颜色空间分离
b,g,r = cv2.split(src)
#通道合并
merge_result = cv2.merge([b,g,r])
#结果显示
cv2.imshow("merge_result", merge_result)
cv2.waitKey(0)
cv2.destroyAllWindows()
```

本案例先使用 split 函数将图像的通道分离，然后使用 merge 函数将分离的通道进行合并，结果如图 3.19 所示。

3.4.3 案例 24：图像旋转

OpenCV 中提供了将图像沿着坐标轴旋转的函数 flip，其定义如下：

```
dst = flip(src, flipCode, dst=None)
```

参数说明如下。

- src：输入图像。

图 3.19

- flipCode：矩阵旋转标志，flipCode 为 0 表示沿着 x 轴旋转，flipCode 为正数也表示沿着 x 轴旋转，flipCode 为负数表示沿着 x 轴和 y 轴旋转。
- dst：输出图像（返回值）。

flip 函数的使用案例如下：

```
import cv2

src = cv2.imread("src.jpg")
#沿 x 轴翻转
flip_x = cv2.flip(src, 0)
#沿 y 轴翻转
flip_y = cv2.flip(src, 1)
#沿 x 轴和 y 轴翻转
flip_xy = cv2.flip(src, -1)
#图像显示
cv2.imshow("flip_x", flip_x)
cv2.imshow("flip_y", flip_y)
cv2.imshow("flip_xy", flip_xy)
cv2.waitKey(0)
cv2.destroyAllWindows()
```

本案例使用的输入图像如图 3.10 所示，旋转后的结果如图 3.20 所示。

图 3.20

另外，OpenCV 中还提供了一种将图像按照角度旋转的函数 rotate，其定义如下：

```
dst = rotate(src, rotateCode, dst=None)
```

参数说明如下。

- src：输入图像。
- rotateCode：矩阵旋转标志，rotateCode 为 ROTATE_90_CLOCKWISE 表示顺时针旋转 90°，rotateCode 为 ROTATE_180 表示顺时针旋转 180°，rotateCode 为 ROTATE_90_COUNTERCLOCKWISE 表示逆时针旋转 90°。
- dst：输出图像（返回值）。

rotate 函数的案例代码如下：

```python
import cv2

src = cv2.imread("src.jpg")
#顺时针旋转 90°
rotate_90 = cv2.rotate(src, cv2.ROTATE_90_CLOCKWISE)
#顺时针旋转 180°
rotate_180 = cv2.rotate(src, cv2.ROTATE_180)
#顺时针旋转 270°，即逆时针旋转 90°
rotate_270 = cv2.rotate(src, cv2.ROTATE_90_COUNTERCLOCKWISE)
#图像显示
cv2.imshow("rotate_90", rotate_90)
cv2.imshow("rotate_180", rotate_180)
cv2.imshow("rotate_270", rotate_270)
cv2.waitKey(0)
cv2.destroyAllWindows()
```

本案例使用的输入图像如图 3.10 所示，旋转后的结果如图 3.21 所示。

图 3.21

3.4.4 案例 25：图像拼接

OpenCV 中提供了用于图像拼接的算法，其中，hconcat 函数是在水平方向上做图像拼接，vconcat 函数是在垂直方向上做图像拼接。

hconcat 函数的定义如下：

```
dst = hconcat(src, dst=None)
```

vconcat 函数的定义如下：

```
dst = vconcat(src, dst=None)
```

两个函数的参数相同，参数说明如下。

- src：输入图像或输入图像组，对于 hconcat 函数，图像组中的图像必须具有相同的高度和深度；对于 vconcat 函数，图像组中的图像必须具有相同的宽度和深度。

- dst：输出图像（返回值）。

水平和垂直两个方向上的图像拼接的案例代码如下：

```python
import cv2

src = cv2.imread("src.jpg")
src1 = cv2.imread("src1.jpg")
#水平拼接图像
hconcat_result = cv2.hconcat([src, src1])
#垂直拼接图像
vconcat_result = cv2.vconcat([src, src1])
#图像显示
cv2.imshow("hconcat_result", hconcat_result)
cv2.imshow("vconcat_result", vconcat_result)
cv2.waitKey(0)
cv2.destroyAllWindows()
```

水平方向上的图像拼接结果如图 3.22 所示，垂直方向上的图像拼接结果如图 3.23 所示。

图 3.22

图 3.23

3.4.5　案例 26：图像边界拓展

OpenCV 中提供了用于图像边界拓展（给图像增加边框）的函数 copyMakeBorder，其定义如下：

```
dst = copyMakeBorder(src, top, bottom, left, right, borderType, dst=None, value=None)
```

参数说明如下。

- src：输入图像。

- top：上边界边框的尺寸（单位为像素，下同）。
- bottom：下边界边框的尺寸。
- left：左边界边框的尺寸。
- right：右边界边框的尺寸。
- borderType：图像边界拓展策略，由 BorderTypes 定义。BorderTypes 的定义如下：

```
enum BorderTypes {
    BORDER_CONSTANT      = 0, //用指定像素值边界"i"
    BORDER_REPLICATE     = 1, //复制边界像素
    BORDER_REFLECT       = 2, //反射复制边界像素
    BORDER_WRAP          = 3, //用另一边的像素补偿填充
    BORDER_REFLECT_101   = 4, //以边界为对称轴反射复制边界
    BORDER_TRANSPARENT   = 5, //透明边界
    BORDER_REFLECT101    = BORDER_REFLECT_101, //与 BORDER_REFLECT_101 相同
    BORDER_DEFAULT       = BORDER_REFLECT_101, //与 BORDER_REFLECT_101 相同
    BORDER_ISOLATED      = 16 //不看 ROI 之外的部分
};
```

- dst：输出图像（返回值）。
- value：borderType 为 BORDER_CONSTANT 时的边框像素值。

图像边界拓展的案例代码如下：

```
import cv2

src = cv2.imread("src.jpg")
#边界拓展，边框类型为 BORDER_CONSTANT
border_constant = cv2.copyMakeBorder(src, 30, 30, 30, 30, cv2.BORDER_CONSTANT, value=88)
cv2.imshow("border_constant", border_constant)
cv2.waitKey(0)
cv2.destroyAllWindows()
```

案例中边界的拓展尺寸为 30 像素，边界拓展策略为常值方式，边界的像素值为 88，案例结果如图 3.24 所示。

若将边界拓展策略设置为 BORDER_REPLICATE，则代码如下：

```
import cv2

src = cv2.imread("src.jpg")
cv2.imshow("src", src)

#边界拓展，边框类型为 BORDER_REPLICATE
border_replicate = cv2.copyMakeBorder(src, 30, 30, 30, 30,
```

图 3.24

```
cv2.BORDER_REPLICATE)
cv2.imshow("border_replicate", border_replicate)
cv2.waitKey(0)
cv2.destroyAllWindows()
```

为了便于对比，将输入图像显示在左边，结果如图 3.25 所示，图中的拓展边界复制了图像边界的像素值。

图 3.25

其他的边界拓展策略此处就不做一一展示了，有兴趣的读者可以自行尝试。

3.4.6　案例 27：傅里叶变换

傅里叶变换在很多领域都有着重要的应用，在图像处理领域常用于图像增强、去噪、压缩编码等。

OpenCV 提供了实现离散傅里叶变换的函数 dft，其定义如下：

```
dst = dft(src, dst=None, flags=None, nonzeroRows=None)
```

参数说明如下。

- src：输入图像。
- dst：输出图像（返回值）。
- flags：变换标志。
- nonzeroRows：设置非零行。

dft 执行傅里叶变换的案例代码如下：

```
import cv2
import numpy as np

img = cv2.imread('src.jpg', 0)
#进行 float32 形式转换
```

```
img_fc1 = np.float32(img)
#执行傅里叶变换操作
dft_img = cv2.dft(img_fc1)
#结果显示
cv2.imshow("dft_img", dft_img)
cv2.waitKey(0)
cv2.destroyAllWindows()
```

傅里叶变换后的结果如图 3.26 所示。

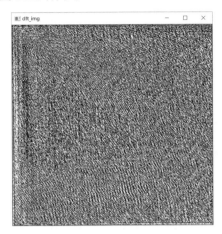

图 3.26

3.5 进阶必备：聊聊图像像素遍历与应用

3.5.1 案例 28：图像像素遍历

数字图像以矩阵的形式在内存中存储，因此，图像像素的遍历也是对矩阵的遍历，最简单的方法就是暴力法，即逐行、逐列、逐通道的遍历图像的每个像素。

本案例使用的输入图像如图 3.27 所示。

图 3.27

使用暴力法遍历，将图像颜色取反的案例代码如下：

```python
import cv2
import time

#颜色取反
def inverse_color(img):
    height = img.shape[0]
    width= img.shape[1]
    channels = img.shape[2]
    print("width : %s, height : %s, channel : %s" % (width, height, channels))

    for row in range(height):              #遍历高
        for col in range(width):           #遍历宽
            for c in range(channels):      #遍历通道
                px = img[row, col, c]
                #像素值取反
                img[row, col, c] = 255-px
    cv2.imshow("inverse_color", img)
    cv2.waitKey(0)

src = cv2.imread("img.jpg")
start = time.clock()
inverse_color(src)
end = time.clock()
print("Time is: %s seconds" % (end - start))
```

颜色取反效果如图 3.28 所示。

图 3.28

图像尺寸与耗时打印信息如下：

```
width : 723, height : 244, channel : 3
Time is: 1.4884088000000002 seconds
```

对于这种并非针对某一个特定像素做处理的操作，使用暴力法遍历太过耗时，而使用切片法访问则更为高效：

```python
import cv2
import time

src = cv2.imread("img.jpg")
```

```
start = time.clock()
#切片法遍历像素
src[:] = 255-src[:]
end = time.clock()
cv2.imshow("inverse_color", src)
cv2.waitKey(0)
print("Time is: %s seconds" % (end - start))
```

耗时打印信息如下：

```
Time is: 0.00101060000000000002 seconds
```

对比两种方法，发现耗时差距很大，对于更大的图像，这两种方式的差距更明显。

3.5.2 案例 29：提取拍照手写签名

生活中，经常遇到需要签名的场景，如果不能亲临现场签名，就需要将文件打印、签名、扫描后发送、再打印。还有一种简单的方式就是签名后传输，将电子签名插入，签名照片如图 3.29 所示。

插入文档后的图像如图 3.30 所示。

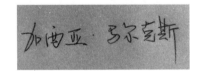
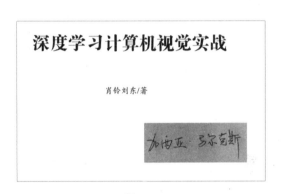

图 3.29　　　　　　　　　　　图 3.30

但是，有背景的签名会显得"很突出"，与原文档格格不入，一眼看去就知道不是手签的。仔细分析可以发现，这种签名图片的前景和背景比较单一均匀，如签名的文字的像素值在 (23,23,16) 附近，而背景的像素值在 (161,162,164) 附近，因此可以直观地判断前景 3 个通道的像素值都低于 100，而背景的像素值都高于 100，对于这种场景，可以简单地通过 OpenCV 工具对像素值的比较进行前景与背景的分离，将背景改为期望的颜色（如纯白色）。案例代码如下：

```
import cv2

#背景消除
def erase_background(img):
    height = img.shape[0]
```

```
        width = img.shape[1]
        channels = img.shape[2]
        print("width : %s, height : %s, channel : %s" % (width, height, channels))

        for row in range(height):            #遍历高
            for col in range(width):         #遍历宽
                for c in range(channels):    #遍历通道
                    #去除背景
                    if img[row, col, c] > 100:
                        #将背景变为纯白色
                        img[row, col, c] = 255
        cv2.imshow("erase_background", img)

src = cv2.imread("img1.jpg")
cv2.imshow("Picture", src)
erase_background(src)
cv2.waitKey(0)
```

去除背景后的结果如图 3.31 所示。

可以将文字的颜色设置为像素值 0，加深笔迹，即在像素遍历时增加如下代码做笔迹修改：

```
                    if img[row, col, c] < 100:
                        #将文字变为纯黑色
                        img[row, col, c] = 0
```

将去除背景并加深笔迹的签名插入文档的结果如图 3.32 所示。

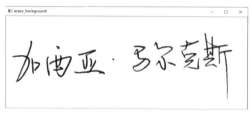

图 3.31   图 3.32

将去除背景的签名插入文档后与原文件完美融合，这种处理方式也可以拓展到给证件照换背景（如将红色背景证件照换为蓝色背景等）。

第 4 章
图像处理模块 imgproc（一）

imgproc 是 OpenCV 最核心的图像处理模块，该模块封装了常用的图像处理算法，这些算法是其他复杂算法的基础，如形态学变换、滤波、边缘检测等。读者在学习时，可以在熟悉本章案例使用方法的基础上逐步了解算法的原理，进一步提升灵活运用 OpenCV 进行图像处理的能力。

4.1 模块导读

imgproc 模块封装了众多的图像处理算法，如图像滤波、几何图像变换、图像绘制、颜色空间变换、直方图、结构分析和形状描述符、运动分析和目标追踪、特征检测、传统算法目标检测、传统算法图像分割等。

imgproc 模块的引用需要包含头文件"opencv2/imgproc.hpp"，该头文件中定义了 OpenCV 图像处理的函数，通过阅读该头文件，读者可以初步了解该模块封装了哪些图像处理功能。该头文件的定义与解读如下。

core 模块是 OpenCV 中的核心模块，因此所有模块的头文件都有对 core.hpp 的引用：

```
#ifndef OPENCV_IMGPROC_HPP
#define OPENCV_IMGPROC_HPP

#include "opencv2/core.hpp"
```

如下为图像处理中的一些标志的定义：

```
namespace cv
```

```
{
enum SpecialFilter {
    FILTER_SCHARR -1
};

//形态学运算类型
enum MorphTypes{
    MORPH_ERODE    = 0,
    MORPH_DILATE   = 1,
    MORPH_OPEN     = 2,
    MORPH_CLOSE    = 3,
    MORPH_GRADIENT = 4,
    MORPH_TOPHAT   = 5,
    MORPH_BLACKHAT = 6,
    MORPH_HITMISS  = 7
};

//结构元素的形状
enum MorphShapes {
    MORPH_RECT    = 0,
    MORPH_CROSS   = 1,
    MORPH_ELLIPSE = 2
};

//! 插值算法
enum InterpolationFlags{
    INTER_NEAREST       = 0,
    INTER_LINEAR        = 1,
    INTER_CUBIC         = 2,
    INTER_AREA          = 3,
    INTER_LANCZOS4      = 4,
    INTER_LINEAR_EXACT  = 5,
    INTER_NEAREST_EXACT = 6,
    INTER_MAX           = 7,
    WARP_FILL_OUTLIERS  = 8,
    WARP_INVERSE_MAP    = 16
};

/** 指定极坐标映射模式*/
enum WarpPolarMode
{
    WARP_POLAR_LINEAR = 0,
    WARP_POLAR_LOG = 256
};

//插值掩模
```

```
enum InterpolationMasks {
    INTER_BITS          = 5,
    INTER_BITS2         = INTER_BITS * 2,
    INTER_TAB_SIZE      = 1 << INTER_BITS,
    INTER_TAB_SIZE2     = INTER_TAB_SIZE * INTER_TAB_SIZE
};

//! 距离变换的距离类型与 M 估计
enum DistanceTypes {
    DIST_USER    = -1,
    DIST_L1      = 1,
    DIST_L2      = 2,
    DIST_C       = 3,
    DIST_L12     = 4,
    DIST_FAIR    = 5,
    DIST_WELSCH  = 6,
    DIST_HUBER   = 7
};

//! 距离变换掩模大小
enum DistanceTransformMasks {
    DIST_MASK_3        = 3, //!< mask=3
    DIST_MASK_5        = 5, //!< mask=5
    DIST_MASK_PRECISE  = 0  //!<
};

//! 阈值操作类型
enum ThresholdTypes {
    THRESH_BINARY      = 0,
    THRESH_BINARY_INV  = 1,
    THRESH_TRUNC       = 2,
    THRESH_TOZERO      = 3,
    THRESH_TOZERO_INV  = 4,
    THRESH_MASK        = 7,
    THRESH_OTSU        = 8,
    THRESH_TRIANGLE    = 16
};

//! 自适应阈值算法类型
enum AdaptiveThresholdTypes {
    ADAPTIVE_THRESH_MEAN_C      = 0,
    ADAPTIVE_THRESH_GAUSSIAN_C  = 1
};

//! GrabCut 算法中的像素类别
enum GrabCutClasses {
```

```cpp
    GC_BGD    = 0,
    GC_FGD    = 1,
    GC_PR_BGD = 2,
    GC_PR_FGD = 3
};

//! GrabCut 算法标志 flags
enum GrabCutModes {
    GC_INIT_WITH_RECT     = 0,
    GC_INIT_WITH_MASK     = 1,
    GC_EVAL               = 2,
    GC_EVAL_FREEZE_MODEL  = 3
};

//! distanceTransform 算法标志 flags
enum DistanceTransformLabelTypes {
    DIST_LABEL_CCOMP  = 0,
    DIST_LABEL_PIXEL  = 1
};

//! floodfill 算法标志 flags
enum FloodFillFlags {
    FLOODFILL_FIXED_RANGE = 1 << 16,
    FLOODFILL_MASK_ONLY   = 1 << 17
};

//! 连通组件统计
enum ConnectedComponentsTypes {
    CC_STAT_LEFT   = 0,
    CC_STAT_TOP    = 1,
    CC_STAT_WIDTH  = 2,
    CC_STAT_HEIGHT = 3,
    CC_STAT_AREA   = 4,
#ifndef CV_DOXYGEN
    CC_STAT_MAX    = 5
#endif
};

//! 连通组件算法
enum ConnectedComponentsAlgorithmsTypes {
    CCL_DEFAULT  = -1,
    CCL_WU       = 0,
    CCL_GRANA    = 1,
    CCL_BOLELLI  = 2,
    CCL_SAUF     = 3,
    CCL_BBDT     = 4,
```

```cpp
    CCL_SPAGHETTI = 5,
};

//! 轮廓检索算法模式
enum RetrievalModes {
    RETR_EXTERNAL  = 0,
    RETR_LIST      = 1,
    RETR_CCOMP     = 2,
    RETR_TREE      = 3,
    RETR_FLOODFILL = 4
};

//! 轮廓逼近算法
enum ContourApproximationModes {
    CHAIN_APPROX_NONE      = 1,
    CHAIN_APPROX_SIMPLE    = 2,
    CHAIN_APPROX_TC89_L1   = 3,
    CHAIN_APPROX_TC89_KCOS = 4
};

/** 形状匹配方式*/
enum ShapeMatchModes {
    CONTOURS_MATCH_I1   =1,
    CONTOURS_MATCH_I2   =2,
    CONTOURS_MATCH_I3   =3
};

//! Hough 变换的模式
enum HoughModes {
    HOUGH_STANDARD      = 0,
    HOUGH_PROBABILISTIC = 1,
    HOUGH_MULTI_SCALE   = 2,
    HOUGH_GRADIENT      = 3,
    HOUGH_GRADIENT_ALT  = 4,
};

//! 线性分割检测器模式
enum LineSegmentDetectorModes {
    LSD_REFINE_NONE= 0,
    LSD_REFINE_STD = 1,
    LSD_REFINE_ADV = 2
};

/**直方图比较方法*/
enum HistCompMethods {
    HISTCMP_CORREL            = 0,
```

```
    HISTCMP_CHISQR          = 1,
    HISTCMP_INTERSECT       = 2,
    HISTCMP_BHATTACHARYYA   = 3,
    HISTCMP_HELLINGER       = HISTCMP_BHATTACHARYYA,
    HISTCMP_CHISQR_ALT      = 4,
    HISTCMP_KL_DIV          = 5
};
```

如下为颜色空间变换方式的标志：

```
enum ColorConversionCodes {
    COLOR_BGR2BGRA       = 0,
    COLOR_RGB2RGBA       = COLOR_BGR2BGRA,

    COLOR_BGRA2BGR       = 1,
    COLOR_RGBA2RGB       = COLOR_BGRA2BGR,

    COLOR_BGR2RGBA       = 2,
    COLOR_RGB2BGRA       = COLOR_BGR2RGBA,

    COLOR_RGBA2BGR       = 3,
    COLOR_BGRA2RGB       = COLOR_RGBA2BGR,

    COLOR_BGR2RGB        = 4,
    COLOR_RGB2BGR        = COLOR_BGR2RGB,

    COLOR_BGRA2RGBA      = 5,
    COLOR_RGBA2BGRA      = COLOR_BGRA2RGBA,

    COLOR_BGR2GRAY       = 6,
    COLOR_RGB2GRAY       = 7,
    COLOR_GRAY2BGR       = 8,
    COLOR_GRAY2RGB       = COLOR_GRAY2BGR,
    COLOR_GRAY2BGRA      = 9,
    COLOR_GRAY2RGBA      = COLOR_GRAY2BGRA,
    COLOR_BGRA2GRAY      = 10,
    COLOR_RGBA2GRAY      = 11,

    COLOR_BGR2BGR565     = 12,
    COLOR_RGB2BGR565     = 13,
    COLOR_BGR5652BGR     = 14,
    COLOR_BGR5652RGB     = 15,
    COLOR_BGRA2BGR565    = 16,
    COLOR_RGBA2BGR565    = 17,
    COLOR_BGR5652BGRA    = 18,
    COLOR_BGR5652RGBA    = 19,
```

```
COLOR_GRAY2BGR565        = 20,
COLOR_BGR5652GRAY        = 21,

COLOR_BGR2BGR555         = 22,
COLOR_RGB2BGR555         = 23,
COLOR_BGR5552BGR         = 24,
COLOR_BGR5552RGB         = 25,
COLOR_BGRA2BGR555        = 26,
COLOR_RGBA2BGR555        = 27,
COLOR_BGR5552BGRA        = 28,
COLOR_BGR5552RGBA        = 29,

COLOR_GRAY2BGR555        = 30,
COLOR_BGR5552GRAY        = 31,

COLOR_BGR2XYZ            = 32,
COLOR_RGB2XYZ            = 33,
COLOR_XYZ2BGR            = 34,
COLOR_XYZ2RGB            = 35,

COLOR_BGR2YCrCb          = 36,
COLOR_RGB2YCrCb          = 37,
COLOR_YCrCb2BGR          = 38,
COLOR_YCrCb2RGB          = 39,

COLOR_BGR2HSV            = 40,
COLOR_RGB2HSV            = 41,

COLOR_BGR2Lab            = 44,
COLOR_RGB2Lab            = 45,

COLOR_BGR2Luv            = 50,
COLOR_RGB2Luv            = 51,
COLOR_BGR2HLS            = 52,
COLOR_RGB2HLS            = 53,

COLOR_HSV2BGR            = 54,
COLOR_HSV2RGB            = 55,

COLOR_Lab2BGR            = 56,
COLOR_Lab2RGB            = 57,
COLOR_Luv2BGR            = 58,
COLOR_Luv2RGB            = 59,
COLOR_HLS2BGR            = 60,
COLOR_HLS2RGB            = 61,

COLOR_BGR2HSV_FULL       = 66,
```

```
    COLOR_RGB2HSV_FULL     = 67,
    COLOR_BGR2HLS_FULL     = 68,
    COLOR_RGB2HLS_FULL     = 69,

    COLOR_HSV2BGR_FULL     = 70,
    COLOR_HSV2RGB_FULL     = 71,
    COLOR_HLS2BGR_FULL     = 72,
    COLOR_HLS2RGB_FULL     = 73,

    COLOR_LBGR2Lab         = 74,
    COLOR_LRGB2Lab         = 75,
    COLOR_LBGR2Luv         = 76,
    COLOR_LRGB2Luv         = 77,

    COLOR_Lab2LBGR         = 78,
    COLOR_Lab2LRGB         = 79,
    COLOR_Luv2LBGR         = 80,
    COLOR_Luv2LRGB         = 81,

    //RGB/BGR 与 YUV 之间转换
    COLOR_BGR2YUV          = 82,
    COLOR_RGB2YUV          = 83,
    COLOR_YUV2BGR          = 84,
    COLOR_YUV2RGB          = 85,

    //YUV 4:2:0 转 RGB
    COLOR_YUV2RGB_NV12     = 90,
    COLOR_YUV2BGR_NV12     = 91,
    COLOR_YUV2RGB_NV21     = 92,
    COLOR_YUV2BGR_NV21     = 93,
    COLOR_YUV420sp2RGB     = COLOR_YUV2RGB_NV21,
    COLOR_YUV420sp2BGR     = COLOR_YUV2BGR_NV21,

    COLOR_YUV2RGBA_NV12    = 94,
    COLOR_YUV2BGRA_NV12    = 95,
    COLOR_YUV2RGBA_NV21    = 96,
    COLOR_YUV2BGRA_NV21    = 97,
    COLOR_YUV420sp2RGBA    = COLOR_YUV2RGBA_NV21,
    COLOR_YUV420sp2BGRA    = COLOR_YUV2BGRA_NV21,

    COLOR_YUV2RGB_YV12     = 98,
    COLOR_YUV2BGR_YV12     = 99,
    COLOR_YUV2RGB_IYUV     = 100,
    COLOR_YUV2BGR_IYUV     = 101,
    COLOR_YUV2RGB_I420     = COLOR_YUV2RGB_IYUV,
    COLOR_YUV2BGR_I420     = COLOR_YUV2BGR_IYUV,
    COLOR_YUV420p2RGB      = COLOR_YUV2RGB_YV12,
```

```
COLOR_YUV420p2BGR     = COLOR_YUV2BGR_YV12,

COLOR_YUV2RGBA_YV12 = 102,
COLOR_YUV2BGRA_YV12 = 103,
COLOR_YUV2RGBA_IYUV = 104,
COLOR_YUV2BGRA_IYUV = 105,
COLOR_YUV2RGBA_I420 = COLOR_YUV2RGBA_IYUV,
COLOR_YUV2BGRA_I420 = COLOR_YUV2BGRA_IYUV,
COLOR_YUV420p2RGBA  = COLOR_YUV2RGBA_YV12,
COLOR_YUV420p2BGRA  = COLOR_YUV2BGRA_YV12,

COLOR_YUV2GRAY_420   = 106,
COLOR_YUV2GRAY_NV21 = COLOR_YUV2GRAY_420,
COLOR_YUV2GRAY_NV12 = COLOR_YUV2GRAY_420,
COLOR_YUV2GRAY_YV12 = COLOR_YUV2GRAY_420,
COLOR_YUV2GRAY_IYUV = COLOR_YUV2GRAY_420,
COLOR_YUV2GRAY_I420 = COLOR_YUV2GRAY_420,
COLOR_YUV420sp2GRAY = COLOR_YUV2GRAY_420,
COLOR_YUV420p2GRAY  = COLOR_YUV2GRAY_420,

//YUV 4:2:2 转 RGB
COLOR_YUV2RGB_UYVY   = 107,
COLOR_YUV2BGR_UYVY   = 108,
//COLOR_YUV2RGB_VYUY = 109,
//COLOR_YUV2BGR_VYUY = 110,
COLOR_YUV2RGB_Y422 = COLOR_YUV2RGB_UYVY,
COLOR_YUV2BGR_Y422 = COLOR_YUV2BGR_UYVY,
COLOR_YUV2RGB_UYNV = COLOR_YUV2RGB_UYVY,
COLOR_YUV2BGR_UYNV = COLOR_YUV2BGR_UYVY,

COLOR_YUV2RGBA_UYVY = 111,
COLOR_YUV2BGRA_UYVY = 112,
//COLOR_YUV2RGBA_VYUY= 113,
//COLOR_YUV2BGRA_VYUY= 114,
COLOR_YUV2RGBA_Y422 = COLOR_YUV2RGBA_UYVY,
COLOR_YUV2BGRA_Y422 = COLOR_YUV2BGRA_UYVY,
COLOR_YUV2RGBA_UYNV = COLOR_YUV2RGBA_UYVY,
COLOR_YUV2BGRA_UYNV = COLOR_YUV2BGRA_UYVY,

COLOR_YUV2RGB_YUY2   = 115,
COLOR_YUV2BGR_YUY2   = 116,
COLOR_YUV2RGB_YVYU   = 117,
COLOR_YUV2BGR_YVYU   = 118,
COLOR_YUV2RGB_YUYV  = COLOR_YUV2RGB_YUY2,
COLOR_YUV2BGR_YUYV  = COLOR_YUV2BGR_YUY2,
COLOR_YUV2RGB_YUNV  = COLOR_YUV2RGB_YUY2,
COLOR_YUV2BGR_YUNV  = COLOR_YUV2BGR_YUY2,
```

```
COLOR_YUV2RGBA_YUY2 = 119,
COLOR_YUV2BGRA_YUY2 = 120,
COLOR_YUV2RGBA_YVYU = 121,
COLOR_YUV2BGRA_YVYU = 122,
COLOR_YUV2RGBA_YUYV = COLOR_YUV2RGBA_YUY2,
COLOR_YUV2BGRA_YUYV = COLOR_YUV2BGRA_YUY2,
COLOR_YUV2RGBA_YUNV = COLOR_YUV2RGBA_YUY2,
COLOR_YUV2BGRA_YUNV = COLOR_YUV2BGRA_YUY2,

COLOR_YUV2GRAY_UYVY = 123,
COLOR_YUV2GRAY_YUY2 = 124,
//CV_YUV2GRAY_VYUY  = CV_YUV2GRAY_UYVY,
COLOR_YUV2GRAY_Y422 = COLOR_YUV2GRAY_UYVY,
COLOR_YUV2GRAY_UYNV = COLOR_YUV2GRAY_UYVY,
COLOR_YUV2GRAY_YVYU = COLOR_YUV2GRAY_YUY2,
COLOR_YUV2GRAY_YUYV = COLOR_YUV2GRAY_YUY2,
COLOR_YUV2GRAY_YUNV = COLOR_YUV2GRAY_YUY2,

//alpha 预乘
COLOR_RGBA2mRGBA    = 125,
COLOR_mRGBA2RGBA    = 126,

//! RGB 转 YUV 4:2:0
COLOR_RGB2YUV_I420  = 127,
COLOR_BGR2YUV_I420  = 128,
COLOR_RGB2YUV_IYUV  = COLOR_RGB2YUV_I420,
COLOR_BGR2YUV_IYUV  = COLOR_BGR2YUV_I420,

COLOR_RGBA2YUV_I420 = 129,
COLOR_BGRA2YUV_I420 = 130,
COLOR_RGBA2YUV_IYUV = COLOR_RGBA2YUV_I420,
COLOR_BGRA2YUV_IYUV = COLOR_BGRA2YUV_I420,
COLOR_RGB2YUV_YV12  = 131,
COLOR_BGR2YUV_YV12  = 132,
COLOR_RGBA2YUV_YV12 = 133,
COLOR_BGRA2YUV_YV12 = 134,

//! 马赛克 (Demosaicing)
COLOR_BayerBG2BGR   = 46,
COLOR_BayerGB2BGR   = 47,
COLOR_BayerRG2BGR   = 48,
COLOR_BayerGR2BGR   = 49,

COLOR_BayerBG2RGB   = COLOR_BayerRG2BGR,
COLOR_BayerGB2RGB   = COLOR_BayerGR2BGR,
COLOR_BayerRG2RGB   = COLOR_BayerBG2BGR,
```

```
    COLOR_BayerGR2RGB      = COLOR_BayerGB2BGR,

    COLOR_BayerBG2GRAY     = 86,
    COLOR_BayerGB2GRAY     = 87,
    COLOR_BayerRG2GRAY     = 88,
    COLOR_BayerGR2GRAY     = 89,

    //! 使用 VNG 去马赛克
    COLOR_BayerBG2BGR_VNG  = 62,
    COLOR_BayerGB2BGR_VNG  = 63,
    COLOR_BayerRG2BGR_VNG  = 64,
    COLOR_BayerGR2BGR_VNG  = 65,

    COLOR_BayerBG2RGB_VNG  = COLOR_BayerRG2BGR_VNG,
    COLOR_BayerGB2RGB_VNG  = COLOR_BayerGR2BGR_VNG,
    COLOR_BayerRG2RGB_VNG  = COLOR_BayerBG2BGR_VNG,
    COLOR_BayerGR2RGB_VNG  = COLOR_BayerGB2BGR_VNG,

    //! 边缘感知马赛克
    COLOR_BayerBG2BGR_EA   = 135,
    COLOR_BayerGB2BGR_EA   = 136,
    COLOR_BayerRG2BGR_EA   = 137,
    COLOR_BayerGR2BGR_EA   = 138,

    COLOR_BayerBG2RGB_EA   = COLOR_BayerRG2BGR_EA,
    COLOR_BayerGB2RGB_EA   = COLOR_BayerGR2BGR_EA,
    COLOR_BayerRG2RGB_EA   = COLOR_BayerBG2BGR_EA,
    COLOR_BayerGR2RGB_EA   = COLOR_BayerGB2BGR_EA,

    //! 带 alpha 通道的马赛克
    COLOR_BayerBG2BGRA     = 139,
    COLOR_BayerGB2BGRA     = 140,
    COLOR_BayerRG2BGRA     = 141,
    COLOR_BayerGR2BGRA     = 142,

    COLOR_BayerBG2RGBA     = COLOR_BayerRG2BGRA,
    COLOR_BayerGB2RGBA     = COLOR_BayerGR2BGRA,
    COLOR_BayerRG2RGBA     = COLOR_BayerBG2BGRA,
    COLOR_BayerGR2RGBA     = COLOR_BayerGB2BGRA,

    COLOR_COLORCVT_MAX     = 143
};
```

如下为图像处理的一些标志定义：

```
//! 矩形之间的相交类型
enum RectanglesIntersectTypes {
    INTERSECT_NONE      = 0, //!< 不相交
```

```
    INTERSECT_PARTIAL    = 1, //!< 局部相交
    INTERSECT_FULL       = 2  //!< 一个被另一个包含
};

/** 线类型*/
enum LineTypes {
    FILLED    = -1,
    LINE_4    = 4,  //!< 4 连通线
    LINE_8    = 8,  //!< 8 连通线
    LINE_AA   = 16 //!< 抗锯齿线
};

/** 只有 Hershey 字体的子集支持*/
enum HersheyFonts {
    FONT_HERSHEY_SIMPLEX          = 0, //!< 正常字号 sans-serif 字体
    FONT_HERSHEY_PLAIN            = 1, //!< 小号 sans-serif 字体
    FONT_HERSHEY_DUPLEX           = 2, //!< 正常字号 sans-serif 字体
    FONT_HERSHEY_COMPLEX          = 3, //!< 正常字号 serif font 字体
    FONT_HERSHEY_TRIPLEX          = 4, //!< 正常字号 serif font 字体
    FONT_HERSHEY_COMPLEX_SMALL    = 5, //!< FONT_HERSHEY_COMPLEX 较小版本
    FONT_HERSHEY_SCRIPT_SIMPLEX   = 6, //!< 手写体
    FONT_HERSHEY_SCRIPT_COMPLEX   = 7, //!< 更加复杂的 FONT_HERSHEY_SCRIPT_SIMPLEX
    FONT_ITALIC                   = 16 //!< italic 字体
};

/** cv::drawMarker 掩模类型的可用设置*/
enum MarkerTypes
{
    MARKER_CROSS          = 0,  //!< 十字线标记
    MARKER_TILTED_CROSS   = 1,  //!< 45°倾斜的十字线标记
    MARKER_STAR           = 2,  //!< 星形标记，十字和斜十字的组合
    MARKER_DIAMOND        = 3,  //!< 钻石标记
    MARKER_SQUARE         = 4,  //!< 正方形标记
    MARKER_TRIANGLE_UP    = 5,  //!< 向上的三角形标记
    MARKER_TRIANGLE_DOWN  = 6  //!< 向下的三角形标记
};
```

如下为 GeneralizedHough 的定义，该类利用广义 Hough 变换在灰度图像中寻找任意模板：

```
class CV_EXPORTS_W GeneralizedHough : public Algorithm
{
public:
    //! 设置检索模板
    CV_WRAP virtual void setTemplate(InputArray templ, Point templCenter = Point(-1, -1)) = 0;
    CV_WRAP virtual void setTemplate(InputArray edges, InputArray dx, InputArray dy, Point
```

```cpp
    templCenter = Point(-1, -1)) = 0;

    //! 在图像中寻找模板
    CV_WRAP virtual void detect(InputArray image, OutputArray positions, OutputArray votes = noArray()) = 0;
    CV_WRAP virtual void detect(InputArray edges, InputArray dx, InputArray dy, OutputArray positions, OutputArray votes = noArray()) = 0;

    //! 设置 Canny 算法下限阈值
    CV_WRAP virtual void setCannyLowThresh(int cannyLowThresh) = 0;
    CV_WRAP virtual int getCannyLowThresh() const = 0;

    //! 设置 Canny 算法上限阈值
    CV_WRAP virtual void setCannyHighThresh(int cannyHighThresh) = 0;
    CV_WRAP virtual int getCannyHighThresh() const = 0;

    //! 设置检测物体的最小中心距离
    CV_WRAP virtual void setMinDist(double minDist) = 0;
    CV_WRAP virtual double getMinDist() const = 0;

    //! 设置累加器分辨率与图像分辨率的反比
    CV_WRAP virtual void setDp(double dp) = 0;
    CV_WRAP virtual double getDp() const = 0;

    //! 设置内部最大 buffer 大小
    CV_WRAP virtual void setMaxBufferSize(int maxBufferSize) = 0;
    CV_WRAP virtual int getMaxBufferSize() const = 0;
};
```

如下为 GeneralizedHoughBallard 的定义，该类利用广义 Hough 变换在灰度图像中寻找任意模板，仅检测位置而不进行平移和旋转：

```cpp
class CV_EXPORTS_W GeneralizedHoughBallard : public GeneralizedHough
{
public:
    //! 设置 R-Table 层级
    CV_WRAP virtual void setLevels(int levels) = 0;
    CV_WRAP virtual int getLevels() const = 0;

    //! 设置检测阶段模板中心的累加器阈值，该值越小，检测到的假位置就越多
    CV_WRAP virtual void setVotesThreshold(int votesThreshold) = 0;
    CV_WRAP virtual int getVotesThreshold() const = 0;
};
```

如下为 GeneralizedHoughGuil 的定义，该类利用广义 Hough 变换在灰度图像中寻找任

意模板，检测位置、平移和旋转：

```
class CV_EXPORTS_W GeneralizedHoughGuil : public GeneralizedHough
{
public:
    //! 设置特征中两点之间的角度差
    CV_WRAP virtual void setXi(double xi) = 0;
    CV_WRAP virtual double getXi() const = 0;

    //! 特征表级别
    CV_WRAP virtual void setLevels(int levels) = 0;
    CV_WRAP virtual int getLevels() const = 0;

    //! 设置最大可视为相等角度的最大差值
    CV_WRAP virtual void setAngleEpsilon(double angleEpsilon) = 0;
    CV_WRAP virtual double getAngleEpsilon() const = 0;

    //! 设置检测的最小旋转角度
    CV_WRAP virtual void setMinAngle(double minAngle) = 0;
    CV_WRAP virtual double getMinAngle() const = 0;

    //! 设置检测的最大旋转角度
    CV_WRAP virtual void setMaxAngle(double maxAngle) = 0;
    CV_WRAP virtual double getMaxAngle() const = 0;

    //! 设置角度步长
    CV_WRAP virtual void setAngleStep(double angleStep) = 0;
    CV_WRAP virtual double getAngleStep() const = 0;

    //! 设置角度阈值
    CV_WRAP virtual void setAngleThresh(int angleThresh) = 0;
    CV_WRAP virtual int getAngleThresh() const = 0;

    //! 设置最小检测范围
    CV_WRAP virtual void setMinScale(double minScale) = 0;
    CV_WRAP virtual double getMinScale() const = 0;

    //! 设置最大检测范围
    CV_WRAP virtual void setMaxScale(double maxScale) = 0;
    CV_WRAP virtual double getMaxScale() const = 0;

    //! 设置范围步长
    CV_WRAP virtual void setScaleStep(double scaleStep) = 0;
    CV_WRAP virtual double getScaleStep() const = 0;

    //! 设置范围投票阈值
```

```cpp
    CV_WRAP virtual void setScaleThresh(int scaleThresh) = 0;
    CV_WRAP virtual int getScaleThresh() const = 0;

    //! 设置位置投票阈值
    CV_WRAP virtual void setPosThresh(int posThresh) = 0;
    CV_WRAP virtual int getPosThresh() const = 0;
};
```

如下为 CLAHE 的定义，该类用于对比度受限自适应直方图均衡化的基类：

```cpp
class CV_EXPORTS_W CLAHE : public Algorithm
{
public:
    /** 使用对比度受限的自适应直方图均衡化来均衡灰度图像的直方图*/
    CV_WRAP virtual void apply(InputArray src, OutputArray dst) = 0;
    CV_WRAP virtual void setClipLimit(double clipLimit) = 0;
    CV_WRAP virtual double getClipLimit() const = 0;
    CV_WRAP virtual void setTilesGridSize(Size tileGridSize) = 0;
    CV_WRAP virtual Size getTilesGridSize() const = 0;
    CV_WRAP virtual void collectGarbage() = 0;
};
```

如下为 Subdiv2D 类的定义：

```cpp
class CV_EXPORTS_W Subdiv2D
{
public:
    /** Subdiv 二维点定位类型*/
    enum { PTLOC_ERROR         = -2,    //!< 定位错误
           PTLOC_OUTSIDE_RECT  = -1,    //!< 定位在 subdivision 边界矩形之外
           PTLOC_INSIDE        = 0,     //!< 定位在某些小平面内部
           PTLOC_VERTEX        = 1,     //!< 定位正好在某个 subdivision 顶点
           PTLOC_ON_EDGE       = 2      //!< 定位在某个边缘
         };

    /** Subdiv2D 边缘类型*/
    enum { NEXT_AROUND_ORG   = 0x00,
           NEXT_AROUND_DST   = 0x22,
           PREV_AROUND_ORG   = 0x11,
           PREV_AROUND_DST   = 0x33,
           NEXT_AROUND_LEFT  = 0x13,
           NEXT_AROUND_RIGHT = 0x31,
           PREV_AROUND_LEFT  = 0x20,
           PREV_AROUND_RIGHT = 0x02
         };

    /** 构造函数*/
    CV_WRAP Subdiv2D();
```

```cpp
    CV_WRAP Subdiv2D(Rect rect);

    CV_WRAP void initDelaunay(Rect rect);
    CV_WRAP int insert(Point2f pt);
    CV_WRAP void insert(const std::vector<Point2f>& ptvec);
    CV_WRAP int locate(Point2f pt, CV_OUT int& edge, CV_OUT int& vertex);
    CV_WRAP int findNearest(Point2f pt, CV_OUT Point2f* nearestPt = 0);
    CV_WRAP void getEdgeList(CV_OUT std::vector<Vec4f>& edgeList) const;
    CV_WRAP void getLeadingEdgeList(CV_OUT std::vector<int>& leadingEdgeList) const;
    CV_WRAP void getTriangleList(CV_OUT std::vector<Vec6f>& triangleList) const;
    CV_WRAP void getVoronoiFacetList(const std::vector<int>& idx,
                                     CV_OUT std::vector<std::vector<Point2f> >& facetList,
                                     CV_OUT std::vector<Point2f>& facetCenters);
    CV_WRAP Point2f getVertex(int vertex, CV_OUT int* firstEdge = 0) const;
    CV_WRAP int getEdge( int edge, int nextEdgeType ) const;
    CV_WRAP int nextEdge(int edge) const;
    CV_WRAP int rotateEdge(int edge, int rotate) const;
    CV_WRAP int symEdge(int edge) const;
    CV_WRAP int edgeDst(int edge, CV_OUT Point2f* dstpt = 0) const;

protected:
    int newEdge();
    void deleteEdge(int edge);
    int newPoint(Point2f pt, bool isvirtual, int firstEdge = 0);
    void deletePoint(int vtx);
    void setEdgePoints( int edge, int orgPt, int dstPt );
    void splice( int edgeA, int edgeB );
    int connectEdges( int edgeA, int edgeB );
    void swapEdges( int edge );
    int isRightOf(Point2f pt, int edge) const;
    void calcVoronoi();
    void clearVoronoi();
    void checkSubdiv() const;

    struct CV_EXPORTS Vertex
    {
        Vertex();
        Vertex(Point2f pt, bool _isvirtual, int _firstEdge=0);
        bool isvirtual() const;
        bool isfree() const;

        int firstEdge;
        int type;
        Point2f pt;
    };

    struct CV_EXPORTS QuadEdge
```

```cpp
{
    QuadEdge();
    QuadEdge(int edgeidx);
    bool isfree() const;

    int next[4];
    int pt[4];
};

//! 存储所有的顶点
std::vector<Vertex> vtx;
//! 存储所有的边缘
std::vector<QuadEdge> qedges;
int freeQEdge;
int freePoint;
bool validGeometry;

int recentEdge;
//! 边界矩形的左上角
Point2f topLeft;
//! 边界矩形的右下角
Point2f bottomRight;
};
```

如下为 LineSegmentDetector 的定义，该类为线性分割检测器：

```cpp
class CV_EXPORTS_W LineSegmentDetector : public Algorithm
{
public:
    /** 输入图像中的检测线*/
    CV_WRAP virtual void detect(InputArray _image, OutputArray _lines,
                    OutputArray width = noArray(), OutputArray prec = noArray(),
                    OutputArray nfa = noArray()) = 0;

    /** 在指定图像上绘制分割线*/
    CV_WRAP virtual void drawSegments(InputOutputArray _image, InputArray lines) = 0;

    /** 以蓝色和红色绘制两组线，计算不重叠（不匹配）的像素*/
    CV_WRAP virtual int compareSegments(const Size& size, InputArray lines1, InputArray lines2,
InputOutputArray _image = noArray()) = 0;

    virtual ~LineSegmentDetector() { }
};
```

如下为图像处理函数的定义：

```cpp
/** 创建指向 LineSegmentDetector 对象的智能指针并对其进行初始化 */
CV_EXPORTS_W Ptr<LineSegmentDetector> createLineSegmentDetector(
```

```
        int _refine = LSD_REFINE_STD, double _scale = 0.8,
        double _sigma_scale = 0.6, double _quant = 2.0, double _ang_th = 22.5,
        double _log_eps = 0, double _density_th = 0.7, int _n_bins = 1024);

/** 返回高斯滤波器系数*/
CV_EXPORTS_W Mat getGaussianKernel( int ksize, double sigma, int ktype = CV_64F );

/** 返回用于计算空间图像导数的滤波器系数*/
CV_EXPORTS_W void getDerivKernels( OutputArray kx, OutputArray ky,
                                   int dx, int dy, int ksize,
                                   bool normalize = false, int ktype = CV_32F );

/** 返回 Gabor 滤波器系数*/
CV_EXPORTS_W Mat getGaborKernel( Size ksize, double sigma, double theta, double lambd,
                                 double gamma, double psi = CV_PI*0.5, int ktype = CV_64F );

//! 返回侵蚀和扩张的"魔法"边界值
static inline Scalar morphologyDefaultBorderValue() { return Scalar::all(DBL_MAX); }

/** 返回用于形态学运算的指定大小和形状的结构元素*/
CV_EXPORTS_W Mat getStructuringElement(int shape, Size ksize, Point anchor = Point(-1,-1));
```

如下为与图像滤波相关的函数（见 4.6 节案例）：

```
/** 中值滤波*/
CV_EXPORTS_W void medianBlur( InputArray src, OutputArray dst, int ksize );

/** 高斯滤波*/
CV_EXPORTS_W void GaussianBlur( InputArray src, OutputArray dst, Size ksize,
                                double sigmaX, double sigmaY = 0,
                                int borderType = BORDER_DEFAULT );

/** 双边滤波*/
CV_EXPORTS_W void bilateralFilter( InputArray src, OutputArray dst, int d,
                                   double sigmaColor, double sigmaSpace,
                                   int borderType = BORDER_DEFAULT );

/** 方框滤波*/
CV_EXPORTS_W void boxFilter( InputArray src, OutputArray dst, int ddepth,
                             Size ksize, Point anchor = Point(-1,-1),
                             bool normalize = true,
                             int borderType = BORDER_DEFAULT );

/** 计算与过滤器重叠的像素值的归一化平方和*/
CV_EXPORTS_W void sqrBoxFilter( InputArray src, OutputArray dst, int ddepth,
                                Size ksize, Point anchor = Point(-1, -1),
                                bool normalize = true,
```

```
                            int borderType = BORDER_DEFAULT );

/** 使用标准化的方框滤波器模糊图像，均值滤波*/
CV_EXPORTS_W void blur( InputArray src, OutputArray dst,
                        Size ksize, Point anchor = Point(-1,-1),
                        int borderType = BORDER_DEFAULT );

/** 计算图像卷积*/
CV_EXPORTS_W void filter2D( InputArray src, OutputArray dst, int ddepth,
                            InputArray kernel, Point anchor = Point(-1,-1),
                            double delta = 0, int borderType = BORDER_DEFAULT );

/** 对图像应用可分离线性滤波*/
CV_EXPORTS_W void sepFilter2D( InputArray src, OutputArray dst, int ddepth,
                               InputArray kernelX, InputArray kernelY,
                               Point anchor = Point(-1,-1),
                               double delta = 0, int borderType = BORDER_DEFAULT );
```

如下为边缘检测相关算子函数的定义（见 4.7 节案例）：

```
/** 使用扩展的 Sobel 算子计算第一、第二或混合图像导数*/
CV_EXPORTS_W void Sobel( InputArray src, OutputArray dst, int ddepth,
                         int dx, int dy, int ksize = 3,
                         double scale = 1, double delta = 0,
                         int borderType = BORDER_DEFAULT );

/** 使用 Sobel 算子计算 x 和 y 上的一阶图像导数*/
CV_EXPORTS_W void spatialGradient( InputArray src, OutputArray dx,
                                   OutputArray dy, int ksize = 3,
                                   int borderType = BORDER_DEFAULT );

/** 使用 Scharr 算子计算第一个 x 或 y 图像导数*/
CV_EXPORTS_W void Scharr( InputArray src, OutputArray dst, int ddepth,
                          int dx, int dy, double scale = 1, double delta = 0,
                          int borderType = BORDER_DEFAULT );

/** 计算图像的拉普拉斯（Laplacian）变换*/
CV_EXPORTS_W void Laplacian( InputArray src, OutputArray dst, int ddepth,
                             int ksize = 1, double scale = 1, double delta = 0,
                             int borderType = BORDER_DEFAULT );

/*Canny 边缘检测*/
CV_EXPORTS_W void Canny( InputArray image, OutputArray edges,
                         double threshold1, double threshold2,
                         int apertureSize = 3, bool L2gradient = false );
```

```
CV_EXPORTS_W void Canny( InputArray dx, InputArray dy,
                         OutputArray edges,
                         double threshold1, double threshold2,
                         bool L2gradient = false );

/** 计算角点检测梯度矩阵的最小特征值*/
CV_EXPORTS_W void cornerMinEigenVal( InputArray src, OutputArray dst,
                                     int blockSize, int ksize = 3,
                                     int borderType = BORDER_DEFAULT );
```

如下为与角点检测相关的函数（见 5.9 节案例）：

```
/** Harris 角点检测*/
CV_EXPORTS_W void cornerHarris( InputArray src, OutputArray dst, int blockSize,
                                int ksize, double k,
                                int borderType = BORDER_DEFAULT );

/** 角点检测中的图像块特征值和特征向量的计算*/
CV_EXPORTS_W void cornerEigenValsAndVecs( InputArray src, OutputArray dst,
                                          int blockSize, int ksize,
                                          int borderType = BORDER_DEFAULT );

/** 计算用于角点检测的特征映射*/
CV_EXPORTS_W void preCornerDetect( InputArray src, OutputArray dst, int ksize,
                                   int borderType = BORDER_DEFAULT );

/** 亚像素角点检测*/
CV_EXPORTS_W void cornerSubPix( InputArray image, InputOutputArray corners,
                                Size winSize, Size zeroZone,
                                TermCriteria criteria );

/** 显著角点检测*/
CV_EXPORTS_W void goodFeaturesToTrack( InputArray image, OutputArray corners,
              int maxCorners, double qualityLevel, double minDistance,
              InputArray mask = noArray(), int blockSize = 3,
              bool useHarrisDetector = false, double k = 0.04 );

CV_EXPORTS_W void goodFeaturesToTrack( InputArray image, OutputArray corners,
              int maxCorners, double qualityLevel, double minDistance,
              InputArray mask, int blockSize,
              int gradientSize, bool useHarrisDetector = false,
              double k = 0.04 );

CV_EXPORTS CV_WRAP_AS(goodFeaturesToTrackWithQuality) void goodFeaturesToTrack(
        InputArray image, OutputArray corners,
        int maxCorners, double qualityLevel, double minDistance,
        InputArray mask, OutputArray cornersQuality, int blockSize = 3,
        int gradientSize = 3, bool useHarrisDetector = false, double k = 0.04);
```

如下为与霍夫变换相关的函数的定义（见 5.1 节案例）：
```
/** 霍夫线变换*/
CV_EXPORTS_W void HoughLines( InputArray image, OutputArray lines,
                              double rho, double theta, int threshold,
                              double srn = 0, double stn = 0,
                              double min_theta = 0, double max_theta = CV_PI );

/** 概率霍夫线变换*/
CV_EXPORTS_W void HoughLinesP( InputArray image, OutputArray lines,
                               double rho, double theta, int threshold,
                               double minLineLength = 0, double maxLineGap = 0 );

/** 使用标准霍夫变换在一组点中查找直线*/
CV_EXPORTS_W void HoughLinesPointSet( InputArray _point, OutputArray _lines,
                                      int lines_max, int threshold,
                                      double min_rho, double max_rho,
                                      double rho_step,
                                      double min_theta, double max_theta,
                                      double theta_step );

/** 霍夫圆变换*/
CV_EXPORTS_W void HoughCircles( InputArray image, OutputArray circles,
                                int method, double dp, double minDist,
                                double param1 = 100, double param2 = 100,
                                int minRadius = 0, int maxRadius = 0 );
```
如下为与图像形态学运算相关的函数（见 4.5 节案例）：
```
/** 使用特定的结构元素腐蚀图像*/
CV_EXPORTS_W void erode( InputArray src, OutputArray dst, InputArray kernel,
                         Point anchor = Point(-1,-1), int iterations = 1,
                         int borderType = BORDER_CONSTANT,
                         const Scalar& borderValue = morphologyDefaultBorderValue() );


/** 使用特定的结构元素膨胀图像*/
CV_EXPORTS_W void dilate( InputArray src, OutputArray dst, InputArray kernel,
                          Point anchor = Point(-1,-1), int iterations = 1,
                          int borderType = BORDER_CONSTANT,
                          const Scalar& borderValue = morphologyDefaultBorderValue() );

/** 形态学运算*/
CV_EXPORTS_W void morphologyEx( InputArray src, OutputArray dst,
                                int op, InputArray kernel,
                                Point anchor = Point(-1,-1), int iterations = 1,
                                int borderType = BORDER_CONSTANT,
                                const Scalar& borderValue = morphologyDefaultBorderValue() );
```

如下为用于图像尺寸调整的函数：
```
CV_EXPORTS_W void resize( InputArray src, OutputArray dst,
                          Size dsize, double fx = 0, double fy = 0,
                          int interpolation = INTER_LINEAR );
```
如下为用于图像仿射变换的函数（见案例 52）：
```
CV_EXPORTS_W void warpAffine( InputArray src, OutputArray dst,
                              InputArray M, Size dsize,
                              int flags = INTER_LINEAR,
                              int borderMode = BORDER_CONSTANT,
                              const Scalar& borderValue = Scalar());
```
如下为用于透视变换的函数（见案例 53）：
```
CV_EXPORTS_W void warpPerspective( InputArray src, OutputArray dst,
                                   InputArray M, Size dsize,
                                   int flags = INTER_LINEAR,
                                   int borderMode = BORDER_CONSTANT,
                                   const Scalar& borderValue = Scalar());
```
如下为用于重映射的函数（见案例 54）：
```
CV_EXPORTS_W void remap( InputArray src, OutputArray dst,
                         InputArray map1, InputArray map2,
                         int interpolation, int borderMode = BORDER_CONSTANT,
                         const Scalar& borderValue = Scalar());
```
如下函数将图像变换 map 从一种表示转换为另一种表示：
```
CV_EXPORTS_W void convertMaps( InputArray map1, InputArray map2,
                               OutputArray dstmap1, OutputArray dstmap2,
                               int dstmap1type, bool nninterpolation = false );
```
如下为计算变换矩阵的函数：
```
/** 计算二维旋转的仿射矩阵*/
CV_EXPORTS_W Mat getRotationMatrix2D(Point2f center, double angle, double scale);

CV_EXPORTS Matx23d getRotationMatrix2D_(Point2f center, double angle, double scale);

inline
Mat getRotationMatrix2D(Point2f center, double angle, double scale)
{
    return Mat(getRotationMatrix2D_(center, angle, scale), true);
}

/** 根据3对对应的点计算仿射变换矩阵*/
CV_EXPORTS Mat getAffineTransform( const Point2f src[], const Point2f dst[] );

/** 逆仿射变换*/
CV_EXPORTS_W void invertAffineTransform( InputArray M, OutputArray iM );
```

/** 根据4对对应的点计算透视变换矩阵*/
CV_EXPORTS_W Mat getPerspectiveTransform(InputArray src, InputArray dst, int solveMethod = DECOMP_LU);
CV_EXPORTS Mat getPerspectiveTransform(const Point2f src[], const Point2f dst[], int solveMethod = DECOMP_LU);

CV_EXPORTS_W Mat getAffineTransform(InputArray src, InputArray dst);

/** 从图像中以亚像素精度检索像素矩形*/
CV_EXPORTS_W void getRectSubPix(InputArray image, Size patchSize,
                                 Point2f center, OutputArray patch, int patchType = -1);

如下为用于坐标空间变换的函数：
/** 将图像重映射到半对数极坐标空间（函数已废弃，结果同warpPolar）*/
CV_EXPORTS_W void logPolar(InputArray src, OutputArray dst,
                            Point2f center, double M, int flags);

/** 将图像重映射到极坐标空间（函数已废弃，结果同warpPolar）*/
CV_EXPORTS_W void linearPolar(InputArray src, OutputArray dst,
                               Point2f center, double maxRadius, int flags);

/** 将图像重映射到极坐标或半对数极坐标空间*/
CV_EXPORTS_W void warpPolar(InputArray src, OutputArray dst, Size dsize,
                            Point2f center, double maxRadius, int flags);

如下为与积分图计算相关的函数：
CV_EXPORTS_W void integral(InputArray src, OutputArray sum, int sdepth = -1);

CV_EXPORTS_AS(integral2) void integral(InputArray src, OutputArray sum,
                                        OutputArray sqsum, int sdepth = -1, int sqdepth = -1);

CV_EXPORTS_AS(integral3) void integral(InputArray src, OutputArray sum,
                                        OutputArray sqsum, OutputArray tilted,
                                        int sdepth = -1, int sqdepth = -1);

如下为用于图像累加等相关操作的函数：
/** 图像累加*/
CV_EXPORTS_W void accumulate(InputArray src, InputOutputArray dst,
                              InputArray mask = noArray());

/** 图像平方累加*/
CV_EXPORTS_W void accumulateSquare(InputArray src, InputOutputArray dst,
                                    InputArray mask = noArray());

/** 将两个输入图像的对应元素乘积添加到累加器图像中*/
CV_EXPORTS_W void accumulateProduct(InputArray src1, InputArray src2,

```
                              InputOutputArray dst, InputArray mask=noArray() );

/** 计算输入图像 src 和累加器 dst 的加权和*/
CV_EXPORTS_W void accumulateWeighted( InputArray src, InputOutputArray dst,
                              double alpha, InputArray mask = noArray() );

/** 该函数用于检测两个图像之间发生的平移偏移*/
CV_EXPORTS_W Point2d phaseCorrelate(InputArray src1, InputArray src2,
                              InputArray window = noArray(), CV_OUT double* response = 0);

/** 此函数用于计算二维汉宁窗系数*/
CV_EXPORTS_W void createHanningWindow(OutputArray dst, Size winSize, int type);
```

如下为与图像阈值化操作相关的函数（见 5.5 节案例）：

```
/** 对每个数组元素应用固定级别的阈值*/
CV_EXPORTS_W double threshold( InputArray src, OutputArray dst,
                              double thresh, double maxVal, int type );

/** 对矩阵应用自适应阈值*/
CV_EXPORTS_W void adaptiveThreshold( InputArray src, OutputArray dst,
                              double maxValue, int adaptiveMethod,
                              int thresholdType, int blockSize, double C );
```

如下为与图像金字塔相关的函数（见 5.6 节案例）：

```
/** 模糊图像并对其进行下采样*/
CV_EXPORTS_W void pyrDown( InputArray src, OutputArray dst,
                              const Size& dstsize = Size(), int borderType = BORDER_DEFAULT );

/** 上采样图像并模糊图像*/
CV_EXPORTS_W void pyrUp( InputArray src, OutputArray dst,
                              const Size& dstsize = Size(), int borderType = BORDER_DEFAULT );

/** 为图像构造高斯金字塔*/
CV_EXPORTS void buildPyramid( InputArray src, OutputArrayOfArrays dst,
                              int maxlevel, int borderType = BORDER_DEFAULT );
```

如下为与直方图相关的函数（见 5.7 节案例）：

```
/** 计算一组数组的直方图*/
CV_EXPORTS void calcHist( const Mat* images, int nimages,
                       const int* channels, InputArray mask,
                       OutputArray hist, int dims, const int* histSize,
                       const float** ranges, bool uniform = true, bool accumulate = false );

CV_EXPORTS void calcHist( const Mat* images, int nimages,
                          const int* channels, InputArray mask,
                          SparseMat& hist, int dims,
                          const int* histSize, const float** ranges,
```

```
                          bool uniform = true, bool accumulate = false );

CV_EXPORTS_W void calcHist( InputArrayOfArrays images,
                            const std::vector<int>& channels,
                            InputArray mask, OutputArray hist,
                            const std::vector<int>& histSize,
                            const std::vector<float>& ranges,
                            bool accumulate = false );

/** 计算直方图的反投影 */
CV_EXPORTS void calcBackProject( const Mat* images, int nimages,
                                 const int* channels, InputArray hist,
                                 OutputArray backProject, const float** ranges,
                                 double scale = 1, bool uniform = true );


CV_EXPORTS void calcBackProject( const Mat* images, int nimages,
                                 const int* channels, const SparseMat& hist,
                                 OutputArray backProject, const float** ranges,
                                 double scale = 1, bool uniform = true );


CV_EXPORTS_W void calcBackProject( InputArrayOfArrays images, const std::vector<int>& channels,
                                   InputArray hist, OutputArray dst,
                                   const std::vector<float>& ranges,
                                   double scale );

/** 直方图对比*/
CV_EXPORTS_W double compareHist( InputArray H1, InputArray H2, int method );

CV_EXPORTS double compareHist( const SparseMat& H1, const SparseMat& H2, int method );

/** 均衡灰度图像的直方图*/
CV_EXPORTS_W void equalizeHist( InputArray src, OutputArray dst );

/** 为 cv::CLAHE 类创建智能指针并初始化*/
CV_EXPORTS_W Ptr<CLAHE> createCLAHE(double clipLimit = 40.0, Size tileGridSize = Size(8, 8));
```

如下为计算 EMD 的函数：

```
CV_EXPORTS float EMD( InputArray signature1, InputArray signature2,
                      int distType, InputArray cost=noArray(),
                      float* lowerBound = 0, OutputArray flow = noArray() );

CV_EXPORTS_AS(EMD) float wrapperEMD( InputArray signature1, InputArray signature2,
                      int distType, InputArray cost=noArray(),
                      CV_IN_OUT Ptr<float> lowerBound = Ptr<float>(), OutputArray flow = noArray() );
```

如下为与图像分割相关的算法的函数（见 5.8 节案例）：
```
/** 使用分水岭算法执行基于标记的图像分割*/
CV_EXPORTS_W void watershed( InputArray image, InputOutputArray markers );

/** 执行图像 meanshift 分割算法的初始化*/
CV_EXPORTS_W void pyrMeanShiftFiltering( InputArray src, OutputArray dst,
                                         double sp, double sr, int maxLevel = 1,
                                         TermCriteria
termcrit=TermCriteria(TermCriteria::MAX_ITER+TermCriteria::EPS,5,1) );

/** 执行 GrabCut 算法*/
CV_EXPORTS_W void grabCut( InputArray img, InputOutputArray mask, Rect rect,
                           InputOutputArray bgdModel, InputOutputArray fgdModel,
                           int iterCount, int mode = GC_EVAL );
```

如下函数用于计算从每个二值图像像素到最近的零像素的近似或精确距离：
```
CV_EXPORTS_AS(distanceTransformWithLabels) void distanceTransform( InputArray src,
                           OutputArray dst,
                           OutputArray labels, int distanceType, int maskSize,
                           int labelType = DIST_LABEL_CCOMP );

CV_EXPORTS_W void distanceTransform( InputArray src, OutputArray dst,
                           int distanceType, int maskSize, int dstType=CV_32F);
```

如下函数用给定的颜色填充连接的组件（见案例 63）：
```
CV_EXPORTS int floodFill( InputOutputArray image,
                          Point seedPoint, Scalar newVal, CV_OUT Rect* rect = 0,
                          Scalar loDiff = Scalar(), Scalar upDiff = Scalar(),
                          int flags = 4 );

CV_EXPORTS_W int floodFill( InputOutputArray image, InputOutputArray mask,
                            Point seedPoint, Scalar newVal, CV_OUT Rect* rect=0,
                            Scalar loDiff = Scalar(), Scalar upDiff = Scalar(),
                            int flags = 4 );
```

如下函数执行两个图像的线性混合操作：
```
CV_EXPORTS_W void blendLinear(InputArray src1, InputArray src2, InputArray weights1, InputArray weights2, OutputArray dst);
```

如下为与颜色空间变换相关的函数（见案例 30）：
```
/** 颜色空间变换*/
CV_EXPORTS_W void cvtColor( InputArray src, OutputArray dst, int code, int dstCn = 0 );

/** 将图像从一个颜色空间变换为另一个颜色空间，其中源图像存储在两个平面中*/
CV_EXPORTS_W void cvtColorTwoPlane( InputArray src1, InputArray src2, OutputArray dst, int code );
```

如下为 demosaicing 函数的定义：
```
CV_EXPORTS_W void demosaicing(InputArray src, OutputArray dst, int code, int dstCn = 0);
```

如下为矩相关操作的函数：
```
/** 计算多边形或光栅化形状的三阶矩*/
CV_EXPORTS_W Moments moments( InputArray array, bool binaryImage = false );

/** 计算 Hu 矩*/
CV_EXPORTS void HuMoments( const Moments& moments, double hu[7] );

CV_EXPORTS_W void HuMoments( const Moments& m, OutputArray hu );
```

如下为与模板匹配相关的操作的函数：
```
//! 模板匹配操作的类型
enum TemplateMatchModes {
    TM_SQDIFF        = 0,
    TM_SQDIFF_NORMED = 1,
    TM_CCORR         = 2,
    TM_CCORR_NORMED  = 3,
    TM_CCOEFF        = 4,
    TM_CCOEFF_NORMED = 5
};

/** 将模板与重叠的图像区域进行比较*/
CV_EXPORTS_W void matchTemplate( InputArray image, InputArray templ,
                     OutputArray result, int method, InputArray mask = noArray() );
```

如下为与连通性分析相关的函数：
```
/** 计算标记为布尔图像的图像连接组件*/
CV_EXPORTS_AS(connectedComponentsWithAlgorithm) int connectedComponents(InputArray image,
OutputArray labels,
                                      int connectivity, int ltype, int ccltype);

CV_EXPORTS_W int connectedComponents(InputArray image, OutputArray labels,
                     int connectivity = 8, int ltype = CV_32S);

/** 计算标记为布尔图像的图像连接组件并为每个标签生成统计输出*/
CV_EXPORTS_AS(connectedComponentsWithStatsWithAlgorithm) int connectedComponentsWithStats(
            InputArray image, OutputArray labels,
            OutputArray stats, OutputArray centroids,
            int connectivity, int ltype, int ccltype);

CV_EXPORTS_W int connectedComponentsWithStats(InputArray image, OutputArray labels,
                                    OutputArray stats, OutputArray centroids,
                                    int connectivity = 8, int ltype = CV_32S);
```

如下为与轮廓操作、图像拟合相关的函数（见 5.10～5.14 节案例）：
```
/** 寻找二值图像的轮廓*/
CV_EXPORTS_W void findContours( InputArray image, OutputArrayOfArrays contours,
                     OutputArray hierarchy, int mode,
```

```
                                int method, Point offset = Point());
CV_EXPORTS void findContours( InputArray image, OutputArrayOfArrays contours,
                              int mode, int method, Point offset = Point());
```

/** 以指定的精度近似多边形曲线*/
```
CV_EXPORTS_W void approxPolyDP( InputArray curve,
                                OutputArray approxCurve,
                                double epsilon, bool closed );
```

/** 计算轮廓周长或曲线长度*/
```
CV_EXPORTS_W double arcLength( InputArray curve, bool closed );
```

/** 计算点集的右上边界矩形或灰度图像的非零像素*/
```
CV_EXPORTS_W Rect boundingRect( InputArray array );
```

/** 计算轮廓面积*/
```
CV_EXPORTS_W double contourArea( InputArray contour, bool oriented = false );
```

/** 查找包含输入二维点集的最小区域的旋转矩形*/
```
CV_EXPORTS_W RotatedRect minAreaRect( InputArray points );
```

/** 查找旋转矩形的4个顶点，用于绘制旋转矩形*/
```
CV_EXPORTS_W void boxPoints(RotatedRect box, OutputArray points);
```

/** 该函数利用迭代算法求出二维点集的最小外接圈*/
```
CV_EXPORTS_W void minEnclosingCircle( InputArray points,
                                      CV_OUT Point2f& center, CV_OUT float& radius );
```

/** 找到一个包围二维点集的面积最小的三角形并返回其面积*/
```
CV_EXPORTS_W double minEnclosingTriangle( InputArray points, CV_OUT OutputArray triangle );
```

/** 形状匹配*/
```
CV_EXPORTS_W double matchShapes( InputArray contour1, InputArray contour2,
                                 int method, double parameter );
```

/** 查找点集的凸包*/
```
CV_EXPORTS_W void convexHull( InputArray points, OutputArray hull,
                              bool clockwise = false, bool returnPoints = true );
```

/** 找出轮廓的凸面缺陷*/
```
CV_EXPORTS_W void convexityDefects( InputArray contour, InputArray convexhull, OutputArray
convexityDefects );
```

/** 测试轮廓凸度*/
```
CV_EXPORTS_W bool isContourConvex( InputArray contour );
```

```
/** @example samples/cpp/intersectExample.cpp
Examples of how intersectConvexConvex works
*/

/** 查找两个凸多边形的交点*/
CV_EXPORTS_W float intersectConvexConvex( InputArray _p1, InputArray _p2,
                                          OutputArray _p12, bool handleNested = true );

/** 围绕一组二维点拟合椭圆*/
CV_EXPORTS_W RotatedRect fitEllipse( InputArray points );

/** AMS 方法围绕一组二维点拟合椭圆*/
CV_EXPORTS_W RotatedRect fitEllipseAMS( InputArray points );


/** 围绕一组二维点拟合椭圆*/
CV_EXPORTS_W RotatedRect fitEllipseDirect( InputArray points );

/** 围绕一组二维点或三维点拟合直线*/
CV_EXPORTS_W void fitLine( InputArray points, OutputArray line, int distType,
                           double param, double reps, double aeps );

/** 执行轮廓点测试操作*/
CV_EXPORTS_W double pointPolygonTest( InputArray contour, Point2f pt, bool measureDist );

/** 找出两个旋转矩形之间是否有交集*/
CV_EXPORTS_W int rotatedRectangleIntersection( const RotatedRect& rect1, const RotatedRect&
rect2, OutputArray intersectingRegion  );
```

如下两个函数的作用是创建智能指针：

```
/** 为 cv::GeneralizedHoughBallard 类创建智能指针并初始化*/
CV_EXPORTS_W Ptr<GeneralizedHoughBallard> createGeneralizedHoughBallard();

/** 为 cv::GeneralizedHoughGuil 类创建智能指针并初始化*/
CV_EXPORTS_W Ptr<GeneralizedHoughGuil> createGeneralizedHoughGuil();
```

如下为 GNU Octave/MATLAB 主要的颜色图类型定义及其相关操作的函数：

```
enum ColormapTypes
{
    COLORMAP_AUTUMN = 0,
    COLORMAP_BONE = 1,
    COLORMAP_JET = 2,
    COLORMAP_WINTER = 3,
    COLORMAP_RAINBOW = 4,
    COLORMAP_OCEAN = 5,
```

```
    COLORMAP_SUMMER = 6,
    COLORMAP_SPRING = 7,
    COLORMAP_COOL = 8,
    COLORMAP_HSV = 9,
    COLORMAP_PINK = 10,
    COLORMAP_HOT = 11,
    COLORMAP_PARULA = 12,
    COLORMAP_MAGMA = 13,
    COLORMAP_INFERNO = 14,
    COLORMAP_PLASMA = 15,
    COLORMAP_VIRIDIS = 16,
    COLORMAP_CIVIDIS = 17,
    COLORMAP_TWILIGHT = 18,
    COLORMAP_TWILIGHT_SHIFTED = 19,
    COLORMAP_TURBO = 20,
    COLORMAP_DEEPGREEN = 21
};

/** 在给定图像上应用 GNU Octave/MATLAB 主要颜色图*/
CV_EXPORTS_W void applyColorMap(InputArray src, OutputArray dst, int colormap);

/** 在给定图像上应用读者颜色图*/
CV_EXPORTS_W void applyColorMap(InputArray src, OutputArray dst, InputArray userColor);

/** OpenCV 创建色彩值 */
#define CV_RGB(r, g, b)  cv::Scalar((b), (g), (r), 0)
```

如下为与图像绘制相关的函数（见 4.4 节案例）：

```
/** 绘制两点连接的线*/
CV_EXPORTS_W void line(InputOutputArray img, Point pt1, Point pt2,
                       const Scalar& color,
                       int thickness = 1, int lineType = LINE_8, int shift = 0);

/** 绘制从第一个点指向第二个点的箭头段*/
CV_EXPORTS_W void arrowedLine(InputOutputArray img, Point pt1, Point pt2,
                       const Scalar& color,
                       int thickness=1, int line_type=8, int shift=0,
                       double tipLength=0.1);

/** 绘制矩形*/
CV_EXPORTS_W void rectangle(InputOutputArray img, Point pt1, Point pt2,
                            const Scalar& color, int thickness = 1,
                            int lineType = LINE_8, int shift = 0);

CV_EXPORTS_W void rectangle(InputOutputArray img, Rect rec,
                            const Scalar& color, int thickness = 1,
                            int lineType = LINE_8, int shift = 0);
```

```
/** 绘制圆*/
CV_EXPORTS_W void circle(InputOutputArray img, Point center, int radius,
                const Scalar& color, int thickness = 1,
                int lineType = LINE_8, int shift = 0);

/** 绘制椭圆*/
CV_EXPORTS_W void ellipse(InputOutputArray img, Point center, Size axes,
                double angle, double startAngle, double endAngle,
                const Scalar& color, int thickness = 1,
                int lineType = LINE_8, int shift = 0);

CV_EXPORTS_W void ellipse(InputOutputArray img, const RotatedRect& box,
                const Scalar& color,
                int thickness = 1, int lineType = LINE_8);

/** 在预定位置绘制标记*/
CV_EXPORTS_W void drawMarker(InputOutputArray img, Point position, const Scalar& color,
                int markerType = MARKER_CROSS, int markerSize=20, int thickness=1,
                int line_type=8);
```

如下为与多边形填充、多边形绘制及轮廓绘制等相关的函数：

```
/** 填充凸多边形*/
CV_EXPORTS void fillConvexPoly(InputOutputArray img, const Point* pts, int npts,
                const Scalar& color, int lineType = LINE_8,
                int shift = 0);

CV_EXPORTS_W void fillConvexPoly(InputOutputArray img, InputArray points,
                const Scalar& color, int lineType = LINE_8,
                int shift = 0);

/** 填充由一个或多个多边形限定的区域*/
CV_EXPORTS void fillPoly(InputOutputArray img, const Point** pts,
                const int* npts, int ncontours,
                const Scalar& color, int lineType = LINE_8, int shift = 0,
                Point offset = Point() );

CV_EXPORTS_W void fillPoly(InputOutputArray img, InputArrayOfArrays pts,
                const Scalar& color, int lineType = LINE_8, int shift = 0,
                Point offset = Point() );

/** 绘制多条多边形曲线*/
CV_EXPORTS void polylines(InputOutputArray img, const Point* const* pts, const int* npts,
                int ncontours, bool isClosed, const Scalar& color,
                int thickness = 1, int lineType = LINE_8, int shift = 0 );

CV_EXPORTS_W void polylines(InputOutputArray img, InputArrayOfArrays pts,
```

```
                         bool isClosed, const Scalar& color,
                         int thickness = 1, int lineType = LINE_8, int shift = 0 );

/** 绘制轮廓或填充轮廓*/
CV_EXPORTS_W void drawContours( InputOutputArray image, InputArrayOfArrays contours,
                              int contourIdx, const Scalar& color,
                              int thickness = 1, int lineType = LINE_8,
                              InputArray hierarchy = noArray(),
                              int maxLevel = INT_MAX, Point offset = Point() );

/** 根据图像矩形剪裁线条*/
CV_EXPORTS bool clipLine(Size imgSize, CV_IN_OUT Point& pt1, CV_IN_OUT Point& pt2);
CV_EXPORTS bool clipLine(Size2l imgSize, CV_IN_OUT Point2l& pt1, CV_IN_OUT Point2l& pt2);
CV_EXPORTS_W bool clipLine(Rect imgRect, CV_OUT CV_IN_OUT Point& pt1, CV_OUT CV_IN_OUT Point& pt2);

/** 用折线近似椭圆弧*/
CV_EXPORTS_W void ellipse2Poly( Point center, Size axes, int angle,
                              int arcStart, int arcEnd, int delta,
                              CV_OUT std::vector<Point>& pts );
CV_EXPORTS void ellipse2Poly(Point2d center, Size2d axes, int angle,
                           int arcStart, int arcEnd, int delta,
                           CV_OUT std::vector<Point2d>& pts);

/** 绘制文字*/
CV_EXPORTS_W void putText( InputOutputArray img, const String& text, Point org,
                         int fontFace, double fontScale, Scalar color,
                         int thickness = 1, int lineType = LINE_8,
                         bool bottomLeftOrigin = false );

/** 计算字符串的宽度和高度*/
CV_EXPORTS_W Size getTextSize(const String& text, int fontFace,
                            double fontScale, int thickness,
                            CV_OUT int* baseLine);


/** 计算用于达到给定高度所需的字体大小*/
CV_EXPORTS_W double getFontScaleFromHeight(const int fontFace,
                                         const int pixelHeight,
                                         const int thickness = 1);
```

如下为 Line 迭代器类及与其相关的操作函数：

```
class CV_EXPORTS LineIterator
{
public:
    LineIterator( const Mat& img, Point pt1, Point pt2,
                int connectivity = 8, bool leftToRight = false )
```

```cpp
    {
        init(&img, Rect(0, 0, img.cols, img.rows), pt1, pt2, connectivity, leftToRight);
        ptmode = false;
    }
    LineIterator( Point pt1, Point pt2,
                  int connectivity = 8, bool leftToRight = false )
    {
        init(0, Rect(std::min(pt1.x, pt2.x),
                     std::min(pt1.y, pt2.y),
                     std::max(pt1.x, pt2.x) - std::min(pt1.x, pt2.x) + 1,
                     std::max(pt1.y, pt2.y) - std::min(pt1.y, pt2.y) + 1),
             pt1, pt2, connectivity, leftToRight);
        ptmode = true;
    }
    LineIterator( Size boundingAreaSize, Point pt1, Point pt2,
                  int connectivity = 8, bool leftToRight = false )
    {
        init(0, Rect(0, 0, boundingAreaSize.width, boundingAreaSize.height),
             pt1, pt2, connectivity, leftToRight);
        ptmode = true;
    }
    LineIterator( Rect boundingAreaRect, Point pt1, Point pt2,
                  int connectivity = 8, bool leftToRight = false )
    {
        init(0, boundingAreaRect, pt1, pt2, connectivity, leftToRight);
        ptmode = true;
    }
    void init(const Mat* img, Rect boundingAreaRect, Point pt1, Point pt2, int connectivity, bool leftToRight);

    uchar* operator *();
    LineIterator& operator ++();
    LineIterator operator ++(int);
    Point pos() const;

    uchar* ptr;
    const uchar* ptr0;
    int step, elemSize;
    int err, count;
    int minusDelta, plusDelta;
    int minusStep, plusStep;
    int minusShift, plusShift;
    Point p;
    bool ptmode;
};

inline
```

```cpp
uchar* LineIterator::operator *()
{
    return ptmode ? 0 : ptr;
}

inline
LineIterator& LineIterator::operator ++()
{
    int mask = err < 0 ? -1 : 0;
    err += minusDelta + (plusDelta & mask);
    if(!ptmode)
    {
        ptr += minusStep + (plusStep & mask);
    }
    else
    {
        p.x += minusShift + (plusShift & mask);
        p.y += minusStep + (plusStep & mask);
    }
    return *this;
}

inline
LineIterator LineIterator::operator ++(int)
{
    LineIterator it = *this;
    ++(*this);
    return it;
}

inline
Point LineIterator::pos() const
{
    if(!ptmode)
    {
        size_t offset = (size_t)(ptr - ptr0);
        int y = (int)(offset/step);
        int x = (int)((offset - (size_t)y*step)/elemSize);
        return Point(x, y);
    }
    return p;
}
} // cv

#include "./imgproc/segmentation.hpp"
#endif
```

4.2 案例 30：颜色空间变换

OpenCV 中提供了用于颜色空间变换的函数，以适应在不同需求中的图像使用要求。颜色空间变换可以调用函数 cvtColor 来实现，其定义如下：

```
dst = cvtColor(src, code, dst=None, dstCn=None)
```

参数说明如下。

- src：输入图像。
- code：颜色空间变换方式，由 ColorConversionCodes 定义（见 4.1 节）。
- dst：输出图像（返回值）。
- dstCn：输出图像通道数，如果将其设置为 0，则根据 src 和 code 自动推导通道数。

本案例使用的输入图像如图 3.10 所示，读取输入图像后，将 BGR 通道的图像变换为灰色图像，案例代码如下：

```
import cv2

#图像读取
src = cv2.imread("src.jpg")
#将 BGR 通道的图像变换为灰色图像
gray_img = cv2.cvtColor(src, cv2.COLOR_BGR2GRAY)
#图像显示
cv2.imshow("gray_img", gray_img)
cv2.waitKey(0)
cv2.destroyAllWindows()
```

imread 函数用来读取图像，其中第一个参数传入的是图像文件路径，第二个参数传入的是图像读取方式标志，默认的读取方式是 IMREAD_COLOR，该方式读取的图像是以 BGR 通道顺序存储的。转为灰度图像后的结果如图 4.1 所示。

使用转换为灰度图像的方式与图像读取时按照灰度图像 IMREAD_GRAYSCALE 读取的结果相同：

```
#灰度图像读取
src = cv2.imread("src.jpg", cv2.IMREAD_GRAYSCALE)
```

OpenCV 中提供的全部颜色空间变换方式都由 ColorConversionCodes 定义，读者可以查阅寻找合适的图像变换方式。

图 4.1

4.3 案例 31：图像尺寸变换

图像尺寸变换即调整图像的大小，OpenCV 提供了用于尺寸变换的函数 resize，其定义如下：

```
dst = resize(src, dsize, dst=None, fx=None, fy=None, interpolation=None)
```

参数说明如下。

- src：输入图像。
- dsize：输出图像尺寸。
- dst：输出图像（返回值）。
- fx：水平方向缩放比例。
- fy：垂直方向缩放比例。
- interpolation：插值方式，由 InterpolationFlags 定义。

插值方式 InterpolationFlags 的定义如下：

```
enum InterpolationFlags{
    INTER_NEAREST           = 0,     #最近邻插值
    INTER_LINEAR            = 1,     #双线性插值
    INTER_CUBIC             = 2,     #双三次插值
    INTER_AREA              = 3,     #不支持
    INTER_LANCZOS4          = 4,     #Lanczos 插值
    INTER_LINEAR_EXACT      = 5,     #按位双线性插值
    INTER_MAX               = 7,     #差值代码 mask
    WARP_FILL_OUTLIERS      = 8,     #是否填充所有目标图像像素的标志
    WARP_INVERSE_MAP        = 16     #是否翻转变换的标志
};
```

本案例使用的输入图像如图 3.10 所示，案例中将图像尺寸变为原始图像尺寸的 1/2，代码如下：

```
import cv2

#图像读取
src = cv2.imread("src.jpg")
#获取图像尺寸
(height, width, channels) = src.shape
print("input image height, width, channels: ", height, width, channels)
#将图像尺寸变为原始图像尺寸的 1/2
resize_result = cv2.resize(src, (height/2, width/2))
(new_height, new_width, new_channels) = resize_result.shape
print("resize image height, width, channels: ", new_height, new_width, new_channels)
```

```
#图像显示
cv2.imshow("resize_result", resize_result)
cv2.waitKey(0)
cv2.destroyAllWindows()
```

resize 函数的第二个参数传入的是新尺寸,该参数的数据类型需要为整型,直接使用 (height/2, width/2) 的方式传入的参数数据类型为浮点型,需要转为整型,即 (int(height/2), int(width/2))。

尺寸变换前后的维度信息为:

```
input image height, width, channels:  512 512 3
resize image height, width, channels:  256 256 3
```

尺寸变换后的图像如图 4.2 所示。

图 4.2

4.4 基本绘制

在图像处理中,经常需要在图像上将处理结果绘制出来,如目标检测中的检测结果可以用矩形框标出来,OpenCV 提供了众多的绘制操作,如在图像上绘制标记,绘制直线、矩形、圆、椭圆、文字等。

4.4.1 案例 32:绘制标记

OpenCV 提供了用于绘制标记的函数 drawMarker,读者可以使用该函数在图像上标记一个点,其定义如下:

```
img = drawMarker(img, position, color, markerType=None, markerSize=None, thickness=None, line_type=None)
```

参数说明如下。

- img：待标记的输入图像（返回值）。
- position：标记绘制的位置，需要传入整型坐标位置。
- color：标记的绘制颜色。
- markerType：标记绘制类型，由 MarkerTypes 定义（见 4.1 节）。
- markerSize：标记的轴向长度。
- thickness：标记线的粗细。
- line_type：线型，由 LineTypes 定义（见 4.1 节）。

本案例在图像中的 5 个不同位置绘制了不同的标记，案例代码如下：

```
import cv2

#图像读取
src = cv2.imread("src.jpg")
#绘制标记
draw_marker = cv2.drawMarker(src, (256, 256), (255, 0, 0), cv2.MARKER_CROSS, thickness=3)
draw_marker = cv2.drawMarker(draw_marker, (50, 50), (0, 255, 0), cv2.MARKER_STAR, thickness=3)
draw_marker = cv2.drawMarker(draw_marker, (400, 400), (0, 0, 255), cv2.MARKER_DIAMOND, thickness=3)
draw_marker = cv2.drawMarker(draw_marker, (50, 400), (255, 255, 0), cv2.MARKER_SQUARE, thickness=3)
draw_marker = cv2.drawMarker(draw_marker, (400, 50), (255, 0, 255), cv2.MARKER_TILTED_CROSS, thickness=3)
#图像显示
cv2.imshow("draw_marker", draw_marker)
cv2.waitKey(0)
cv2.destroyAllWindows()
```

绘制标记后的结果如图 4.3 所示。

图 4.3

4.4.2 案例 33：绘制直线

OpenCV 提供的直线绘制函数为 line，其定义如下：

```
img = line(img, pt1, pt2, color, thickness=None, lineType=None, shift=None)
```

参数说明如下。

- img：待绘制的输入图像（返回值）。
- pt1：绘制直线的第一个点。
- pt2：绘制直线的第二个点。
- color：绘制的直线的颜色。
- thickness：绘制的直线的粗细。
- lineType：线型，由 LineTypes 定义（见 4.1 节）。
- shift：点坐标中的小数位数。

本案例在点(50, 50)和点(450, 450)之间绘制一条粗细为 3 像素的红色直线，代码如下：

```python
import cv2

#图像读取
src = cv2.imread("src.jpg")
#绘制直线
draw_line = cv2.line(src, (50, 50), (450, 450), (0, 0, 255), thickness=3)
#图像显示
cv2.imshow("draw_line", draw_line)
cv2.waitKey(0)
cv2.destroyAllWindows()
```

绘制的结果如图 4.4 所示。

图 4.4

4.4.3 案例34：绘制矩形

OpenCV 提供的用于绘制矩形的函数为 rectangle，其定义如下：

```
img = rectangle(img, pt1, pt2, color, thickness=None, lineType=None, shift=None)
```

参数说明如下。

- img：待绘制的输入图像（返回值）。
- pt1：矩形的顶点。
- pt2：与 pt1 相对的矩形的顶点。
- color：绘制矩形的线的颜色。
- thickness：绘制线的粗细。
- lineType：线型，由 LineTypes 定义（见 4.1 节）。
- shift：点坐标中的小数位数。

本节案例在顶点(50,50)和顶点(450,450)之间绘制一个粗细为 3 像素的红色矩形框，绘制代码如下：

```python
import cv2

#图像读取
src = cv2.imread("src.jpg")
#绘制矩形
draw_rectangle = cv2.rectangle(src, (50, 50), (450, 450), (0, 0, 255), thickness=3)
#图像显示
cv2.imshow("draw_line", draw_rectangle)
cv2.waitKey(0)
cv2.destroyAllWindows()
```

绘制结果如图 4.5 所示。

图 4.5

4.4.4 案例 35：绘制圆

OpenCV 提供的用于绘制圆的函数为 circle，其定义如下：
```
img = circle(img, center, radius, color, thickness=None, lineType=None, shift=None)
```
参数说明如下。

- img：待绘制的输入图像（返回值）。
- center：圆心坐标。
- radius：圆的半径。
- color：绘制圆的线的颜色。
- thickness：绘制线的粗细。
- lineType：线型，由 LineTypes 定义（见 4.1 节）。
- shift：点坐标中的小数位数。

本案例在圆心位置(256, 256)处绘制一个半径为 200 像素、线粗细为 3 像素的红色圆，并标记圆心位置，绘制代码如下：

```
import cv2

#图像读取
src = cv2.imread("src.jpg")
#绘制圆
draw_circle = cv2.circle(src, (256, 256), 200, (0, 0, 255), thickness=3)
#标记圆心
draw_circle = cv2.drawMarker(draw_circle, (256, 256), (255, 0, 0), cv2.MARKER_CROSS, thickness=3)
#图像显示
cv2.imshow("draw_circle", draw_circle)
cv2.waitKey(0)
cv2.destroyAllWindows()
```

绘制结果如图 4.6 所示。

图 4.6

4.4.5 案例 36：绘制椭圆

OpenCV 提供的用于绘制椭圆的函数为 ellipse，其定义如下：

```
img = ellipse(img, center, axes, angle, startAngle, endAngle, color, thickness=None, lineType=None, shift=None)
```

参数说明如下。

- img：待绘制的输入图像（返回值）。
- center：椭圆圆心坐标。
- axes：主轴尺寸的一半。
- angle：椭圆旋转角度。
- startAngle：椭圆圆弧的起始角度。
- endAngle：椭圆圆弧的终止角度。
- color：绘制椭圆的线的颜色。
- thickness：绘制线的粗细。
- lineType：线型，由 LineTypes 定义（见 4.1 节）。
- shift：点坐标中的小数位数。

本案例选择椭圆的圆心位置为(256, 256)，绘制的椭圆的长、短轴的长度的一半分别为 250 像素、150 像素，旋转角度为 45°，圆弧起始角度为 0°，终止角度为 360°，线颜色为红色，线粗细为 3 像素，并标记了椭圆圆心位置，绘制代码如下：

```
import cv2

#图像读取
src = cv2.imread("src.jpg")
#绘制椭圆
draw_ellipse = cv2.ellipse(src, (256, 256), (250, 150), 45, 0, 360, (0, 0, 255), thickness=3)
#标记圆心
draw_ellipse = cv2.drawMarker(draw_ellipse, (256, 256), (255, 0, 0), cv2.MARKER_CROSS, thickness=3)
#图像显示
cv2.imshow("draw_ellipse", draw_ellipse)
cv2.waitKey(0)
cv2.destroyAllWindows()
```

绘制结果如图 4.7 所示。

图 4.7

4.4.6 案例 37：绘制文字

在图像处理中，经常需要对图像增加一些说明性的文字，OpenCV 提供了用于绘制文字的函数 putText，其定义如下：

```
img = putText(img, text, org, fontFace, fontScale, color, thickness=None, lineType=None, bottomLeftOrigin=None)
```

参数说明如下。
- img：待绘制的输入图像（返回值）。
- text：待绘制的文字。
- org：文字在图像中绘制区域的左下角位置。
- fontFace：字体，由 HersheyFonts 定义（见 4.1 节）。
- fontScale：基于设定字号大小的缩放因子。
- color：绘制文字的线的颜色。
- thickness：绘制文字的线的粗细。
- lineType：线型，由 LineTypes 定义（见 4.1 节）。
- bottomLeftOrigin：如果将其设为 true，则图像数据原点位于左下角；否则位于左上角。

本案例给输入图像添加说明文字的代码如下：

```
import cv2

#图像读取
src = cv2.imread("src.jpg")
#绘制文字
text_content = "This is Lena"
```

```
draw_text = cv2.putText(src, text_content, (50, 50), cv2.FONT_ITALIC, 1, (255, 0, 0), thickness=2)
#图像显示
cv2.imshow("draw_text", draw_text)
cv2.waitKey(0)
cv2.destroyAllWindows()
```

案例中绘制的文字内容由 text_content 定义，文字的左下角坐标为(50,50)，字体为 FONT_ITALIC，缩放因子为 1，颜色为蓝色，绘制文字的线粗细为 2 像素。文字绘制的结果如图 4.8 所示。

图 4.8

4.5 形态学运算

形态学运算包括腐蚀、膨胀、开运算、闭运算、形态学梯度、顶帽运算、底帽运算 7 种，其中，膨胀与腐蚀是最常用的两种形态学运算方法，它们在消除噪声、元素分割和连接中有着广泛的应用；其他形态学运算方法多是在这两种运算方法的基础上组合而成的。形态学运算在图像去噪、图像分割等方面有着广泛的应用。

4.5.1 案例 38：腐蚀

腐蚀运算计算核覆盖范围内的局部最小值，OpenCV 提供的用于腐蚀运算的函数为 erode，其定义如下：

```
dst = erode(src, kernel, dst=None, anchor=None, iterations=None, borderType=None, borderValue=None)
```

参数说明如下。
- src：输入图像。
- kernel：用于腐蚀运算的核结构。

- dst：输出图像（返回值）。
- anchor：锚点位置，默认是 kernel 对应区域的中心位置。
- iterations：应用腐蚀操作迭代的次数。
- borderType：边界模式，由 BorderTypes 定义（见 3.4.5 节）。
- borderValue：边界模式为 BORDER_CONSTANT 时的边界像素值。

本案例使用的输入图像如图 4.9 所示。

图 4.9

腐蚀运算的案例代码如下：

```
import cv2

#读取图像
src = cv2.imread("cvbook.jpg")
#定义 3×3（单位为像素，书本涉及图像的单位均为像素）的腐蚀运算矩形核结构
element = cv2.getStructuringElement(cv2.MORPH_RECT, (3, 3))
#腐蚀运算
erode_img = cv2.erode(src, element)
#图像显示
cv2.imshow("erode_img", erode_img)
cv2.waitKey(0)
cv2.destroyAllWindows()
```

为了使效果更加明显，选用了 3×3 的腐蚀运算矩形核结构。腐蚀运算的结果如图 4.10 所示。

图 4.10

腐蚀运算选取核结构中的最小像素值作为锚点处的像素值，因此，对于图 4.9 中的文字，经过腐蚀运算，会让黑色文字区域变大，文字会变粗。

4.5.2 案例 39：膨胀

与腐蚀运算相反，膨胀运算计算的是核覆盖范围内的局部最大值，OpenCV 提供的用于膨胀运算的函数为 dilate，其定义如下：

```
dst = dilate(src, kernel, dst=None, anchor=None, iterations=None, borderType=None, borderValue=None)
```

参数说明如下。

- src：输入图像。
- kernel：用于膨胀运算的核结构。
- dst：输出图像（返回值）。
- anchor：锚点位置，默认是 kernel 对应区域的中心位置。
- iterations：应用膨胀操作迭代的次数。
- borderType：边界模式，由 BorderTypes 定义（见 3.4.5 节）。
- borderValue：边界模式为 BORDER_CONSTANT 时的边界像素值。

本案例使用的输入图像如图 4.9 所示，膨胀运算的案例代码如下：

```python
import cv2

#读取图像
src = cv2.imread("logo.jpg")
#定义 3×3 的膨胀运算矩形核结构
element = cv2.getStructuringElement(cv2.MORPH_RECT, (3, 3))
#膨胀运算
dilate_img = cv2.dilate(src, element)
#图像显示
cv2.imshow("dilate_img", dilate_img)
cv2.waitKey(0)
cv2.destroyAllWindows()
```

膨胀运算的结果如图 4.11 所示。

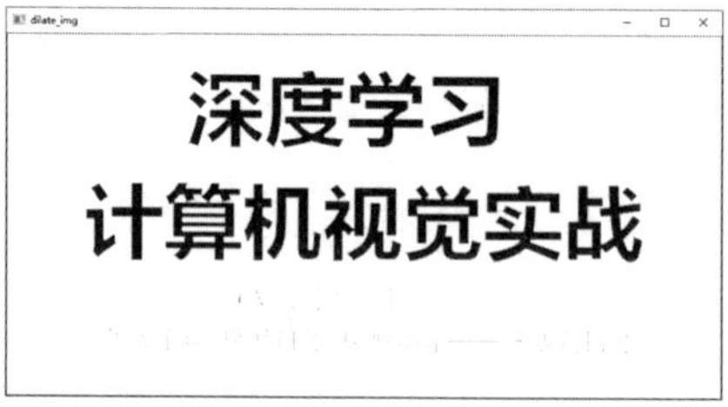

图 4.11

膨胀运算选取核结构中的最大像素值作为锚点处的像素值，因此，对于图 4.9 中的文字，经过膨胀运算，黑色的文字区域变小，文字会变细，而下面较小的文字则几乎被磨灭。

对比两种运算可以发现，膨胀运算会使明亮区域扩张，腐蚀运算会使阴暗区域扩张，因此，膨胀运算可以用于填充凹面，腐蚀运算可以用于消除凸起。

提示：膨胀和腐蚀操作计算的是核覆盖范围内的局部最大值和最小值，可以在强度图像上运算，但是图像的形态学运算通常在阈值化后的二值图像上进行。

4.5.3　案例 40：其他形态学运算

OpenCV 提供了用于形态学运算的函数 morphologyEx，其定义如下：

```
dst = morphologyEx(src, op, kernel, dst=None, anchor=None, iterations=None, borderType=None, borderValue=None)
```

参数说明如下。

- src：输入图像。
- op：形态学运算类型，由 MorphTypes 定义（见 4.1 节）。
- kernel：用于形态学运算的核结构。
- dst：输出图像（返回值）。
- anchor：锚点位置，默认是 kernel 对应区域的中心位置。
- iterations：应用形态学操作迭代的次数。
- borderType：边界模式，由 BorderTypes 定义（见 3.4.5 节）。
- borderValue：边界模式为 BORDER_CONSTANT 时的边界像素值。

几种形态学运算的计算方法与用途如表 4.1 所示。

表 4.1

形态学运算	计算方法	用途
膨胀	核覆盖范围内的局部最大值	用于发现连通分支等
腐蚀	核覆盖范围内的局部最小值	用于消除斑点噪声等
开运算	先腐蚀后膨胀	对二值图像中的区域进行计数
闭运算	先膨胀后腐蚀	用于消除较小异常值
形态学梯度	膨胀减腐蚀	寻找图像边界
顶帽运算	原始图像减开运算	提取明亮区域
底帽运算	闭运算减原始图像	提取阴暗区域

本案例使用的输入图像如图 4.9 所示，腐蚀和膨胀操作在前面做了介绍，本节不再赘述，其他的形态学运算的案例代码如下：

```
import cv2
```

```python
#图像读取
src = cv2.imread("cvbook.jpg")
#获取形态学运算结构
element = cv2.getStructuringElement(cv2.MORPH_RECT, (3,3))

#开运算
open_img = cv2.morphologyEx(src, cv2.MORPH_OPEN, element)
#结果如图 4.12 所示
cv2.imshow("open_img", open_img)

#闭运算
close_img = cv2.morphologyEx(src, cv2.MORPH_CLOSE, element)
#结果如图 4.13 所示
cv2.imshow("close_img", close_img)

#形态学梯度运算
grad_img = cv2.morphologyEx(src, cv2.MORPH_GRADIENT, element)
#结果如图 4.14 所示
cv2.imshow("grad_img", grad_img)

#顶帽运算
tophat_img = cv2.morphologyEx(src, cv2.MORPH_TOPHAT, element)
#结果如图 4.15 所示
cv2.imshow("tophat_img", tophat_img)

#底帽运算
blackhat_img = cv2.morphologyEx(src, cv2.MORPH_BLACKHAT, element)
#结果如图 4.16 所示
cv2.imshow("blackhat_img", blackhat_img)

cv2.waitKey(0)
cv2.destroyAllWindows()
```

开运算的计算方法为先腐蚀后膨胀，执行结果如图 4.12 所示。

闭运算的计算方法为先膨胀后腐蚀，执行结果如图 4.13 所示。

图 4.12

图 4.13

形态学梯度运算的计算方法为膨胀结果减去腐蚀结果,执行结果如图 4.14 所示。

顶帽运算的计算方法为原始图像减去开运算的结果,执行结果如图 4.15 所示。

图 4.14

图 4.15

底帽运算的计算方法为闭运算减去原始图像的结果,执行结果如图 4.16 所示。

图 4.16

4.6 图像滤波

图像滤波也称图像平滑或图像模糊,是为了减少噪声和伪影的一种图像处理操作。根据原始图像与核函数计算方式的不同,图像滤波分为线性滤波和非线性滤波。

线性滤波包括方框滤波、均值滤波和高斯滤波,通过线性运算得到目标图像。

非线性滤波包括中值滤波和双边滤波。其中,中值滤波选取 kernel 区域中的中值作为锚点处的值;双边滤波是一种联合滤波方式,对于其具体原理,读者可以自行查阅资料。

4.6.1 案例 41:方框滤波

方框滤波的滤波器是矩形的,滤波器中所有的元素值均相等。OpenCV 提供了方框滤波函数 boxFilter,其定义如下:

```
dst = boxFilter(src, ddepth, ksize, dst=None, anchor=None, normalize=None, borderType=None)
```

参数说明如下。

- src：输入图像。
- ddepth：处理后的目标图像的深度，若为-1，则深度与原始图像的深度相同。
- ksize：滤波运算的核尺寸。
- dst：输出图像（返回值）。
- anchor：锚点。
- normalize：核是否需要被归一化处理。
- borderType：边界模式，由 BorderTypes 定义（见 3.4.5 节）。

为了便于对比滤波效果，本案例将图 4.9 中的图像添加了椒盐噪声。添加噪声后的图像如图 4.17 所示。

图 4.17

方框滤波的案例代码如下：

```
import cv2

#读取图像
src = cv2.imread("noise.jpg")
#方框滤波
box_img = cv2.boxFilter(src, -1, (3,3))
#图像显示
cv2.imshow("box_img", box_img)
cv2.waitKey(0)
cv2.destroyAllWindows()
```

方框滤波的结果如图 4.18 所示。

图 4.18

4.6.2 案例 42：均值滤波

均值滤波是方框滤波的特殊形式，均值滤波的输出图像的深度和输入图像的深度一致，而方框滤波可以以归一化的形式调用（如 3×3 的滤波器，在归一化调用方式中，滤波器的每个元素值为 1/9；在非归一化调用中，每个元素值为 1），且输出图像深度可以控制。OpenCV 提供了均值滤波函数 blur，其定义如下：

```
dst = blur(src, ksize, dst=None, anchor=None, borderType=None)
```

参数说明如下。
- src：输入图像。
- ksize：滤波运算的核尺寸。
- dst：输出图像（返回值）。
- anchor：锚点。
- borderType：边界模式，由 BorderTypes 定义（见 3.4.5 节）。

均值滤波的案例代码如下：

```
import cv2

#读取图像
src = cv2.imread("src.jpg")
#均值滤波
blur_img = cv2.blur(src, (3,3))
#图像显示
cv2.imshow("blur_img", blur_img)
cv2.waitKey(0)
cv2.destroyAllWindows()
```

本案例使用的输入图像如图 4.17 所示，均值滤波的结果如图 4.19 所示。

图 4.19

4.6.3 案例 43：高斯滤波

高斯滤波是最有用的一种滤波方式，OpenCV 为 3×3、5×5、7×7 这 3 种常见的高斯核提供了性能上的优化。OpenCV 提供了高斯滤波函数 GaussianBlur，其定义如下：

```
dst = GaussianBlur(src, ksize, sigmaX, dst=None, sigmaY=None, borderType=None)
```

参数说明如下。
- src：输入图像。
- ksize：滤波运算的核尺寸。
- sigmaX：高斯核在 X 方向上的 sigma 值。
- dst：输出图像（返回值）。
- sigmaY：高斯核在 Y 方向上的 sigma 值。
- borderType：边界模式，由 BorderTypes 定义（见 3.4.5 节）。

高斯滤波的案例代码如下：

```
import cv2

#读取图像
src = cv2.imread("src.jpg")
#高斯滤波
gaussian_img = cv2.GaussianBlur(src, (3,3), 0.8)
#图像显示
cv2.imshow("gaussian_img", gaussian_img)
cv2.waitKey(0)
cv2.destroyAllWindows()
```

本案例使用的输入图像如图 4.17 所示，高斯滤波的结果如图 4.20 所示。

图 4.20

4.6.4 案例 44：双边滤波

高斯滤波可以较好地减弱噪声并保留小信号，但是边缘信息损失比较严重，而双边滤波则能较好地保留边缘信息。OpenCV 提供了双边滤波函数 bilateralFilter，其定义如下：

```
dst = bilateralFilter(src, d, sigmaColor, sigmaSpace, dst=None, borderType=None)
```

参数说明如下。
- src：输入图像。
- d：滤波过程中每个像素邻域的直径范围。
- sigmaColor：颜色空间滤波器的 sigma 值。
- sigmaSpace：坐标空间滤波器的 sigma 值。
- dst：输出图像（返回值）。
- borderType：边界模式，由 BorderTypes 定义（见 3.4.5 节）。

双边滤波的案例代码如下：

```python
import cv2

#读取图像
src = cv2.imread("noise.jpg")
#双边滤波
bilateral_img = cv2.bilateralFilter(src, 0, 50, 50)
#图像显示
cv2.imshow("bilateral_img", bilateral_img)
cv2.waitKey(0)
cv2.destroyAllWindows()
```

本案例使用的输入图像如图 4.17 所示，双边滤波的结果如图 4.21 所示。

图 4.21

4.6.5 案例 45：中值滤波

中值滤波是将每个像素替换为核覆盖范围内的中值像素，对消除像素异常值有显著效果。OpenCV 提供了中值滤波函数 medianBlur，其定义如下：

```
dst = medianBlur(src, ksize, dst=None)
```

参数说明如下。
- src：输入图像。
- ksize：滤波核大小。
- dst：输出图像（返回值）。

中值滤波的案例代码如下：

```
import cv2

#读取图像
src = cv2.imread("noise.jpg")
#中值滤波，ksize=3
median_img = cv2.medianBlur(src, 3)
#图像显示
cv2.imshow("median_img", median_img)
cv2.waitKey(0)
cv2.destroyAllWindows()
```

本案例使用的输入图像如图 4.17 所示，中值滤波的结果如图 4.22 所示。

图 4.22

如图 4.22 所示，中值滤波较好地滤除了图中的噪点，但是也导致图中较小文字的笔画断裂，因为中值滤波选取了核范围内的中值作为滤波结果。

4.7 边缘检测

边缘是图像中非常重要的特征，在图像识别、目标检测等领域发挥着重要的作用。OpenCV 封装了常用的边缘检测算法，如 Sobel、Scharr、Laplacian 和 Canny 边缘检测算法，本节介绍这些算法的使用方法。

4.7.1 案例 46：Sobel 边缘检测

OpenCV 提供了用于 Sobel 边缘检测的函数 Sobel，其定义如下：

dst = Sobel(src, ddepth, dx, dy, dst=None, ksize=None, scale=None, delta=None, borderType=None)

参数说明如下。

- src：输入图像。
- ddepth：输出图像的深度，若将其设置为-1，则深度与输入图像的深度相同。
- dx：计算 x 方向的导数。
- dy：计算 y 方向的导数。
- dst：输出图像（返回值）。
- ksize：核的大小。
- scale：梯度计算结果的放大比例。
- delta：在图像存储前，可以将像素值增加 delta 数值，其默认值为 0。
- borderType：边界模式，由 BorderTypes 定义（见 3.4.5 节）。

本案例调用 Sobel 函数进行边缘检测分为 3 步：第一步，计算 x 方向的边缘；第二步，计算 y 方向的边缘；第三步，将 x 和 y 方向的边缘叠加形成图像的边缘。

本案例使用的输入图像如图 3.10 所示，案例代码如下：

```
import cv2

src = cv2.imread("src.jpg")
#高斯滤波
src = cv2.GaussianBlur(src, (3, 3), 0)
#转为灰度图像
gray = cv2.cvtColor(src, cv2.COLOR_BGR2GRAY)
#求 x 方向的边缘
sobel_gradx = cv2.Sobel(gray, -1, 1, 0)
#求 y 方向的边缘
sobel_grady = cv2.Sobel(gray, -1, 0, 1)
#边缘合并
sobel_grad = cv2.addWeighted(sobel_gradx, 0.5, sobel_grady, 0.5, 0)

#图像显示
cv2.imshow("sobel_gradx", sobel_gradx)
cv2.imshow("sobel_grady", sobel_grady)
cv2.imshow("sobel_grad", sobel_grad)
cv2.waitKey(0)
cv2.destroyAllWindows()
```

计算 x 方向的边缘结果如图 4.23 所示。

计算 y 方向的边缘结果如图 4.24 所示。

图 4.23

图 4.24

合并后的边缘结果如图 4.25 所示。

Sobel 边缘检测也可以同时计算 x 和 y 方向的边缘，即将 dx 和 dy 同时设置为 1，案例代码如下：

```
import cv2

src = cv2.imread("src.jpg")
#高斯滤波
src = cv2.GaussianBlur(src, (3, 3), 0)
#转为灰度图像
gray = cv2.cvtColor(src, cv2.COLOR_BGR2GRAY)
#Sobel 边缘检测，同时计算 x 和 y 方向的边缘
sobel_grad = cv2.Sobel(gray, -1, 1, 1)
cv2.imshow("sobel_grad", sobel_grad)
cv2.waitKey(0)
cv2.destroyAllWindows()
```

计算结果如图 4.26 所示。

图 4.25

图 4.26

4.7.2 案例 47：Scharr 边缘检测

Sobel 边缘检测不是很精准，另外一种对 Sobel 算法进行改进的算法为 Scharr 算法，该算法的精度比 Sobel 算法的精度高且检测速度相当。

OpenCV 提供了用于 Scharr 边缘检测的函数 Scharr，其定义如下：

```
dst = Scharr(src, ddepth, dx, dy, dst=None, scale=None, delta=None, borderType=None)
```

参数说明如下。
- src：输入图像。
- ddepth：输出图像的深度，若将其设置为-1，则深度与输入图像的深度相同。
- dx：计算 x 方向的导数。
- dy：计算 y 方向的导数。
- dst：输出图像（返回值）。
- scale：梯度计算结果的放大比例。
- delta：在图像存储前，可以将像素值增加 delta 数值，其默认值为 0。
- borderType：边界模式，由 BorderTypes 定义（见 3.4.5 节）。

Scharr 算法的核大小为 3，因此，和 Sobel 函数相比，它不用设置 ksize，其他参数均相同。本案例调用 Scharr 函数进行边缘检测同样分为 3 步：第一步，计算 x 方向的边缘；第二步，计算 y 方向的边缘；第三步，将 x 和 y 方向的边缘叠加形成图像的边缘。

案例代码如下：

```python
import cv2

src = cv2.imread("src.jpg")
#高斯滤波
src = cv2.GaussianBlur(src, (3, 3), 0)
#转为灰度图像
gray = cv2.cvtColor(src, cv2.COLOR_BGR2GRAY)

#求 x 方向的边缘
scharr_gradx = cv2.Scharr(gray, -1, 1, 0)
#求 y 方向的边缘
scharr_grady = cv2.Scharr(gray, -1, 0, 1)
#边缘合并
scharr_grad = cv2.addWeighted(scharr_gradx, 0.5, scharr_grady, 0.5, 0)

#图像显示
cv2.imshow("scharr_gradx", scharr_gradx)
cv2.imshow("scharr_grady", scharr_grady)
cv2.imshow("scharr_grad", scharr_grad)
cv2.waitKey(0)
cv2.destroyAllWindows()
```

计算 x 方向的边缘结果如图 4.27 所示，计算 y 方向的边缘结果如图 4.28 所示。

图 4.27　　　　　　　　　　　　　　图 4.28

边缘合并的结果如图 4.29 所示。

图 4.29

注意：cv2.Scharr 函数不支持同时计算 x 和 y 方向的边缘，将 dx 和 dy 同时设置为 1 将会报错。

4.7.3 案例 48：Laplacian 边缘检测

Sobel 边缘检测计算的是一阶梯度，而 Laplacian 边缘检测计算的则是二阶梯度，OpenCV 提供了用于 Laplacian 边缘检测的函数 Laplacian，其定义如下：

```
dst = Laplacian(src, ddepth, dst=None, ksize=None, scale=None, delta=None, borderType=None)
```

参数说明如下。
- src：输入图像。
- ddepth：输出图像的深度，若将其设置为-1，则深度与输入图像的深度相同。
- dst：输出图像（返回值）。
- ksize：用于计算二阶导数滤波器的核大小。
- scale：梯度计算结果的放大比例。
- delta：在图像存储前，可以将像素值增加 delta 数值，其默认值为 0。
- borderType：边界模式，由 BorderTypes 定义（见 3.4.5 节）。

Laplacian 边缘检测的案例代码如下：

```
import cv2

src = cv2.imread("src.jpg")
#高斯滤波
src = cv2.GaussianBlur(src, (3, 3), 0)
#转为灰度图像
```

```
gray = cv2.cvtColor(src, cv2.COLOR_BGR2GRAY)
#Laplacian 边缘检测
laplacian_grad = cv2.Laplacian(gray, -1)
#图像显示
cv2.imshow("laplacian_grad", laplacian_grad)
cv2.waitKey(0)
cv2.destroyAllWindows()
```

Laplacian 边缘检测结果如图 4.30 所示。

图 4.30

4.7.4 案例 49：Canny 边缘检测

Canny 边缘检测是传统图像处理边缘检测中效果较好的算法，该算法先计算 x 和 y 方向的一阶导数，然后由它们组合为 4 个方向导数，而方向导数则是（局部最大值的点）组成边缘的候选项。Canny 算法采用了两个阈值，如果像素的梯度大于较大阈值，则接受其为轮廓；如果小于较小值，则舍弃。对于介于阈值之间的像素，如果连接到大于阈值的像素则接受；否则就舍弃，算法建议的阈值介于 2:1 与 3:1 之间。OpenCV 提供了用于 Canny 边缘检测的函数 Canny，其定义如下：

```
edges = Canny(image, threshold1, threshold2, edges=None, apertureSize=None, L2gradient=None)
```

参数说明如下。
- image：输入图像。
- threshold1：阈值 1。

- threshold2：阈值 2。
- edges：输出的边缘图像（返回值）。
- apertureSize：Sobel 核大小。
- L2gradient：是否使用 L2 梯度。

提示：针对阈值参数 threshold1 和 threshold2，用户只需设置高阈值和低阈值即可，不用关心设置顺序，算法会自动选择高阈值和低阈值。如果阈值设置得较小，则会有更多的边缘被检测到。

Canny 边缘检测的案例代码如下：

```python
import cv2

src = cv2.imread("src.jpg")
#高斯滤波
src = cv2.GaussianBlur(src, (3, 3), 0)
#转为灰度图像
gray = cv2.cvtColor(src, cv2.COLOR_BGR2GRAY)
#Canny 边缘检测
canny_grad = cv2.Canny(gray, 70, 160)
#图像显示
cv2.imshow("canny_grad", canny_grad)
cv2.waitKey(0)
cv2.destroyAllWindows()
```

Canny 边缘检测的结果如图 4.31 所示。

图 4.31

4.8 进阶必备：聊聊颜色模型

颜色模型也称为颜色空间，是用一组数值来描述颜色的数学模型。在图像处理中，颜色模型的选择非常重要，常用的颜色模型有 RGB、HSV 等 6 种。

1. RGB 颜色模型

RGB（Red Green Blue，红绿蓝）颜色模型是依据人眼识别的颜色定义的，采用加法混色法，即将红、绿、蓝 3 种颜色按比例混合叠加产生新的颜色。RGB 是目前应用最为广泛的一种颜色模型，也是工业界最常用的一种颜色标准。

RGB 3 种颜色的取值均为 0～255，即每个通道有 256 种颜色值，可以表示大约 1677 万种颜色，如白色表示为 RGB(255, 255, 255)、黑色表示为 RGB(0, 0, 0)。RGBA 是在 RGB 的基础上增加了阿尔法通道来实现透明效果的。

RGB 颜色模型是最常用的彩色图像和彩色视频的表示方法，但因为它难以数字化调整图像细节，所以在某些场合使用受限。

2. CMY/CMYK 颜色模型

CMY（Cyan-Mag）采用减法混色法，即采用反光的色彩模式，通过光的反射来显示颜色。

CMYK 是在 CMY 的基础上增加了 Key Plate Black（黑色墨盒），CMY 颜色混合会导致亮度降低，而在 CMYK 中加入黑色是因为由青、品红、黄 3 种颜色构成的黑色不够纯粹。

CMYK 是彩色印刷时常用的模式，通过三原色加上黑色 4 种颜色的混合来实现彩色印刷。

3. HSI/HSL 颜色模型

HSL（Hue Saturation Lightness／Luminance，色相饱和度亮度）也称为 HSI（Hue Saturation Intensity，色相饱和度光强）。HSL 颜色模型是为了更好地进行数字化处理颜色而提出的。

HSL 的 H（色相）分量代表人眼能感知的颜色范围，所有的颜色都分布在平面色相环上，每个角度代表一种颜色，其中六大主色为红（360°/0°）、黄（60°）、绿（120°）、青（180°）、蓝（240°）、洋红（300°）。在不改变光感的情况下，旋转色相环可以改变颜色。S（饱和度）分量表示色彩的饱和度，其值为 0%～100%，描述了在色相和明度相同的条件下色彩纯度的变化，S 值越大，颜色中的灰色越少，颜色看起来越鲜艳，当饱和度从 0%变化到 100%时，颜色呈现为从灰色变化到纯色。L 分量表示色彩的明度，用于控制色彩的明暗变化，其值为 0%～

100%，L 值越小，色彩越接近黑色，看起来也就越暗；L 值越大，越接近白色，看起来也就越亮。

4. HSV/HSB 颜色模型

HSV（Hue Saturation Value，色相饱和度色调）也称为 HSB（Hue Saturation Brightness，色相饱和度明度），该颜色模型的提出也是为了更好地进行数字化处理颜色，常用于台式机图形程序的颜色表示。与 RGB 加法混色和 CMYK 减法混色相比，HSV/HSB 在颜色描述上显得更加自然直观，在艺术中常用。

HSV 的 H（色相）分量的含义与 HSL 相同，S（饱和度）分量表示色彩的纯度，值为 0%～100%，S 为 0%时表示灰色，白、黑和其他灰色色彩没有 S 分量，S 最大时表示为每一色相具有的最纯色光。V（亮度）分量表示色彩的亮度，值为 0%～100%，V 为 0 表示黑色，亮度最大时色彩最鲜明。

5. Lab 颜色模型

Lab 颜色模型是由 CIE（国际照明委员会）制定的一种色彩模式，其色彩空间比 RGB 的色彩空间大，可以表示自然界中的任何颜色。相比于 RGB 和 CMYK，Lab 颜色模型不依赖于设备色彩特性。Lab 颜色模型取坐标 Lab，其中，L 表示亮度；a 为正数代表红色，a 为负数代表绿色；b 为正数代表黄色，b 为负数代表蓝色。

6. YUV/YCbCr 颜色模型

YUV 是一种常用的颜色编码方法，是通过亮度—色差来描述颜色的颜色模型。其中，Y 表示亮度，U 和 V 表示色度。色度的作用是描述图像色彩及饱和度，用于指定像素的颜色。色度是由两个互相独立的信号组成的，被称为 UV、PbPr 或 CbCr，它们的差异是由不同的编码格式产生的，但实际上概念基本相同。

第 5 章
图像处理模块 imgproc（二）

本章讲解 imgproc 模块中较为复杂的变换与算法，如霍夫变换、角点检测等。

5.1 霍夫变换

霍夫变换是图像处理中识别基本形状（如线、圆）的方法之一，OpenCV 封装了利用霍夫变换检测直线和圆的算法。

5.1.1 案例 50：霍夫线变换

OpenCV 支持 3 种霍夫线变换，即标准霍夫变换、多尺度霍夫变换（HoughLines 函数）和累积概率霍夫变换（HoughLinesP 函数）。

HoughLines 函数的定义如下：

```
lines = HoughLines(image, rho, theta, threshold, lines=None, srn=None, stn=None, min_theta=None, max_theta=None)
```

参数说明如下。
- image：输入图像，图像需要为 8 位单通道二值图像。
- rho：距离分辨率，单位为像素。
- theta：角度分辨率，单位为弧度。
- threshold：累加平面的阈值参数。
- lines：检测到的直线（返回值）。

- srn：对于多尺度霍夫变换，它是距离分辨率 rho 的除数距离。
- stn：对于多尺度霍夫变换，它是角度分辨率 theta 的除数距离。
- min_theta：检查线条的最小角度，介于 0 与 max_theta 之间。
- max_theta：检查线条的最大角度，介于 min_theta 和 CV_PI 之间。

本节案例使用的输入图像如图 5.1 所示。

图 5.1

使用标准霍夫变换进行直线检测的案例代码如下：

```
import cv2
import numpy as np

img = cv2.imread('chess_board.jpg')
draw_lines = np.zeros(img.shape[:], dtype=np.uint8)
gray = cv2.cvtColor(img, cv2.COLOR_BGR2GRAY)
#Canny 边缘检测
edges = cv2.Canny(gray, 50, 150)
#标准霍夫变换
lines = cv2.HoughLines(edges, 1, np.pi / 180, 150)

#绘制检测到的直线
for line in lines:
    rho, theta = line[0]
    a = np.cos(theta)
    b = np.sin(theta)
    x0 = a * rho
    y0 = b * rho
    x1 = int(x0 + 1000 * (-b))
    y1 = int(y0 + 1000 * (a))
    x2 = int(x0 - 1000 * (-b))
    y2 = int(y0 - 1000 * (a))
    cv2.line(draw_lines, (x1, y1), (x2, y2), (255, 255, 255))

#图像显示
```

```
cv2.imshow("draw_lines", draw_lines)
cv2.waitKey(0)
cv2.destroyAllWindows()
```

标准霍夫变换检测到的直线绘制结果如图 5.2 所示。

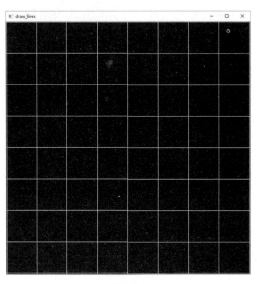

图 5.2

累积概率霍夫变换函数 HoughLinesP 的定义如下：

```
lines = HoughLinesP(image, rho, theta, threshold, lines=None, minLineLength=None,
maxLineGap=None)
```

参数说明如下。

- image：输入图像，图像需要为 8 位单通道二值图像。
- rho：距离分辨率，单位为像素。
- theta：角度分辨率，单位为弧度。
- threshold：累加平面的阈值参数。
- lines：检测到的直线（返回值）。
- minLineLength：小的线段长度。
- maxLineGap：同一行两点连接起来的最大距离。

累积概率霍夫变换案例代码如下：

```
import cv2
import numpy as np

img = cv2.imread('chess_board.jpg')
draw_lines = np.zeros(img.shape[:], dtype=np.uint8)
```

```
gray = cv2.cvtColor(img, cv2.COLOR_BGR2GRAY)
#Canny 边缘检测
edges = cv2.Canny(gray, 50, 150)
#累积概率霍夫变换
lines = cv2.HoughLinesP(edges, 1, np.pi / 180, 90, minLineLength=50, maxLineGap=10)
#绘制检测到的直线
for line in lines:
    x1, y1, x2, y2 = line[0]
    cv2.line(draw_lines, (x1, y1), (x2, y2), (255, 255, 255), 1, lineType = cv2.LINE_AA)

#图像显示
cv2.imshow("draw_lines", draw_lines)
cv2.waitKey(0)
cv2.destroyAllWindows()
```

检测到的直线绘制结果如图 5.3 所示。

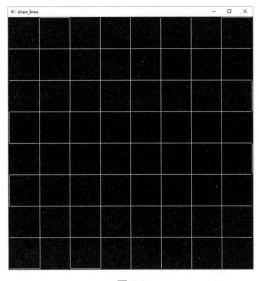

图 5.3

HoughLines 函数中的 threshold 参数设置得越小，就有越多的直线被检测到。HoughLinesP 函数中的 minLineLength 和 maxLineGap 设置得越小，被检测到的直线也会越多。读者可以根据图像的先验知识设置合适的参数，这样能够达到更加精确的检测效果。

5.1.2 案例 51：霍夫圆变换

OpenCV 提供了用于霍夫圆变换的函数 HoughCircles，其定义如下：

```
circles = HoughCircles(image, method, dp, minDist, circles=None, param1=None, param2=None,
minRadius=None, maxRadius=None)
```

参数说明如下。

- image：输入图像，需要传入 8 位单通道二值图像。
- method：检测方法。
- dp：累加器分辨率与图像分辨率的反比。
- minDist：圆心之间的最小距离。
- circles：找到的圆（返回值）。
- param1：Canny 边缘检测算法的高阈值，低阈值是它的 1/2。
- param2：累加器阈值，表示确定一个圆的阈值。
- minRadius：待检测圆的最小半径。
- maxRadius：待检测圆的最大半径。

本案例使用的输入图像如图 5.4 所示。

图 5.4

霍夫圆变换的案例代码如下：

```
import cv2
import numpy as np

img = cv2.imread('circle_src.jpg')
#创建纯黑色图像
draw_circle = np.zeros(img.shape[:], dtype=np.uint8)
gray = cv2.cvtColor(img, cv2.COLOR_BGR2GRAY)
#Canny 边缘检测
edges = cv2.Canny(gray, 50, 150)
#霍夫圆变换
circles = cv2.HoughCircles(edges, cv2.HOUGH_GRADIENT, 1, 100)
#浮点数转整型数
circles = np.int0(np.around(circles))

#将检测到的圆绘制出来
for i in circles[0, :]:
    cv2.circle(draw_circle, (i[0], i[1]), i[2], (255, 255, 255), 2)
#图像显示
cv2.imshow("edges", edges)
cv2.imshow("draw_circle", draw_circle)
```

```
cv2.waitKey(0)
cv2.destroyAllWindows()
```

使用 Canny 算法进行边缘检测的结果如图 5.5 所示，使用霍夫圆变换检测到的圆的绘制结果如图 5.6 所示。

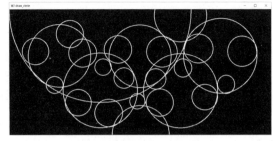

图 5.5　　　　　　　　　　　　　　图 5.6

图 5.6 所示的检测效果不尽如人意，有很多不是期望的圆被检测并绘制出来。本案例中设置的参数 dp=1，即累加器与输入图像具有相同的分辨率；设置的 minDist 为 100 像素，即检测到的圆心之间的最小距离为 100 像素。根据图 5.5 中的圆的信息，可以调整输入参数，设置待检测圆的最小半径 minRadius 为 130 像素，最大半径 maxRadius 为 180 像素：

```
circles = cv2.HoughCircles(edges, cv2.HOUGH_GRADIENT, 1, 100, param2=30, minRadius=130,
maxRadius=180)
```

设置新参数后的检测结果如图 5.7 所示，可见，检测的结果比较准确。

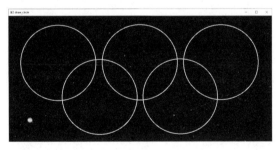

图 5.7

5.2　案例 52：仿射变换

仿射变换是实现图像旋转、平移和缩放的常见操作，仿射变换后的图像具有平行性与平直性，即能够很好地保持原有图像中的直线关系与平行关系。OpenCV 提供了用于仿射变换的函数 warpAffine，其定义如下：

```
dst = warpAffine(src, M, dsize, dst=None, flags=None, borderMode=None, borderValue=None)
```

参数说明如下。

- src：输入图像。
- M：2×3 的变换矩阵。
- dsize：输出图像的尺寸。
- dst：输出图像（返回值）。
- flags：插值方式，由 InterpolationFlags 定义（见 4.3 节）。
- borderMode：边界模式，由 BorderTypes 定义（见 3.4.5 节）。
- borderValue：边界模式为 BORDER_CONSTANT 时的边界值。

仿射变换需要通过函数 getRotationMatrix2D 生成变换矩阵，该函数的定义如下：

```
retval = getRotationMatrix2D(center, angle, scale)
```

参数说明如下。

- center：输入图像的旋转中心。
- angle：旋转角度，正数值为逆时针旋转。
- scale：缩放系数。
- retval：生成的旋转矩阵（返回值）。

如果设置旋转角度 angle 为 0°，缩放系数为 0.6，就可以实现图像的缩放，案例代码如下：

```python
import cv2

img = cv2.imread('src.jpg')
rows, cols, channels = img.shape
#获取仿射变换矩阵
M = cv2.getRotationMatrix2D((cols/2, rows/2), 0, 0.6)
#仿射变换
affine_img = cv2.warpAffine(img, M, (cols, rows))
#图像显示
cv2.imshow("affine_img", affine_img)
cv2.waitKey(0)
cv2.destroyAllWindows()
```

缩放的结果如图 5.8 所示。

如果设置旋转角度 angle 为 45°，即逆时针旋转 45°，缩放系数为 0.6，就可以实现图像的旋转和缩放，变换矩阵的生成代码如下：

```
M = cv2.getRotationMatrix2D((cols/2, rows/2), 45, 0.6)
```

旋转缩放的结果如图 5.9 所示。

图 5.8   图 5.9

5.3 案例 53：透视变换

透视变换是在二维平面获得接近真实三维物体的视觉效果的一种算法，OpenCV 提供了用于透视变换的函数 warpPerspective，其定义如下：

`dst = warpPerspective(src, M, dsize, dst=None, flags=None, borderMode=None, borderValue=None)`

参数含义说明如下。

- src：输入图像。
- M：3×3 的变换矩阵。
- dsize：输出图像的尺寸。
- dst：输出图像（返回值）。
- flags：插值方式，由 InterpolationFlags 定义。
- borderMode：边界模式，由 BorderTypes 定义（见 3.4.5 节）。
- borderValue：边界模式为 BORDER_CONSTANT 时的边界值。

插值方式 InterpolationFlags 的定义如下：

```
enum InterpolationFlags{
    INTER_NEAREST       = 0,          #最近邻插值
    INTER_LINEAR        = 1,          #双线性插值
    INTER_CUBIC         = 2,          #双三次插值
    INTER_AREA          = 3,          #不支持
    INTER_LANCZOS4      = 4,          #Lanczos 插值
    INTER_LINEAR_EXACT  = 5,          #按位双线性插值
```

```
    INTER_MAX            = 7,        #差值代码 mask
    WARP_FILL_OUTLIERS   = 8,        #是否填充所有目标图像像素的标志
    WARP_INVERSE_MAP     = 16        #是否翻转变换的标志
};
```

3×3 的变换矩阵可以通过函数 getPerspectiveTransform 获取，其定义如下：

```
retval = getPerspectiveTransform(src, dst, solveMethod=None)
```

函数参数说明如下。

- src：输入图像中四边形顶点的坐标。
- dst：输出图像中相应四边形顶点的坐标。
- solveMethod：矩阵分解类型。
- retval：生成的透视变换矩阵（返回值）。

使用 warpPerspective 函数进行透视变换的案例代码如下：

```python
import cv2
import numpy as np

img = cv2.imread('src.jpg')
height, width, channel = img.shape
#设置原始图像中四边形顶点的坐标
pSrc = np.float32([[0, 0], [width-1, 0], [width-1, height-1], [0, height-1]])
#设置目标图像中相应四边形顶点的坐标
pDst = np.float32([[width*0.02, height*0.1], [width*0.8, width*0.2], [width*0.9, height*0.8], [width*0.2, height*0.8]])
#生成 3×3 的透视变换矩阵
M = cv2.getPerspectiveTransform(pSrc, pDst)
#透视变换
dst = cv2.warpPerspective(img, M, (height, width))
cv2.imshow("perspective_result", dst)
cv2.waitKey(0)
cv2.destroyAllWindows()
```

透视变换的结果如图 5.10 所示。

图 5.10

5.4 案例 54：重映射

重映射是指将输入图像与输出图像的像素按照一定的对应关系进行映射变换，OpenCV 中进行重映射的函数为 remap，其定义如下：

```
dst = remap(src, map1, map2, interpolation, dst=None, borderMode=None, borderValue=None)
```

参数说明如下。

- src：输入图像，需要传入单通道 8 位整型图像或浮点型图像。
- map1：第一个 map，对应(x,y)的变化或仅 x 值的变化。
- map2：第二个 map，对应 y 值的变化。
- interpolation：插值方式，由 InterpolationFlags 定义（见 4.3 节）。
- dst：输出图像（返回值）。
- borderMode：边界模式，由 BorderTypes 定义（见 3.4.5 节）。
- borderValue：当边界模式为 BORDER_CONSTANT 时的边界值。

本案例实现输入图像在水平方向上的翻转，实现的案例代码如下：

```python
import cv2
import numpy as np

img = cv2.imread("src.jpg", 0)
mapx = np.zeros_like(img, dtype=np.float32)
mapy = np.zeros_like(img, dtype=np.float32)

#水平翻转
for i in range(mapx.shape[1]):
    mapx[:, i:i+1] = mapx.shape[1] - i - 1

#垂直方向保持不变
for j in range(mapx.shape[0]):
    mapy[j:j+1, :] = j

#重映射
remap_x = cv2.remap(img, mapx, mapy, cv2.INTER_NEAREST)
cv2.imshow("src_img", img)
cv2.imshow("remap_x", remap_x)
cv2.waitKey(0)
cv2.destroyAllWindows()
```

水平翻转的结果如图 5.11 所示，其中，左图为输入图像，右图为水平翻转后的图像。

图 5.11

如果想实现图像沿着水平和垂直方向都翻转，就可以修改对应的映射关系：

```
#水平翻转
for i in range(mapx.shape[1]):
    mapx[:, i:i+1] = mapx.shape[1] - i - 1

#垂直翻转
for j in range(mapx.shape[0]):
    mapy[j:j+1, :] = mapx.shape[0] - j - 1

#重映射
remap_img = cv2.remap(img, mapx, mapy, cv2.INTER_NEAREST)
cv2.imshow("src_img", img)
cv2.imshow("remap_x", remap_img)
cv2.waitKey(0)
cv2.destroyAllWindows()
```

翻转的结果如图 5.12 所示，其中，左图为输入图像，右图为翻转后的图像。

图 5.12

5.5 阈值化

在图像处理过程中，经常需要根据像素值高于或低于一个像素值做出决定，将高于或低于这一像素值的像素置零，同时其他像素保持不变，这种操作可以通过阈值化来实现。OpenCV 提供了基本阈值化和自适应阈值化两种方式。

5.5.1 案例 55：基本阈值化

OpenCV 中用于执行基本阈值化操作的函数为 threshold，其定义如下：

```
dst = threshold(src, thresh, maxVal, type, dst=None)
```

参数说明如下。
- src：输入图像。
- thresh：设定的阈值。
- maxVal：参数 type 为 THRESH_BINARY 或 THRESH_BINARY_INV 时的最大值。
- type：阈值类型，由 ThresholdTypes 定义（见 4.1 节）。
- dst：输出图像（返回值）。

可以通过设置阈值类型 type 来计算不同类型的阈值。将阈值类型设置为 THRESH_BINARY（二值阈值）的案例代码如下：

```
import cv2

src = cv2.imread("src.jpg", cv2.IMREAD_GRAYSCALE)
#应用二值阈值
_, thresh_bin = cv2.threshold(src, 128, 255, cv2.THRESH_BINARY)
cv2.imshow("thresh_bin", thresh_bin)
cv2.waitKey(0)
cv2.destroyAllWindows()
```

将阈值类型设置为 THRESH_BINARY，阈值化后，输入图像中像素值高于阈值的被设置为最大值 maxVal，低于阈值的被设置为 0。二值阈值化后的结果如图 5.13 所示。

读者可以设置由 ThresholdTypes 定义的其他阈值类型，如果将阈值类型 type 设置为 THRESH_BINARY_INV，则阈值化后，输入图像中像素值高于阈值的被设置为 0，低于阈值的被设置为 maxVal。THRESH_BINARY_INV 阈值化后的结果如图 5.14 所示。

图 5.13

图 5.14

如果将阈值类型 type 设置为 THRESH_TRUNC，则阈值化后，输入图像中像素值高于阈值的被设置为阈值，低于阈值的被设置为 0。THRESH_TRUNC 阈值化后的结果如图 5.15 所示。

图 5.15

其他阈值化类型就不做一一展示了，有兴趣的读者可以自行修改阈值类型 type 进行尝试。

5.5.2 案例 56：自适应阈值化

基本阈值化需要读者手动设置阈值进行阈值化操作，阈值的设置需要尝试选出最佳值，而自适应阈值化则可以变化阈值而完成阈值化操作。OpenCV 提供了用于进行自适应阈值化操作的函数 adaptiveThreshold，其定义如下：

```
dst = adaptiveThreshold(src, maxValue, adaptiveMethod, thresholdType, blockSize, C, dst=None)
```

参数说明如下。

- src：输入图像，需要传入单通道图像。
- maxValue：分配给满足条件的像素的非零像素值。
- adaptiveMethod：自适应算法，由 AdaptiveThresholdTypes 定义（见 4.1 节）。
- thresholdType：阈值类型。
- blockSize：计算阈值的邻域尺寸。
- C：减去平均值或加权平均值后的常数值。
- dst：输出图像（返回值）。

自适应阈值化可以使用的自适应算法包括两种：平均法（ADAPTIVE_THRESH_MEAN_C）和高斯法（ADAPTIVE_THRESH_GAUSSIAN_C）。这两种算法的自适应阈值化的案例代码如下：

```
import cv2

#读取灰度图像
src = cv2.imread("src.jpg", cv2.IMREAD_GRAYSCALE)
#平均法自适应阈值化
thresh_mean = cv2.adaptiveThreshold(src, 255, cv2.ADAPTIVE_THRESH_MEAN_C, cv2.THRESH_BINARY, 5, 0)
#高斯法自适应阈值化
thresh_gaussian = cv2.adaptiveThreshold(src, 255, cv2.ADAPTIVE_THRESH_GAUSSIAN_C, cv2.THRESH_BINARY, 5, 0)
#图像显示
cv2.imshow("thresh_mean", thresh_mean)
cv2.imshow("thresh_gaussian", thresh_gaussian)
cv2.waitKey(0)
cv2.destroyAllWindows()
```

平均法自适应阈值化的输出结果如图 5.16 所示，高斯法自适应阈值化的输出结果如图 5.17 所示。

图 5.16　　　　　　　　　　　　　　　　图 5.17

5.6　图像金字塔

图像金字塔是来源于同一图像不同分辨率的图像组成的集合，在图像缩放或图像分割中有较多应用。图像金字塔由采样完成，向下采样形成高斯金字塔，向上采样形成拉普拉斯金字塔。

 注意：高斯金字塔下采样生成过程不是拉普拉斯金字塔上采样生成过程的逆，因为在生成高斯金字塔的过程中存在信息的丢失现象，这些丢失的信息在上采样中不可恢复。

5.6.1　案例 57：高斯金字塔

OpenCV 提供了用于生成高斯金字塔的函数 pyrDown，其定义如下：

```
dst = pyrDown(src, dst=None, dstsize=None, borderType=None)
```

参数说明如下。

- src：输入图像。
- dst：输出图像（返回值）。
- dstsize：输出图像的尺寸。
- borderType：边界模式，由 BorderTypes 定义（见 3.4.5 节）。

本案例使用的输入图像如图 3.10 所示，下采样 1 次的案例代码如下：

```
import cv2
```

```
src = cv2.imread("src.jpg")
#图像下采样
pyrdown1 = cv2.pyrDown(src)
#图像显示
cv2.imshow("src", src)
cv2.imshow("pyrdown1", pyrdown1)
cv2.waitKey(0)
cv2.destroyAllWindows()
```

执行 1 次下采样的结果如图 5.18 所示。

图 5.18

执行 3 次下采样的案例代码如下:

```
import cv2

src = cv2.imread("src.jpg")
#3 次下采样
pyrdown1 = cv2.pyrDown(src)
pyrdown2 = cv2.pyrDown(pyrdown1)
pyrdown3 = cv2.pyrDown(pyrdown2)
#图像显示
cv2.imshow("src", src)
cv2.imshow("pyrdown1", pyrdown1)
cv2.imshow("pyrdown2", pyrdown2)
cv2.imshow("pyrdown3", pyrdown3)
cv2.waitKey(0)
cv2.destroyAllWindows()
```

执行 3 次下采样得到的高斯金字塔图像如图 5.19 所示。

图 5.19

5.6.2 案例 58：拉普拉斯金字塔

OpenCV 提供了用于生成拉普拉斯金字塔的函数 pyrUp，其定义如下：

dst = pyrUp(src, dst=None, dstsize=None, borderType=None)

参数说明如下。

- src：输入图像。
- dst：输出图像（返回值）。
- dstsize：输出图像的尺寸。
- borderType：边界模式，由 BorderTypes 定义（见 3.4.5 节）。

进行 3 次上采样，建立拉普拉斯金字塔的案例代码如下：

```
import cv2

src = cv2.imread("src_pyrup.jpg")
#3次上采样
pyrup1 = cv2.pyrUp(src)
pyrup2 = cv2.pyrUp(pyrup1)
pyrup3 = cv2.pyrUp(pyrup2)
#图像显示
cv2.imshow("src", src)
cv2.imshow("pyrup1", pyrup1)
cv2.imshow("pyrup2", pyrup2)
cv2.imshow("pyrup3", pyrup3)
cv2.waitKey(0)
cv2.destroyAllWindows()
```

执行 3 次上采样形成的拉普拉斯金字塔图像如图 5.20 所示。

图 5.20

5.7 直方图

直方图是一种揭示数据分布的统计特性的工具,将数据归入预先定义的不同组中,并对每个组进行计数,用以表示数据的统计分布信息。在图像处理中,可以使用直方图统计图像中的亮度或像素灰度值,也可以对提取的图像特征进行计数形成直方图。它是计算机视觉中进行图像特征分析的一种重要工具。

5.7.1 案例 59:直方图计算

OpenCV 提供的用于直方图计算的函数为 calcHist,其定义如下:
```
hist = calcHist(images, channels, mask, histSize, ranges, hist=None, accumulate=None)
```
参数说明如下。
- images:输入图像。
- channels:用于计算直方图的通道列表。
- mask:掩模,计算掩模内的直方图。
- histSize:直方图分成的区间数。
- ranges:统计像素值的区间。
- hist:输出的直方图数组(返回值)。
- accumulate:多幅图像是否累积计算像素值个数。

本案例使用的输入图像如图 3.10 所示，直方图计算并绘制的案例代码如下：

```python
import cv2
import numpy as np

img = cv2.imread('src.jpg')
#计算通道 0 直方图
hist = cv2.calcHist([img], [0], None, [256], [0,255])
#获取直方图中的最大值和最小值
minVal, maxVal, minLoc, maxLoc = cv2.minMaxLoc(hist)
#创建纯白色图像用于绘制直方图
hist_img = np.zeros([256, 256], np.uint8)
hist_img[:] = 255
#直方图绘制
for i in range(256):
    #将直方图中的像素统计数值归一化到[0, 256]区间
    norm_value = int(hist[i] * 256 / maxVal)
    #以黑色线绘制每个像素值的数量
    cv2.line(hist_img, (i, 256), (i, 256 - norm_value), [0, 0, 0])
cv2.imshow("hist_img", hist_img)
cv2.waitKey(0)
cv2.destroyAllWindows()
```

计算直方图并绘制的结果如图 5.21 所示。

对于亮度较暗的图像，直方图数据会比较集中在左侧区域，如图 5.22 所示。

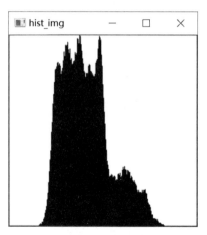

图 5.21

图 5.22

对于亮度较亮的图像，直方图数据会比较集中在右侧区域，如图 5.23 所示。

图 5.23

5.7.2 案例 60：直方图均衡化

直方图均衡化是一种增强图像对比度的方法，主要思想是让一幅图像的直方图均匀分布于 0 到 255 的范围内，从而增强对比度。

OpenCV 提供了用于直方图均衡化的函数 equalizeHist，其定义如下：

```
dst = equalizeHist(src, dst=None)
```

参数说明如下。

- src：输入图像。
- dst：均衡化后的输出图像（返回值）。

直方图均衡化的案例代码如下：

```python
import cv2

src = cv2.imread('src1.jpg', 0)
#直方图均衡化
equalize_img = cv2.equalizeHist(src)
#图像显示
cv2.imshow("src_img", src)
cv2.imshow("equalize_img", equalize_img)
cv2.waitKey(0)
cv2.destroyAllWindows()
```

均衡化之后的图像与原始图像的对比如图 5.24 所示。

均衡化之后的图像直方图如图 5.25 所示，图中像素值均匀分布在 0 到 255 之间。

图 5.24

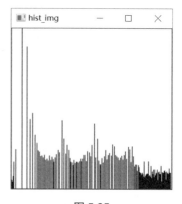
图 5.25

5.8 传统图像分割

图像分割是计算机视觉研究中非常活跃的一个领域,即使是深度学习技术非常先进的今天,也没有一种通用有效的方法。虽然如此,但经过长期研究,已经形成了一些比较稳定的图像分割算法,本节就讲解 OpenCV 中封装的常用图像分割算法。

5.8.1 案例 61:分水岭算法

在没有背景模板可用的情况下,分水岭算法首先计算强度图像的梯度(如查找轮廓),形成的线条组成了山脉或岭,没有纹理的地方形成盆地或山谷;然后从指定的点向盆地灌水,当图像被灌满时,所有有标记的区域就被分割开了,这就是分水岭算法的分割思想。

OpenCV 中提供了分水岭算法函数 watershed,其定义如下:

```
markers = watershed(image, markers)
```

参数说明如下。

- image:输入图像,需要传入 8 位 3 通道图像。
- markers:与输入图像大小相同的标记图像(返回值)。

本案例使用的案例代码来源于 OpenCV 源码中的样例,位置为"samples/cpp/watershed.cpp",为了便于理解,对源码做了一些调整,如增加中文注释。案例中使用鼠标标记分割目标(标记点即注水点),然后使用分水岭算法进行图像分割,案例代码如下:

```
#include <opencv2/core/utility.hpp>
#include "opencv2/imgproc.hpp"
```

```cpp
#include "opencv2/imgcodecs.hpp"
#include "opencv2/highgui.hpp"

#include <cstdio>
#include <iostream>

using namespace cv;
using namespace std;

//帮助说明
static void help(char** argv)
{
    //第一个参数传入的是输入图像，默认是 fruits.jpg
    cout << "\nThis program demonstrates the famous watershed segmentation algorithm in OpenCV: watershed()\n"
        "Usage:\n" << argv[0] << " [image_name -- default is fruits.jpg]\n" << endl;

    //操作键：按住鼠标左键，拖动选取前景；按 Esc 键，退出操作；按 R 键，图像复原；按 W 或空格键，执行操作
    cout << "Hot keys: \n"
        "\tESC - quit the program\n"
        "\tr - restore the original image\n"
        "\tw or SPACE - run watershed segmentation algorithm\n"
        "\t\t(before running it, *roughly* mark the areas to segment on the image)\n"
        "\t\t(before that, roughly outline several markers on the image)\n";
}
Mat markerMask, img;
Point prevPt(-1, -1);

//鼠标事件
static void onMouse(int event, int x, int y, int flags, void*)
{
    //若图像有问题或鼠标当前点在图像外，则返回不处理
    if (x < 0 || x >= img.cols || y < 0 || y >= img.rows)
        return;
    //鼠标左键抬起
    if (event == EVENT_LBUTTONUP || !(flags & EVENT_FLAG_LBUTTON))
        prevPt = Point(-1, -1);
    //鼠标左键按下
    else if (event == EVENT_LBUTTONDOWN)
        prevPt = Point(x, y);
    //鼠标移动
    else if (event == EVENT_MOUSEMOVE && (flags & EVENT_FLAG_LBUTTON))
    {
        Point pt(x, y);
        if (prevPt.x < 0)
            prevPt = pt;
```

```cpp
        //绘制标记线
        line(markerMask, prevPt, pt, Scalar::all(255), 5, 8, 0);
        line(img, prevPt, pt, Scalar::all(255), 5, 8, 0);
        prevPt = pt;
        imshow("image", img);
    }
}

//执行主函数, 程序入口
int main(int argc, char** argv)
{
    //参数解析
    cv::CommandLineParser parser(argc, argv, "{help h | | }{ @input | fruits.jpg | }");
    if (parser.has("help"))
    {
        help(argv);
        return 0;
    }
    //获取输入图像文件名
    string filename = samples::findFile(parser.get<string>("@input"));
    Mat img0 = imread(filename, 1), imgGray;

    if (img0.empty())
    {
        cout << "Couldn't open image ";
        help(argv);
        return 0;
    }
    //创建图像窗口
    namedWindow("image", 1);

    img0.copyTo(img);
    cvtColor(img, markerMask, COLOR_BGR2GRAY);
    cvtColor(markerMask, imgGray, COLOR_GRAY2BGR);
    markerMask = Scalar::all(0);
    imshow("image", img);
    //鼠标事件回调函数
    setMouseCallback("image", onMouse, 0);

    for (;;)
    {
        char c = (char)waitKey(0);

        //按 Esc 键, 退出
        if (c == 27)
            break;
```

```cpp
        //按 R 键,图像复原
        if (c == 'r')
        {
            markerMask = Scalar::all(0);
            img0.copyTo(img);
            imshow("image", img);
        }

        //按 W 或空格键,执行算法
        if (c == 'w' || c == ' ')
        {
            int i, j, compCount = 0;
            vector<vector<Point> > contours;
            vector<Vec4i> hierarchy;

            //查找轮廓
            findContours(markerMask, contours, hierarchy, RETR_CCOMP, CHAIN_APPROX_SIMPLE);

            if (contours.empty())
                continue;
            Mat markers(markerMask.size(), CV_32S);
            markers = Scalar::all(0);
            int idx = 0;
            //绘制轮廓
            for (; idx >= 0; idx = hierarchy[idx][0], compCount++)
                drawContours(markers, contours, idx, Scalar::all(compCount + 100), -1, 8, hierarchy, INT_MAX);

            if (compCount == 0)
                continue;

            vector<Vec3b> colorTab;
            for (i = 0; i < compCount; i++)
            {
                int b = theRNG().uniform(0, 255);
                int g = theRNG().uniform(0, 255);
                int r = theRNG().uniform(0, 255);

                colorTab.push_back(Vec3b((uchar)b, (uchar)g, (uchar)r));
            }

            double t = (double)getTickCount();
            imwrite("markers.jpg", markers);
            //执行分水岭算法
            watershed(img0, markers);
```

```cpp
        t = (double)getTickCount() - t;
        //打印处理时间
        printf("execution time = %gms\n", t * 1000. / getTickFrequency());

        Mat wshed(markers.size(), CV_8UC3);
        cout << wshed.at<int>(0, 0) << endl;

        // 绘制分割结果图像，对不同分割区域着色
        for (i = 0; i < markers.rows; i++) {
            for (j = 0; j < markers.cols; j++)
            {
                int index = markers.at<int>(i, j);
                if (index == -1)
                    wshed.at<Vec3b>(i, j) = Vec3b(255, 255, 255);
                else if (index <= 0 || index > compCount)
                    wshed.at<Vec3b>(i, j) = Vec3b(0, 0, 0);
                else
                    wshed.at<Vec3b>(i, j) = colorTab[index - 1];
            }
        }

        //显示分水岭算法的分割线
        imshow("watershed", wshed);
        wshed = wshed * 0.5 + imgGray * 0.5;
        //显示分水岭算法的结果
        imshow("watershed_result", wshed);
    }
}

return 0;
}
```

本案例使用的输入图像如图 5.26 所示。

在输入图像中，用鼠标左键标识待分割的目标，如图 5.27 所示。

图 5.26

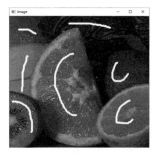
图 5.27

按下 W 或空格键，执行分割算法，分割结果如图 5.28 所示。
分割结果在输入图像中绘制的结果如图 5.29 所示。

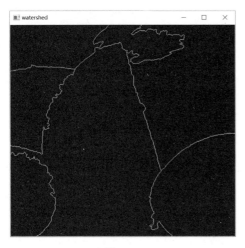

图 5.28

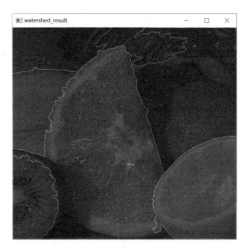

图 5.29

5.8.2 案例 62：GrabCut 算法

GrabCut 是一种交互式前景提取算法，OpenCV 允许读者在待分割的图像周围提供矩形框，矩形框之外的部分属于背景，此时不用指定前景。读者也可以使用一个全局掩膜，将图像的像素点分为确定前景、确定背景，以及疑似前景和疑似背景，这样，确定区域将被算法用于对疑似区域进行分割。

OpenCV 中提供了 GrabCut 算法函数 grabCut，其定义如下：
`mask, bgdModel, fgdModel = grabCut(img, mask, rect, bgdModel, fgdModel, iterCount, mode=None)`
参数说明如下。

- img：输入图像。
- mask：掩模（返回值）。
- rect：包含分割对象的感兴趣区域。
- bgdModel：背景模型的临时数组（返回值）。
- fgdModel：前景模型的临时数组（返回值）。
- iterCount：算法迭代次数。
- mode：处理模式，由 GrabCutModes 定义。

本案例实现了一个交互式的图像分割，用鼠标左键选取确定前景，用鼠标右键选取确定背景，GrabCut 算法根据确定前景和确定背景进行图像分割。

提示：图像窗口操作和鼠标事件在模块 highgui 中定义，该部分内容将在第 6 章讲解。

GrabCut 算法图像分割案例代码如下：

```python
import cv2
import numpy as np

#绘制前景/背景标识线标志
drawing = False

#定义 GrabCut 类，作用是设置一些参数
class GrabCut:
    def __init__(self, t_img):
        self.img = t_img
        self.img_raw = img.copy()
        self.img_width = img.shape[0]
        self.img_height = img.shape[1]
        self.img_show = self.img.copy()
        self.img_gc = self.img.copy()
        self.img_gc = cv2.GaussianBlur(self.img_gc, (3, 3), 0)
        self.lb_up = False
        self.rb_up = False
        self.lb_down = False
        self.rb_down = False
        self.mask = np.full(self.img.shape[:2], 2, dtype=np.uint8)
        self.firt_choose = True

#鼠标操作的回调函数
def mouse_event(event, x, y, flags, param):
    global drawing, last_point, start_point
    #鼠标左键按下，开始标识前景
    if event == cv2.EVENT_LBUTTONDOWN:
        drawing = True
        #设置鼠标按下的起始点
        last_point = (x, y)
        start_point = last_point
        param.lb_down = True
    #鼠标右键按下，开始标识背景
    elif event == cv2.EVENT_RBUTTONDOWN:
        #要先标识前景，否则无法分割
        if param.firt_choose:
            print("Please select foreground first!")
            return
        drawing = True
```

```python
            last_point = (x, y)
            start_point = last_point
            param.rb_down = True
        #鼠标移动，绘制标识前景和背景的线
        elif event == cv2.EVENT_MOUSEMOVE:
            if drawing:
                #鼠标左键按下的绘制
                if param.lb_down:
                    cv2.line(param.img_show, last_point, (x,y), (0, 0, 255), 2, -1)
                    cv2.rectangle(param.mask, last_point, (x, y), 1, -1, 4)
                #鼠标右键按下的绘制
                if param.rb_down:
                    cv2.line(param.img_show, last_point, (x, y), (255, 0, 0), 2, -1)
                    cv2.rectangle(param.mask, last_point, (x, y), 0, -1, 4)
                last_point = (x, y)
        #左键释放，结束标识前景
        elif event == cv2.EVENT_LBUTTONUP:
            drawing = False
            param.lb_up = True
            param.lb_down = False
            cv2.line(param.img_show, last_point, (x,y), (0, 0, 255), 2, -1)
            #如果是第一次标识，则切换状态
            if param.firt_choose:
                param.firt_choose = False
            cv2.rectangle(param.mask, last_point, (x,y), 1, -1, 4)
        #右键释放，结束标识背景
        elif event == cv2.EVENT_RBUTTONUP:
            #如果首先标识背景，则不做处理
            if param.firt_choose:
                return
            drawing = False
            param.rb_up = True
            param.rb_down = False
            cv2.line(param.img_show, last_point, (x,y), (255, 0, 0), 2, -1)
            cv2.rectangle(param.mask, last_point, (x,y), 0, -1, 4)

#执行操作
def process(img):
    if img is None:
        print('Can not read image correct!')
        return
    g_img = GrabCut(img)

    cv2.namedWindow('image')
    #定义鼠标的回调函数
```

```python
        cv2.setMouseCallback('image', mouse_event, g_img)
        while (True):
            cv2.imshow('image', g_img.img_show)
            #当鼠标左键或右键抬起时,按照标识执行GrabCut算法
            if g_img.lb_up or g_img.rb_up:
                g_img.lb_up = False
                g_img.rb_up = False
                #背景model
                bgdModel = np.zeros((1, 65), np.float64)
                #前景model
                fgdModel = np.zeros((1, 65), np.float64)

                rect = (1, 1, g_img.img.shape[1], g_img.img.shape[0])
                mask = g_img.mask
                g_img.img_gc = g_img.img.copy()
                #执行GrabCut算法
                cv2.grabCut(g_img.img_gc, mask, rect, bgdModel, fgdModel, 5, cv2.GC_INIT_WITH_MASK)
                #0 和 2 做背景
                mask2 = np.where((mask == 2) | (mask == 0), 0, 1).astype('uint8')
                #使用蒙版获取前景区域
                g_img.img_gc = g_img.img_gc * mask2[:, :, np.newaxis]
                cv2.imshow('Grabcut_result', g_img.img_gc)

            #按下Esc键,退出
            if cv2.waitKey(20) == 27:
                break

if __name__ == '__main__':
    img = cv2.imread("./src.jpg")
    process(img)
```

第一次选取前景的操作如图 5.30 所示。

图 5.31 为分割后的结果,可见,图中人物被有效地抠取出来了。

图 5.30

图 5.31

继续使用鼠标右键选取背景区域，如图 5.32 所示。

抠取人像的结果如图 5.33 所示。

图 5.32

图 5.33

如果分割效果不佳，则可以继续选取前景或背景，多次迭代，以进行更加精细化的分割。

5.8.3 案例 63：漫水填充算法

漫水填充算法是一种常用的填充算法，常被用于生成进行进一步图像分析的标记图像或掩模图像。漫水填充的结果是一个单连通域，算法思想为选取一个填充的种子点，与该点相似的点（像素值差异在指定范围内）被填充为同一种颜色。

OpenCV 中提供了漫水填充算法函数 floodFill，其定义如下：

```
retval, image, mask, rect = floodFill(image, mask, seedPoint, newVal, loDiff=None, upDiff=None, flags=None)
```

参数说明如下。

- image：输入图像（返回值）。
- mask：掩模（返回值）。
- seedPoint：注水点（填充算法的起始点）。
- newVal：填充的新的颜色值。
- loDiff：注水点邻域中的像素点与注水点像素值差值的下限。
- upDiff：注水点邻域中的像素点与注水点像素值差值的上限。
- flags：操作标志符。
- rect：将要重绘区域的最小边界矩形区域（返回值）。

漫水填充算法的案例代码如下：

```
import cv2
import numpy as np
```

```
src = cv2.imread("fruits.jpg")
#创建 mask，尺寸需要比输入图像大 2 像素
mask = np.zeros([src.shape[0]+2, src.shape[1]+2], np.uint8)
#执行漫水填充算法，填充颜色为(128,128,128)
cv2.floodFill(src, mask, (160, 240), (128,128,128), (10,10,10), (10,10,10))    #执行填充算法
cv2.imshow("floodFill_Result", src)
cv2.waitKey(0)
cv2.destroyAllWindows()
```

执行漫水填充算法后的分割结果如图 5.34 所示。

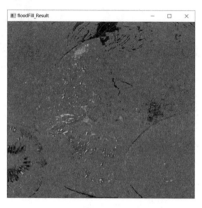

图 5.34

5.9 角点检测

角点是含有丰富局部信息的一小块图像，在具有相同场景或目标的另一幅图像中，角点能够以一种相同的不变形式进行表示。它是一种重要的局部特征，在目标追踪中有重要作用。

5.9.1 案例 64：Harris 角点检测

OpenCV 中提供了 Harris 角点检测算法函数 cornerHarris，其定义如下：
```
dst = cornerHarris(src, blockSize, ksize, k, dst=None, borderType=None)
```
参数说明如下。
- src：输入图像。
- blockSize：邻域大小。
- ksize：Sobel 算子的孔径大小。
- k：Harris 参数。
- dst：输出图像（返回值）。

- borderType：边界模式，由 BorderTypes 定义（见 3.4.5 节）。

案例代码如下：

```
import cv2
import numpy as np

#图像输入并转为灰度图像
img = cv2.imread("logo.jpg")
gray = cv2.cvtColor(img,cv2.COLOR_BGR2GRAY)
gray = np.float32(gray)

#输入图像必须是 float32，最后一个参数在 0.04 到 0.05 之间
dst = cv2.cornerHarris(gray, 2, 3, 0.04)

#对结果做膨胀操作（可以不用执行该操作）
dst = cv2.dilate(dst,None)

#将超过阈值的点标记为红色，不同图像的阈值不同
img[dst>0.01*dst.max()] = [0,0,255]

#图像显示
cv2.imshow('harris-result', img)
cv2.waitKey(0)
cv2.destroyAllWindows()
```

检测的角点绘制结果如图 5.35 所示。

图 5.35

5.9.2 案例 65：Shi-Tomasi 角点检测

OpenCV 中提供了 Shi-Tomasi 角点检测算法函数 goodFeaturesToTrack，其定义如下：

```
corners = goodFeaturesToTrack(image, maxCorners, qualityLevel, minDistance, corners=None,
mask=None, blockSize=None, useHarrisDetector=None, k=None)
```

参数说明如下。

- image：输入图像，需要传入单通道 8 位或浮点型图像。
- maxCorners：角点的最大数量。
- qualityLevel：可接受的角点检测的最小特征值。
- minDistance：角点之间的最小距离。
- corners：检测到的角点向量（返回值）。
- mask：感兴趣区域。
- blockSize：邻域大小。
- useHarrisDetector：是否使用 Harris 角点检测。
- k：用于设置 Hessian 自相关矩阵行列式的相对权重的权重系数。

Shi-Tomasi 角点检测案例代码如下：

```
import numpy as np
import cv2

#以灰度图像的形式读取原始图像
img = cv2.imread('cvbook.jpg', 0)
#角点检测，设置检测的最大数量为200
corners = cv2.goodFeaturesToTrack(img, 200, 0.01, 10)
#在输入图像上绘制检测到的角点
corners = np.int0(corners)
#在原始图像上绘制角点
src_img = cv2.imread('cvbook.jpg')
for i in corners:
    x, y = i.ravel()
    cv2.circle(src_img, (x, y), 5, (0, 0, 255), -1)
#图像显示
cv2.imshow('shi-tomasi', src_img)
cv2.waitKey(0)
cv2.destroyAllWindows()
```

角点检测结果如图 5.36 所示。

图 5.36

5.9.3 案例 66：亚像素角点检测

如果用户需要更高的分辨率或实数坐标的特征，则可以使用亚像素角点检测算法。亚像素角点检测广泛用于相机校正、跟踪三维重建等场景。OpenCV 中提供了亚像素角点检测算法函数 cornerSubPix，其定义如下：

```
corners = cornerSubPix(image, corners, winSize, zeroZone, criteria)
```

参数说明如下。

- image：输入图像，需要传入单通道 8 位或浮点型图像。
- corners：输出角点的精确坐标（返回值）。
- winSize：搜索窗口尺寸的一半。
- zeroZone：死区尺寸的一半。
- criteria：角点计算迭代终止条件。

本案例对比了 Shi-Tomasi 角点检测算法与亚像素角点检测算法。案例代码如下：

```
import cv2
import numpy as np
#以灰度图像模式读取图像
src = cv2.imread('logo.jpg', 0)

#Shi-Tomasi 角点检测参数
max_corners = 100
quality_level = 0.01
min_dist = 50
block_size = 3
use_harris = False
k = 0.04
#图像拷贝，用于绘制角点
copy = np.copy(src)
copy1 = np.copy(src)

#Shi-Tomasi 角点检测
corners = cv2.goodFeaturesToTrack(src,
                                  max_corners,
                                  quality_level,
                                  min_dist, None,
                                  blockSize=block_size,
                                  useHarrisDetector = use_harris,
                                  k = k)
#保存绘制角点的图像
radius = 8
for i in range(corners.shape[0]):
    cv2.circle(copy,
```

```
            (corners[i, 0, 0], corners[i, 0, 1]),
            radius,
            (0, 0, 255))
cv2.imshow("shi_tomasi", copy)
#坐标打印
print("Shi-Tomasi 角点坐标")
for i in range(corners.shape[0]):
    print("坐标", i, ":(", corners[i, 0, 0], ",", corners[i, 0, 1], ")")

#亚像素角点检测
win_size = (5, 5)
zero_zone = (-1, -1)
criteria = (cv2.TERM_CRITERIA_EPS + cv2.TermCriteria_COUNT, 40, 0.001)
corners = cv2.cornerSubPix(src, corners, win_size, zero_zone, criteria)
#保存绘制角点的图像
for i in range(corners.shape[0]):
    cv2.circle(copy1, (corners[i, 0, 0], corners[i, 0, 1]), radius,
            (0, 0, 255))
cv2.imshow("subpixel_result", copy1)
#坐标打印
print("亚像素角点坐标")
for i in range(corners.shape[0]):
    print("坐标", i, ":(", corners[i, 0, 0], ",", corners[i, 0, 1], ")")

cv2.waitKey(0)
cv2.destroyAllWindows()
```

角点检测结果对比如图 5.37 所示，部分角点坐标结果对比如图 5.38 所示。

图 5.37　　　　　　　　　　　图 5.38

由坐标结果可以发现，亚像素角点检测可以精确到亚像素级别，精度更高。

5.10 图像轮廓

Canny 边缘检测算法可以很好地检测出目标的边缘，但是没有将这些边缘作为一个整体进行处理。而轮廓是一系列连续的点组成的曲线，可以表达物体的基本外形，在目标识别中有着重要的作用。

5.10.1 案例 67：轮廓查找

OpenCV 中提供了轮廓查找函数 findContours，该函数处理的图像可以是 Canny 边缘检测的结果，也可以是阈值化得到的图像，其定义如下：

```
contours, hierarchy = findContours(image, mode, method, contours=None, hierarchy=None, offset=None)
```

参数说明如下。

- image：输入图像，需要传入二值图像。
- mode：轮廓查找的模式，由 RetrievalModes 定义（见 4.1 节）。
- method：轮廓近似方式，由 ContourApproximationModes 定义（见 4.1 节）。
- contours：检测到的轮廓的坐标点（返回值）。
- hierarchy：轮廓层次（返回值）。
- offset：每个轮廓点移动的偏移量。

本案例使用的原始图像如图 5.39 所示。

轮廓查找的案例代码如下：

图 5.39

```
import cv2
import numpy as np

#读取图像，转为灰度图像
img = cv2.imread('contour_src.jpg')
gray = cv2.cvtColor(img, cv2.COLOR_BGR2GRAY)
#阈值化
ret, thresh = cv2.threshold(gray, 0, 255, cv2.THRESH_BINARY+cv2.THRESH_OTSU)
#寻找二值图像的轮廓
contours, hierarchy = cv2.findContours(thresh, cv2.RETR_TREE, cv2.CHAIN_APPROX_SIMPLE)
#打印轮廓数量
print("Contour Numbers:", len(contours))
#打印层次关系
print("hierarchy\n", hierarchy)
```

执行后的打印信息如下：

```
Contour Numbers: 4
hierarchy
[[[-1 -1  1 -1]
  [-1 -1  2  0]
  [-1 -1  3  1]
  [-1 -1 -1  2]]]
```

由此可见，轮廓数量为 4 个，包括五边形、圆形、四边形和整个图像最外围的轮廓。

hierarchy 为轮廓的层次关系，每个轮廓中的 4 个元素的含义对应为[下一个轮廓索引,上一个轮廓索引,第一个子轮廓索引,父轮廓索引]，如果对应的关系不存在，则对应位置的值为−1。hierarchy 与参数 mode 有关，不同的 mode 对应的 hierarchy 不同。

5.10.2　案例 68：轮廓绘制

对于查找到的轮廓，可以使用 OpenCV 中提供的轮廓绘制函数 drawContours 进行绘制，其定义如下：

```
image = drawContours(image, contours, contourIdx, color, thickness=None, lineType=None, hierarchy=None, maxLevel=None, offset=None)
```

参数说明如下。

- image：输入图像，在该图像上绘制轮廓（返回值）。
- contours：查找到的轮廓。
- contourIdx：绘制的轮廓编号，若全部绘制，则编号为负数。
- color：绘制颜色。
- thickness：绘制线的粗细。
- lineType：绘制线的线型，由 LineTypes 定义（见 4.1 节）。
- hierarchy：绘制轮廓的层次。
- maxLevel：轮廓绘制级别。
- offset：绘制轮廓偏移。

轮廓查找与轮廓绘制的完整案例代码如下：

```python
import cv2
import numpy as np

#读取图像，转为灰度图像
img = cv2.imread('contour_src.jpg')
gray = cv2.cvtColor(img, cv2.COLOR_BGR2GRAY)
#阈值化
ret, thresh = cv2.threshold(gray, 0, 255, cv2.THRESH_BINARY+cv2.THRESH_OTSU)
#寻找二值图像的轮廓
contours, hierarchy = cv2.findContours(thresh, cv2.RETR_TREE, cv2.CHAIN_APPROX_SIMPLE)
```

```python
#打印轮廓数量
print("Contour Numbers:", len(contours))
#打印层次关系
print("hierarchy\n", hierarchy)
#创建白底图像
contours_img = np.zeros(img.shape[:], dtype=np.uint8)
contours_img[:] = 255
#绘制轮廓
cv2.drawContours(contours_img, contours, -1, (0, 0, 255), 2)
cv2.imshow('contours_result', contours_img)
cv2.waitKey(0)
cv2.destroyAllWindows()
```

本案例中绘制了所有的轮廓，绘制结果如图 5.40 所示。

如果想绘制某个轮廓，则传入对应的轮廓编号即可，本案例中包括 4 个轮廓，从外到内的轮廓编号分别为 0、1、2、3，轮廓 2 的绘制结果如图 5.41 所示。

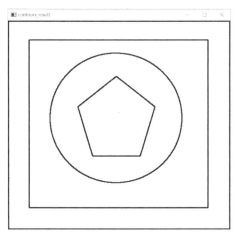 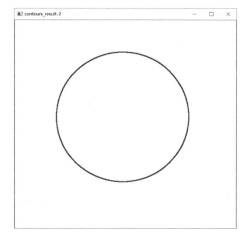

图 5.40　　　　　　　　　　　　　　图 5.41

OpenCV 中提供了轮廓面积的计算函数 contourArea，其定义如下：

```
retval = contourArea(contour, oriented=None)
```

参数说明如下。

- contour：待计算面积的轮廓。
- oriented：带方向的面积标志。如果此值为 true，则函数将根据轮廓方向（顺时针或逆时针）返回带符号的面积值，这样就可以通过获取某个面积的符号来确定轮廓的方向。默认情况下，该参数值为 false，这意味着返回面积的绝对值。
- retval：计算的面积结果。

在轮廓绘制案例代码的基础上，计算面积的案例代码如下：

```
area = cv2.contourArea(contours_list[1])
print("Contour 1 Area is: ", area)
```

计算结果输出如下：

```
Contour 1 Area is:  219110.0
```

5.11 轮廓包裹

在图像中识别到目标后，经常需要绘制目标的包围框，本节案例在轮廓查找后，对某个轮廓进行绘制包围矩形框或最小外接矩形，或者外接圆的操作。

5.11.1 案例 69：矩形边框

OpenCV 中提供了用于返回包围输入点集合的矩形边框的函数 boundingRect，其定义如下：

```
retval = boundingRect(array)
```

参数说明如下。
- array：输入灰度图像或二维点集。
- retval：返回的包围矩形（返回值）。

本案例检测输入图像的轮廓，对检测到的轮廓绘制外接包围矩形。本案例使用的输入图像如图 5.42 所示。

案例代码如下：

```
import cv2
import numpy as np


#读取图像，转为灰度图像
img = cv2.imread('bounding_src.jpg')
gray = cv2.cvtColor(img, cv2.COLOR_BGR2GRAY)
#阈值化
ret, thresh = cv2.threshold(gray, 0, 255, cv2.THRESH_BINARY+cv2.THRESH_OTSU)
#寻找二值图像的轮廓
contours, hierarchy = cv2.findContours(thresh, cv2.RETR_TREE, cv2.CHAIN_APPROX_SIMPLE)
#创建白底图像
contours_img = np.zeros(img.shape[:], dtype=np.uint8)
contours_img[:] = 255
#绘制轮廓
cv2.drawContours(contours_img, contours, 3, (0, 0, 255), 2)
```

```
#返回包围矩形
rect = cv2.boundingRect(contours[3])
print("rect: ", rect)
#绘制包围矩形
cv2.rectangle(contours_img, rect, (128, 128, 128))
cv2.imshow('boundingRect', contours_img)
cv2.waitKey(0)
cv2.destroyAllWindows()
```

案例中绘制索引为 3 的轮廓，并找到其外接矩形，返回的包围矩形信息打印如下：

```
rect:  (163, 165, 298, 367)
```

索引为 3 的轮廓及其外接矩形的绘制结果如图 5.43 所示。

图 5.42

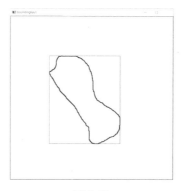

图 5.43

5.11.2　案例 70：最小外接矩形

OpenCV 中提供了用于返回最小外接矩形的函数 minAreaRect，其定义如下：

```
retval = minAreaRect(points)
```

参数说明如下。

- points：输入的二维点集。
- retval：返回的最小外接矩形（返回值）。

本案例检测输入图像的轮廓，然后对检测到的轮廓绘制最小外接矩形。本案例使用的输入图像如图 5.42 所示。案例代码如下：

```
import cv2
import numpy as np

#读取图像，转为灰度图像
img = cv2.imread('bounding_src.jpg')
gray = cv2.cvtColor(img, cv2.COLOR_BGR2GRAY)
#阈值化
```

```
ret, thresh = cv2.threshold(gray, 0, 255, cv2.THRESH_BINARY+cv2.THRESH_OTSU)
#寻找二值图像的轮廓
contours, hierarchy = cv2.findContours(thresh, cv2.RETR_TREE, cv2.CHAIN_APPROX_SIMPLE)
#创建白底图像
contours_img = np.zeros(img.shape[:], dtype=np.uint8)
contours_img[:] = 255
#绘制轮廓
cv2.drawContours(contours_img, contours, 3, (0, 0, 255), 2)

#返回最小外接矩形
min_rect = cv2.minAreaRect(contours[3])
print("min_rect: ", min_rect)
#获取最小外接矩形的4个顶点坐标
points = cv2.boxPoints(min_rect)
print("min_rect points:\n", points)
box_rect = np.int0(points)
#绘制包围矩形
cv2.drawContours(contours_img, [box_rect], 0, (128, 128, 128), 1)
cv2.imshow('minAreaRect', contours_img)
cv2.waitKey(0)
cv2.destroyAllWindows()
```

案例中绘制索引为 3 的轮廓，并找到其最小外接矩形，最小外接矩形返回结果与其 4 个顶点的坐标打印如下：

```
min_rect:  ((315.33319091796875, 335.85968017578125), (405.6804504394531, 159.4354248046875), 58.50078201293945)
min_rect points:
 [[140.64197 205.9863 ]
 [276.58417 121.68338]
 [488.0244  466.73306]
 [352.0822  550.036  ]]
```

索引为 3 的轮廓及其最小外接矩形的绘制结果如图 5.44 所示。

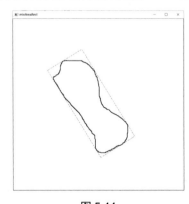

图 5.44

5.11.3 案例 71：最小外接圆

OpenCV 中提供了用于返回最小外接圆的函数 minEnclosingCircle，其定义如下：

`center, radius= minEnclosingCircle(points)`

参数说明如下。

- points：输入的二维点集。
- center：返回的最小外接圆的圆心（返回值）。
- radius：返回的最小外接圆的半径（返回值）。

本案例检测输入图像的轮廓，然后对检测到的轮廓绘制最小外接圆。本案例使用的输入图像如图 5.42 所示。案例代码如下：

```
import cv2
import numpy as np

#读取图像，转为灰度图像
img = cv2.imread('bounding_src.jpg')
gray = cv2.cvtColor(img, cv2.COLOR_BGR2GRAY)
#阈值化
ret, thresh = cv2.threshold(gray, 0, 255, cv2.THRESH_BINARY+cv2.THRESH_OTSU)
#寻找二值图像的轮廓
contours, hierarchy = cv2.findContours(thresh, cv2.RETR_TREE, cv2.CHAIN_APPROX_SIMPLE)
#创建白底图像
contours_img = np.zeros(img.shape[:], dtype=np.uint8)
contours_img[:] = 255
#绘制轮廓
cv2.drawContours(contours_img, contours, 3, (0, 0, 255), 2)

#返回最小外接圆
min_circle = cv2.minEnclosingCircle(contours[3])
print("min_circle: ", min_circle)
#获取绘制外接圆的圆心与半径
center_x = int(min_circle[0][0])
center_y = int(min_circle[0][0])
radius = int(min_circle[1])
#绘制最小外接圆
cv2.circle(contours_img, (center_x, center_y), radius, (128, 128, 128), 1)
cv2.imshow('min_circle', contours_img)
cv2.waitKey(0)
cv2.destroyAllWindows()
```

寻找最小外接圆的结果如下（返回的结果类型为 tuple 类型数据，其中第一个元素为圆心坐标，第二个元素为半径值）：

`min_circle:  ((315.9704895019531, 335.1665954589844), 202.88648986816406)`

绘制的最小外接圆的结果如图 5.45 所示。

图 5.45

5.12 案例 72：多边形填充

OpenCV 中提供了用于凸多边形填充的函数 fillConvexPoly，其定义如下：
`img = fillConvexPoly(img, points, color, lineType=None, shift=None)`

参数说明如下。

- img：输入图像（返回值）。
- points：多边形的顶点。
- color：填充的颜色。
- lineType：线型，由 LineTypes 定义（见 4.1 节）。
- shift：顶点坐标中的小数位数。

本案例自定义了一个凸多边形，并对其进行白色填充。案例代码如下：

```
import cv2
import numpy as np

#创建黑底空白图像
image = np.zeros((512, 512, 3), np.uint8)
#定义多边形的顶点
poly = np.array([[300, 50], [50, 250], [300,300], [170, 450], [320, 450], [470, 250]])
#凸多边形填充
cv2.fillConvexPoly(image, poly, (255, 255, 255))
#图像显示
cv2.imshow("image", image)
cv2.waitKey(0)
```

```
cv2.destroyAllWindows()
```
凸多边形填充结果如图 5.46 所示。

图 5.46

OpenCV 中提供了用于任意多边形填充的函数 fillPoly，该函数可以同时填充多个多边形，其定义如下：

```
img = fillPoly(img, pts, color, lineType=None, shift=None, offset=None)
```

参数说明如下。

- img：输入图像（返回值）。
- pts：多边形数组，其中每个多边形均表示为顶点数组。
- color：填充的颜色。
- lineType：线型，由 LineTypes 定义（见 4.1 节）。
- shift：顶点坐标中的小数位数。
- offset：轮廓所有点的可选偏移。

本案例自定义了一个凸多边形和一个三角形，并对其进行白色填充。案例代码如下：

```
import cv2
import numpy as np

#创建黑底空白图像
image = np.zeros((512, 512, 3), np.uint8)
#定义多边形的顶点
poly1 = np.array([[200, 50], [50, 250], [100, 450], [200, 450], [250, 250]])
poly2 = np.array([[300, 80], [450, 250], [330,380]])
#凸多边形填充
cv2.fillPoly(image, [poly1, poly2], (255, 255, 255))
#图像显示
cv2.imshow("image", image)
```

```
cv2.waitKey(0)
cv2.destroyAllWindows()
```

填充结果如图 5.47 所示。

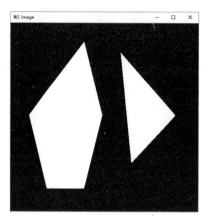

图 5.47

5.13 图像拟合

在图像处理中，经常需要拟合出一组点集的直线或椭圆，或者多边形，OpenCV 中提供了用于对点集进行拟合的函数。

5.13.1 案例 73：直线拟合

OpenCV 中提供了直线拟合函数 fitLine，其定义如下：

```
line = fitLine(points, distType, param, reps, aeps, line=None)
```

参数说明如下。

- points：输入的二维或三维点集。
- distType：距离类型，由 DistanceTypes 定义（见 4.1 节）。
- param：距离参数，若该值为 0，则函数会自动选择最优值。
- reps：径向精度。
- aeps：角度精度。
- line：拟合的直线参数（返回值）。

本案例检测输入图像的轮廓，对检测到的轮廓选取其中一个轮廓做直线拟合。本案例使用的输入图像如图 5.42 所示，案例代码如下：

```
import cv2
```

```python
import numpy as np
#读取图像，转为灰度图像
img = cv2.imread('bounding_src.jpg')
gray = cv2.cvtColor(img, cv2.COLOR_BGR2GRAY)
#阈值化
ret, thresh = cv2.threshold(gray, 0, 255, cv2.THRESH_BINARY+cv2.THRESH_OTSU)
#寻找二值图像的轮廓
contours, hierarchy = cv2.findContours(thresh, cv2.RETR_TREE, cv2.CHAIN_APPROX_SIMPLE)
#创建白底图像
fitline_img = np.zeros(img.shape[:], dtype=np.uint8)
fitline_img[:] = 255
#绘制轮廓
cv2.drawContours(fitline_img, contours, 3, (0, 0, 0), 2)
#直线拟合
line_para = cv2.fitLine(contours[3], cv2.DIST_L2, 0, 0.01, 0.01)
print("line_para: ",line_para)
cv2.line(fitline_img, (int(line_para[0]), int(line_para[1])), (int(line_para[2]), int(line_para[3])), (0,0,255))
#拟合结果显示
cv2.imshow('fitLine', fitline_img)
cv2.waitKey(0)
cv2.destroyAllWindows()
```

直线拟合结果如图 5.48 所示。

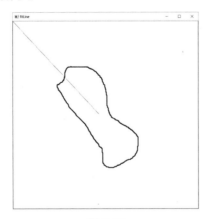

图 5.48

读者可以根据拟合直线返回的参数将图 5.48 中的拟合直线延长。

5.13.2 案例 74：椭圆拟合

OpenCV 中提供了椭圆拟合函数 fitEllipse，其定义如下：

```
retval = fitEllipse(points)
```

参数说明如下。

- points：输入的二维点集。
- retval：拟合的椭圆参数（返回值）。

本案例检测输入图像的轮廓，对检测到的轮廓选取其中一个轮廓做椭圆拟合。本案例使用的输入图像如图 5.42 所示，案例代码如下：

```
import cv2
import numpy as np
#读取图像，转为灰度图像
img = cv2.imread('bounding_src.jpg')
gray = cv2.cvtColor(img, cv2.COLOR_BGR2GRAY)
#阈值化
ret, thresh = cv2.threshold(gray, 0, 255, cv2.THRESH_BINARY+cv2.THRESH_OTSU)
#寻找二值图像的轮廓
contours, hierarchy = cv2.findContours(thresh, cv2.RETR_TREE, cv2.CHAIN_APPROX_SIMPLE)
#创建白底图像
fitellipse_img = np.zeros(img.shape[:], dtype=np.uint8)
fitellipse_img[:] = 255
#绘制轮廓
cv2.drawContours(fitellipse_img, contours, 3, (0, 0, 0), 2)
#椭圆拟合
ellipse_para = cv2.fitEllipse(contours[3])
print("ellipse_para: ",ellipse_para)
cv2.ellipse(fitellipse_img, ellipse_para, (0,0,255))
#拟合结果显示
cv2.imshow('fitellipse_img', fitellipse_img)
cv2.waitKey(0)
cv2.destroyAllWindows()
```

椭圆拟合结果如图 5.49 所示。

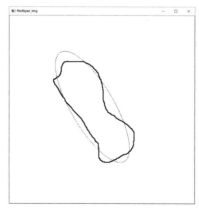

图 5.49

5.13.3 案例 75：多边形拟合

OpenCV 中提供了多边形拟合（或称多边形近似）函数 approxPolyDP，其定义如下：

`approxCurve = approxPolyDP(curve, epsilon, closed, approxCurve=None)`

参数说明如下。

- curve：输入的二维点集。
- epsilon：拟合精度。
- closed：拟合曲线是否闭合。
- approxCurve：拟合结果（返回值）。

本案例检测输入图像的轮廓，对检测到的轮廓选取其中一个轮廓做多边形拟合。本案例使用的输入图像如图 5.42 所示，案例代码如下：

```
import cv2
import numpy as np
#读取图像，转为灰度图像
img = cv2.imread('bounding_src.jpg')
gray = cv2.cvtColor(img, cv2.COLOR_BGR2GRAY)
#阈值化
ret, thresh = cv2.threshold(gray, 0, 255, cv2.THRESH_BINARY+cv2.THRESH_OTSU)
#寻找二值图像的轮廓
contours, hierarchy = cv2.findContours(thresh, cv2.RETR_TREE, cv2.CHAIN_APPROX_SIMPLE)
#创建白底图像
polydp_img = np.zeros(img.shape[:], dtype=np.uint8)
polydp_img[:] = 255
#绘制轮廓
cv2.drawContours(polydp_img, contours, 3, (0, 0, 0), 2)
#多边形拟合
polydp = cv2.approxPolyDP(contours[3], 5, True)
#绘制拟合结果
cv2.drawContours(polydp_img, [polydp], 0, (0, 0, 255), 2)
#拟合结果显示
cv2.imshow('polydp_img', polydp_img)
cv2.waitKey(0)
cv2.destroyAllWindows()
```

本案例使用的拟合精度为 5 像素，拟合的曲线会闭合，结果如图 5.50 所示。

如果使用的拟合精度为 50 像素，则拟合结果如图 5.51 所示。

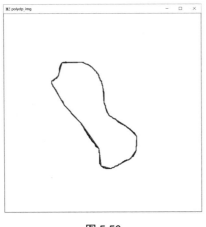
图 5.50

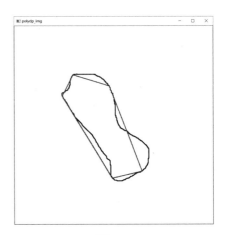
图 5.51

5.14 案例 76：凸包检测

凸包是由轮廓上的点组成的包裹轮廓的多边形，该多边形的每一处都必须是凸的（多边形上任意连续 3 个点的内角小于 180°），可以用于检测一个点是否落在一个凸多边形内部。OpenCV 中提供了用于凸包检测的函数 convexHull，其定义如下：

```
hull = convexHull(points, hull=None, clockwise=None, returnPoints=None)
```

参数说明如下。

- points：输入的二维点集。
- clockwise：方向标志，若该值为 True，则为顺时针方向。
- returnPoints：操作标志，对于图像矩阵，若该值为 True，则返回凸包点；否则返回凸包点的索引。
- hull：输出凸包参数（返回值）。

本案例检测输入图像的轮廓，对检测到的轮廓选取其中一个轮廓做凸包检测。本案例使用的输入图像如图 5.42 所示，案例代码如下：

```python
import cv2
import numpy as np
#读取图像，转为灰度图像
img = cv2.imread('bounding_src.jpg')
gray = cv2.cvtColor(img, cv2.COLOR_BGR2GRAY)
#阈值化
ret, thresh = cv2.threshold(gray, 0, 255, cv2.THRESH_BINARY+cv2.THRESH_OTSU)
#寻找二值图像的轮廓
```

```
contours, hierarchy = cv2.findContours(thresh, cv2.RETR_TREE, cv2.CHAIN_APPROX_SIMPLE)
#创建白底图像
hull_img = np.zeros(img.shape[:], dtype=np.uint8)
hull_img[:] = 255
#绘制轮廓
cv2.drawContours(hull_img, contours, 3, (0, 0, 0), 2)
#检测凸包
hull = cv2.convexHull(contours[3])
#绘制凸包结果
cv2.polylines(hull_img, [hull], True, (0, 0, 255), 2)
#拟合结果显示
cv2.imshow('hull_img', hull_img)
cv2.waitKey(0)
cv2.destroyAllWindows()
```

凸包检测并绘制的结果如图 5.52 所示。

图 5.52

5.15 进阶必备：图像处理算法概述

图像处理模块 imgproc 是 OpenCV 中最重要的图像处理功能模块，因为内容很多，所以将该模块分两章进行讲述。本节对模块中的图像处理算法做一个总结，如表 5.1 所示。

表 5.1

算法专题	子专题	算法名称	OpenCV 函数
滤波与卷积	平滑滤波	均值滤波	blur
		方框滤波	boxFilter
		高斯滤波	GaussianBlur
		中值滤波	medianBlur
		双边滤波	bilateralFilter
	阈值化	基本阈值化	threshold
		自适应阈值化	adaptiveThreshold
	图像形态学	膨胀	dilate
		腐蚀	erode
		开运算	morphologyEx
		闭运算	
		形态学梯度	
		顶帽运算	
		底帽运算	
	边缘检测	Sobel 算法	Sobel
		Scharr 算法	Scharr
		Laplacian 算法	Laplacian
		Canny 算法	Canny
图像变换	图像金字塔	高斯金字塔	pyrDown
		拉普拉斯金字塔	pyrUp
	仿射变换	仿射变换	warpAffine
	透视变换	透视变换	warpPerspective
	重映射	重映射	remap
图像分析	霍夫变换	霍夫线变换	HoughLines
		霍夫圆变换	HoughCircles
	图像分割	漫水填充算法	floodFill
		分水岭算法	watershed
		GrabCut 算法	grabCut
		Mean-Shift 分割	pyrMeanShiftFiltering
直方图	直方图表示	创建直方图	calcHist
	直方图操作	直方图归一化	normalize
		直方图二值化	threshold
		直方图比较	compareHist
轮廓	轮廓操作	轮廓查找	findContours
		轮廓绘制	drawContours
	轮廓分析	多边形拟合	approxPolyDP
	轮廓匹配	矩	Moments 类
特征点	角点检测	Harris 角点检测	cornerHarris
		Shi-Tomasi 角点检测	goodFeaturesToTrack
		亚像素角点检测	cornerSubPix

第 6 章
可视化模块 highgui

在图像处理中,处理结果的展示是必不可少的一个环节,highgui 模块提供了图像显示功能,用户与显示的图像之间通过鼠标、键盘等进行实时交互的功能,在图像显示窗口添加进度条进行参数动态调整的功能等。

6.1 模块导读

尽管 OpenCV 可以被 Qt 等 UI 框架使用,或者在没有 UI 的情况下被使用,但是有时用户想将结果可视化,如果引入第三方 UI 框架,则比较复杂,因此,OpenCV 提供了 highgui 模块,为读者提供结果快速可视化功能。该模块提供的接口可以创建和操作窗口、展示图像,还可以添加进度条、处理鼠标事件和键盘命令。

highgui 模块需要通过包含头文件#include "opencv2/highgui.hpp"来引入,模块说明如下:

```
#ifndef OPENCV_HIGHGUI_HPP
#define OPENCV_HIGHGUI_HPP

#include "opencv2/core.hpp"
#ifdef HAVE_OPENCV_IMGCODECS
#include "opencv2/imgcodecs.hpp"
#endif
#ifdef HAVE_OPENCV_VIDEOIO
#include "opencv2/videoio.hpp"
#endif
```

```
namespace cv
{
```
如下为与定义的枚举类型的可视化操作相关的标志定义：
```
//! cv::namedWindow 标志
enum WindowFlags {
       WINDOW_NORMAL     = 0x00000000, //!< 正常窗口
       WINDOW_AUTOSIZE   = 0x00000001, //!< 窗口大小不可调整
       WINDOW_OPENGL     = 0x00001000, //!< OpenGL 支持的窗口
       WINDOW_FULLSCREEN = 1,          //!< 窗口全屏
       WINDOW_FREERATIO  = 0x00000100, //!< 图像尽可能地扩展（没有比率约束）
       WINDOW_KEEPRATIO  = 0x00000000, //!< 保证图像比率的显示窗口
       WINDOW_GUI_EXPANDED=0x00000000, //!< 状态条和工具条
       WINDOW_GUI_NORMAL = 0x00000010, //!< 旧风格
};

//! cv::setWindowProperty 和 cv::getWindowProperty 标志
enum WindowPropertyFlags {
       WND_PROP_FULLSCREEN   = 0, //!< 全屏属性
       WND_PROP_AUTOSIZE     = 1, //!< 自动大小属性
       WND_PROP_ASPECT_RATIO = 2, //!< 窗口纵横比
       WND_PROP_OPENGL       = 3, //!< OpenGL 支持
       WND_PROP_VISIBLE      = 4, //!< 检查窗口是否存在且可用
       WND_PROP_TOPMOST      = 5, //!< 切换普通窗口是否为最顶层的属性
       WND_PROP_VSYNC        = 6  //!< 开/关 VSYNC 功能
};

//! 鼠标事件类型
enum MouseEventTypes {
       EVENT_MOUSEMOVE      = 0,  //!< 表明鼠标移到窗口上
       EVENT_LBUTTONDOWN    = 1,  //!< 鼠标左键按下
       EVENT_RBUTTONDOWN    = 2,  //!< 鼠标右键按下
       EVENT_MBUTTONDOWN    = 3,  //!< 鼠标中键按下
       EVENT_LBUTTONUP      = 4,  //!< 鼠标左键释放
       EVENT_RBUTTONUP      = 5,  //!< 鼠标右键释放
       EVENT_MBUTTONUP      = 6,  //!< 鼠标中键释放
       EVENT_LBUTTONDBLCLK  = 7,  //!< 鼠标左键双击
       EVENT_RBUTTONDBLCLK  = 8,  //!< 鼠标右键双击
       EVENT_MBUTTONDBLCLK  = 9,  //!< 鼠标中键双击
       EVENT_MOUSEWHEEL     = 10, //!< 正值和负值分别表示向前和向后滚动
       EVENT_MOUSEHWHEEL    = 11  //!< 正值和负值分别表示向右和向左滚动
};

//! 鼠标事件标志
```

```
enum MouseEventFlags {
    EVENT_FLAG_LBUTTON     = 1,  //!< 鼠标左键按下
    EVENT_FLAG_RBUTTON     = 2,  //!< 鼠标右键按下
    EVENT_FLAG_MBUTTON     = 4,  //!< 鼠标中键按下
    EVENT_FLAG_CTRLKEY     = 8,  //!< Ctrl 键按下
    EVENT_FLAG_SHIFTKEY    = 16, //!< Shift 键按下
    EVENT_FLAG_ALTKEY      = 32  //!< Alt 键按下
};

//! Qt 字体粗细
enum QtFontWeights {
    QT_FONT_LIGHT      = 25, //!< 粗细为 25 像素
    QT_FONT_NORMAL     = 50, //!< 粗细为 50 像素
    QT_FONT_DEMIBOLD   = 63, //!< 粗细为 63 像素
    QT_FONT_BOLD       = 75, //!< 粗细为 75 像素
    QT_FONT_BLACK      = 87  //!< 粗细为 87 像素
};

//! Qt 字体风格
enum QtFontStyles {
    QT_STYLE_NORMAL    = 0, //!< 正常字体
    QT_STYLE_ITALIC    = 1, //!< 粗体
    QT_STYLE_OBLIQUE   = 2  //!< 斜体
};

//! Qt 按键类型
enum QtButtonTypes {
    QT_PUSH_BUTTON     = 0,    //!< 按压按键
    QT_CHECKBOX        = 1,    //!< 复选框按键
    QT_RADIOBOX        = 2,    //!< 单选按键
    QT_NEW_BUTTONBAR   = 1024  //!< 需要创建一个新的按钮栏的按键
};
```

如下为与可视化中的鼠标操作（见案例 82）、进度条控制（见案例 83）相关的回调函数：

```
/** 鼠标事件回调函数，参见 cv::setMouseCallback*/
typedef void (*MouseCallback)(int event, int x, int y, int flags, void* userdata);

/** 进度条回调函数，参见 cv::createTrackbar*/
typedef void (*TrackbarCallback)(int pos, void* userdata);

/** 每帧图像调用的回调函数，参见 cv::setOpenGlDrawCallback*/
typedef void (*OpenGlDrawCallback)(void* userdata);

/** 按键创建回调，参见 cv::createButton*/
typedef void (*ButtonCallback)(int state, void* userdata);
```

如下是与窗口创建和销毁相关的函数（见案例 77）：
```
/** 创建窗口*/
CV_EXPORTS_W void namedWindow(const String& winname, int flags = WINDOW_AUTOSIZE);

/** 销毁指定窗口*/
CV_EXPORTS_W void destroyWindow(const String& winname);

/** 销毁所有窗口*/
CV_EXPORTS_W void destroyAllWindows();

CV_EXPORTS_W int startWindowThread();
```
如下为与按键操作相关的函数（见 6.4 节案例）：
```
/** 等待操作，类似 waitKey*/
CV_EXPORTS_W int waitKeyEx(int delay = 0);

/** 等待按键操作*/
CV_EXPORTS_W int waitKey(int delay = 0);

/** 在不等待的情况下轮询按键事件，返回被按下的按键的代码*/
CV_EXPORTS_W int pollKey();
```
如下为图像显示函数（见案例 79）：
```
/** 在指定窗口显示图像*/
CV_EXPORTS_W void imshow(const String& winname, InputArray mat);
```
如下为与窗口操作相关的函数（见案例 78）：
```
/** 调整窗口到指定大小*/
CV_EXPORTS_W void resizeWindow(const String& winname, int width, int height);
CV_EXPORTS_W void resizeWindow(const String& winname, const cv::Size& size);

/** 移动窗口到指定位置*/
CV_EXPORTS_W void moveWindow(const String& winname, int x, int y);

/** 动态调整窗口参数*/
CV_EXPORTS_W void setWindowProperty(const String& winname, int prop_id, double prop_value);

/** 更新窗口标题*/
CV_EXPORTS_W void setWindowTitle(const String& winname, const String& title);

/** 获取窗口参数*/
CV_EXPORTS_W double getWindowProperty(const String& winname, int prop_id);

/** 获取窗口中的图像矩形*/
CV_EXPORTS_W Rect getWindowImageRect(const String& winname);
```

```
/** 为指定窗口设置鼠标句柄*/
CV_EXPORTS void setMouseCallback(const String& winname, MouseCallback onMouse, void* userdata
= 0);

/** 获取鼠标滚轮运动增量*/
CV_EXPORTS int getMouseWheelDelta(int flags);
```

如下为用于设置感兴趣区域的函数（见案例 80）:

```
/** 允许读者在指定图像上选择感兴趣区域*/
CV_EXPORTS_W Rect selectROI(const String& windowName, InputArray img, bool showCrosshair = true,
bool fromCenter = false);

CV_EXPORTS_W Rect selectROI(InputArray img, bool showCrosshair = true, bool fromCenter = false);

/** 允许读者在指定图像上选取多个感兴趣区域*/
CV_EXPORTS_W void selectROIs(const String& windowName, InputArray img,
                             CV_OUT std::vector<Rect>& boundingBoxes,
                             bool showCrosshair = true, bool fromCenter = false);
```

如下为与进度条操作相关的函数（见案例 83）:

```
/** 创建进度条并附加到指定窗口*/
CV_EXPORTS int createTrackbar(const String& trackbarname, const String& winname,
                              int* value, int count,
                              TrackbarCallback onChange = 0,
                              void* userdata = 0);

/** 获取进度条位置*/
CV_EXPORTS_W int getTrackbarPos(const String& trackbarname, const String& winname);

/** 设置进度条位置*/
CV_EXPORTS_W void setTrackbarPos(const String& trackbarname, const String& winname, int pos);

/** 设置进度条到最大位置*/
CV_EXPORTS_W void setTrackbarMax(const String& trackbarname, const String& winname, int maxVal);

/** 设置进度条到最小位置*/
CV_EXPORTS_W void setTrackbarMin(const String& trackbarname, const String& winname, int minVal);
```

如下为与 OpenGL 相关的操作函数：

```
/** 在指定窗口中显示 OpenGL 二维纹理*/
CV_EXPORTS void imshow(const String& winname, const ogl::Texture2D& tex);

/** 调用回调函数在显示的图像上绘制*/
CV_EXPORTS void setOpenGlDrawCallback(const String& winname, OpenGlDrawCallback onOpenGlDraw,
void* userdata = 0);

/** 将指定的窗口设置为当前 OpenGL 背景*/
```

```
CV_EXPORTS void setOpenGlContext(const String& winname);

/** 强制窗口刷新*/
CV_EXPORTS void updateWindow(const String& winname);
```

如下为 OpenCV 与 Qt 相关的操作函数：

```
/** Qt 字体类，仅 Qt 可用*/
struct QtFont
{
    const char* nameFont;
    Scalar      color;
    int         font_face;
    const int*  ascii;
    const int*  greek;
    const int*  cyrillic;
    float       hscale, vscale;
    float       shear;
    int         thickness;
    float       dx;
    int         line_type;
};

/** 创建字体，在图像上绘制文字*/
CV_EXPORTS QtFont fontQt(const String& nameFont, int pointSize = -1,
                    Scalar color = Scalar::all(0), int weight = QT_FONT_NORMAL,
                    int style = QT_STYLE_NORMAL, int spacing = 0);

/** 在图像上绘制文字*/
CV_EXPORTS void addText( const Mat& img, const String& text, Point org, const QtFont& font);
CV_EXPORTS_W void addText(const Mat& img, const String& text, Point org,
                    const String& nameFont, int pointSize = -1,
                    Scalar color = Scalar::all(0), int weight = QT_FONT_NORMAL,
                    int style = QT_STYLE_NORMAL, int spacing = 0);

/** 在指定时间段内，将窗口图像上的文本显示为覆盖内容*/
CV_EXPORTS_W void displayOverlay(const String& winname, const String& text, int delayms = 0);

/** 在指定的时间段内，在窗口状态栏上显示文本*/
CV_EXPORTS_W void displayStatusBar(const String& winname, const String& text, int delayms = 0);

/** 保存指定窗口参数*/
CV_EXPORTS void saveWindowParameters(const String& windowName);

/** 加载指定窗口参数*/
CV_EXPORTS void loadWindowParameters(const String& windowName);
```

```
CV_EXPORTS  int startLoop(int (*pt2Func)(int argc, char *argv[]), int argc, char* argv[]);

CV_EXPORTS  void stopLoop();

/** 将按键附加到控制面板中*/
CV_EXPORTS int createButton( const String& bar_name, ButtonCallback on_change,
                             void* userdata = 0, int type = QT_PUSH_BUTTON,
                             bool initial_button_state = false);
} // cv

#endif
```

6.2 图像窗口

6.2.1 案例 77：创建与销毁窗口

OpenCV 中提供了用于创建图像窗口的函数 namedWindow，其定义如下：

`namedWindow(winname, flags=None)`

参数说明如下。

- winname：窗口名。
- flags：窗口标志，由 WindowFlags 定义（见 6.1 节）。

namedWindow 函数的返回值为 None。

OpenCV 中提供了用于销毁图像窗口的函数 destroyWindow，其定义如下：

`destroyWindow(winname)`

参数说明如下。

- winname：窗口名。

destroyWindow 函数的返回值为 None。

另外，OpenCV 中还提供了用于销毁所有图像窗口的函数 destroyAllWindows，其定义如下：

`destroyAllWindows()`

destroyAllWindows 函数没有任何参数，返回值为 None。

窗口创建与销毁的案例代码如下：

```
import cv2

#创建窗口，窗口名称为 image_window，窗口标志为 WINDOW_NORMAL
cv2.namedWindow("image_window", cv2.WINDOW_NORMAL)
```

```
#窗口销毁
cv2.destroyWindow("image_window")
#销毁所有图像窗口
#cv2.destroyAllWindows()
```

6.2.2 案例78：图像窗口操作

OpenCV 中提供了用于设置窗口标题的函数 setWindowTitle，其定义如下：

```
setWindowTitle(winname, title)
```

参数说明如下。

- winname：窗口名。
- title：窗口标题。

setWindowTitle 函数的返回值为 None。

OpenCV 中提供了用于调整窗口尺寸的函数 resizeWindow，其定义如下：

```
resizeWindow(winname, width, height)
```

参数说明如下。

- winname：窗口名。
- width：新窗口的宽度。
- height：新窗口的高度。

resizeWindow 函数的返回值为 None。

OpenCV 中提供了用于移动窗口的函数 moveWindow，其定义如下：

```
moveWindow(winname, x, y)
```

参数说明如下。

- winname：窗口名。
- x：新位置的 x 坐标。
- y：新位置的 y 坐标。

moveWindow 函数的返回值为 None。

窗口操作的案例代码如下：

```
import cv2

#创建窗口，窗口名称为 image_window，窗口标志为 WINDOW_NORMAL
cv2.namedWindow("image_window", cv2.WINDOW_NORMAL)
#读取图像
image = cv2.imread("src.jpg")
#图像显示
cv2.imshow("image_window", image)
```

```
#设置窗口标题
cv2.setWindowTitle("image_window", "lena")
#等待5s
cv2.waitKey(10000)
#调整窗口大小
cv2.resizeWindow("image_window", 100, 100)
#窗口移动
cv2.moveWindow("image_window", 1000, 1000)
#等待操作
cv2.waitKey(0)
#窗口销毁
cv2.destroyWindow("image_window")
```

本案例使用的输入图像为图 3.10,图像显示如图 6.1 所示。

设置窗口标题并调整窗口大小后的结果如图 6.2 所示。

图 6.1

图 6.2

6.3 图像操作

6.3.1 案例 79:图像显示

图像可视化中最重要的一个函数就是图像显示函数,OpenCV 中提供了用于图像显示的函数 imshow,其定义如下:

```
imshow(winname, mat)
```

参数说明如下。

- winname：窗口名称。
- mat：待显示图像。

imshow 函数的返回值为 None。

imshow 函数用于在指定窗口中显示图像，案例代码如下：

```
import cv2

#读取图像
image = cv2.imread("src.jpg")
#图像显示
cv2.imshow("image_window", image)
#等待操作
cv2.waitKey(0)
#窗口销毁
cv2.destroyWindow("image_window")
```

对 imshow 函数的 3 点说明如下。

- imshow 中的图像显示窗口如果没有提前创建，则会自动创建。
- imshow 后必须紧跟 waitKey（见 6.4 节）函数，图像才会显示。
- 在使用 imshow 进行图像显示时，在 Windows 系统中，使用 Ctrl+C 组合键，会将图像拷贝到剪贴板上；使用 Ctrl+S 组合键，会弹出图像保存窗口。

6.3.2 案例 80：选取感兴趣区域

在图像处理中，如果对某个目标对象比较感兴趣，则可以将其设置为感兴趣区域（ROI）。OpenCV 中提供了用于在显示的图像窗口上选取感兴趣区域的函数 selectROI，其定义如下：

```
retval = selectROI(windowName, img, showCrosshair=None, fromCenter=None)
```

参数说明如下。

- windowName：窗口名称。
- img：待选取的图像。
- showCrosshair：若该值为 true，则显示矩形十字线。
- fromCenter：若该值为 true，则选择的中心将匹配鼠标的初始位置。
- retval：选取的感兴趣区域（返回值）。

图像显示后，可以调用函数 selectROI 在图像上选择感兴趣区域，选择完成后，使用空格键或 Enter 键完成选择，按 C 键取消选择。

选择感兴趣区域的案例代码如下：

```
import cv2
```

```
#读取图像
image = cv2.imread("src.jpg")
#图像显示
cv2.imshow("image", image)
#选择感兴趣区域
roi = cv2.selectROI("image", image)
#打印感兴趣区域矩形
print(roi)
#裁剪感兴趣区域
roi_img = image[roi[1]:(roi[1]+roi[3]), roi[0]:(roi[0]+roi[2]), :]
#感兴趣区域显示
cv2.imshow("roi_img", roi_img)
cv2.waitKey(0)
cv2.destroyWindow("image_window")
```

选择的感兴趣区域如图 6.3 所示。

打印的感兴趣区域的矩形值如下：

(108, 66, 274, 429)

裁剪后的感兴趣区域如图 6.4 所示。

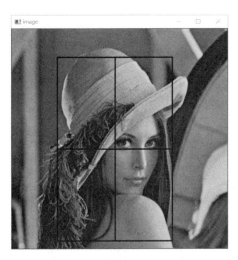

图 6.3

图 6.4

6.4 案例 81：键盘操作

OpenCV 中提供了键盘操作函数 waitKey，其定义如下：

```
retval = waitKey(delay=None)
```

参数说明如下。

- delay：延迟等待的时间，单位为 ms，当将该值设置为 0 时，表示无限期等待直到键盘操作。
- retval：返回操作的按键（返回值）。

函数 waitKey 无限期地（当参数为 0 时）等待一个按键事件，或者等待延迟毫秒（当参数为正数时）。由于操作系统在切换线程之间有一个最短的时间间隔，因此函数不会完全只等待 delay（ms），它至少会等待 delay（ms），这取决于此时计算机上运行的其他内容。它返回被按下的按键的编码，如果在指定的时间过去之前没有按下任何键，则返回-1。

函数 waitKey 是 highgui 模块中唯一可以获取和处理事件的方法，因此，需要定期调用它以正常地进行事件处理，除非在处理事件的环境中已经使用了 highgui。仅当至少创建了一个 highgui 窗口且该窗口处于活动状态时，该函数才起作用。如果有几个 highgui 窗口，则其中任何一个都可以处于活动状态。

图像显示函数 imshow 后必须跟有 waitKey 函数，否则图像无法正常显示。

waitKey 使用的案例代码如下：

```python
import cv2

#读取图像
image = cv2.imread("src.jpg")
#图像显示
cv2.imshow("image", image)
#按键等待
key = cv2.waitKey(0)
print("Press key is:", key)
cv2.destroyWindow("image")
```

按下 G 键，显示窗口关闭，按键按下打印信息如下：

```
Press key is: 103
```

6.5 案例 82：鼠标操作

OpenCV 中用于设置鼠标事件回调的函数为 setMouseCallback，其定义如下：

```
setMouseCallback(windowName, onMouse, param=None)
```

参数说明如下。

- windowName：窗口名。
- onMouse：鼠标事件回调函数。
- param：参数。

setMouseCallback 函数的返回值为 None。

OpenCV 中的鼠标事件可以通过如下代码查询：

```
import cv2
events = [ev for ev in dir(cv2) if 'EVENT' in ev]
print( events )         #将鼠标事件显示出来
```

鼠标事件打印结果如下：

```
['EVENT_FLAG_ALTKEY', 'EVENT_FLAG_CTRLKEY', 'EVENT_FLAG_LBUTTON', 'EVENT_FLAG_MBUTTON',
'EVENT_FLAG_RBUTTON', 'EVENT_FLAG_SHIFTKEY', 'EVENT_LBUTTONDBLCLK', 'EVENT_LBUTTONDOWN',
'EVENT_LBUTTONUP', 'EVENT_MBUTTONDBLCLK', 'EVENT_MBUTTONDOWN', 'EVENT_MBUTTONUP',
'EVENT_MOUSEHWHEEL', 'EVENT_MOUSEMOVE', 'EVENT_MOUSEWHEEL', 'EVENT_RBUTTONDBLCLK',
'EVENT_RBUTTONDOWN', 'EVENT_RBUTTONUP']
```

在 6.1 节中提到，鼠标事件类型由 MouseEventTypes 定义，鼠标事件标志由 MouseEventFlags 定义。

本节案例通过不同的鼠标操作进行图形绘制，双击可以绘制圆，鼠标左键按下拖动绘制红色线，鼠标右键按下拖动绘制蓝色线，案例代码如下：

```
import cv2
import numpy as np

#设置鼠标起始点
start_point, end_point= (0,0), (0,0)
#设置鼠标左键和右键按下/抬起标志
lb_down, lb_up, rb_down, rb_up = False, False, False, False
#鼠标回调函数
def mouse_event(event, x, y, flags, param):
    global end_point, start_point, lb_down, lb_up, rb_down, rb_up
    #双击
    if event == cv2.EVENT_LBUTTONDBLCLK:
        cv2.circle(img, (x, y), 100, (0, 255, 0), -1)
    #左键按下
    elif event == cv2.EVENT_LBUTTONDOWN:
        #设置鼠标按下的起始点
        end_point = (x, y)
        start_point = end_point
        lb_down = True
    #右键按下
    elif event == cv2.EVENT_RBUTTONDOWN:
        end_point = (x, y)
        start_point = end_point
        rb_down = True
    #鼠标移动，绘制线
    elif event == cv2.EVENT_MOUSEMOVE:
        #鼠标左键按下绘制红色线
```

```
        if lb_down:
            cv2.line(img, end_point, (x,y), (0, 0, 255), 2, -1)
        #鼠标右键按下绘制蓝色线
        if rb_down:
            cv2.line(img, end_point, (x, y), (255, 0, 0), 2, -1)
        end_point = (x, y)
    #左键释放
    elif event == cv2.EVENT_LBUTTONUP:
        lb_up = True
        lb_down = False
        cv2.line(img, end_point, (x,y), (0, 0, 255), 2, -1)
    #右键释放
    elif event == cv2.EVENT_RBUTTONUP:
        rb_up = True
        rb_down = False
        cv2.line(img, end_point, (x,y), (255, 0, 0), 2, -1)

#创建一个黑色的图像 img；创建一个窗口，并命名为 image
img = np.zeros((512, 512, 3), np.uint8)
cv2.namedWindow('image')
#设置回调
cv2.setMouseCallback('image', mouse_event)
while (1):
    cv2.imshow('image', img)
    #按 Esc 键退出
    if cv2.waitKey(10) == 27:
        break
cv2.destroyAllWindows()
```

鼠标绘制结果如图 6.5 所示。

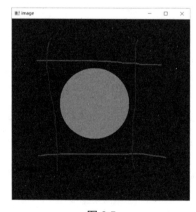

图 6.5

6.6 案例 83：进度条操作

OpenCV 中提供了用于创建进度条的函数 createTrackbar，其定义如下：

```
createTrackbar(trackbarName, windowName, value, count, onChange)
```

参数说明如下。

- trackbarName：进度条名称。
- windowName：窗口名。
- value：进度条初始位置值。
- count：进度条最大值。
- onChange：进度条滑动回调函数。

createTrackbar 函数的返回值为 None。

OpenCV 中提供了用于获取进度条位置的函数 getTrackbarPos，其定义如下：

```
retval = getTrackbarPos(trackbarname, winname)
```

参数说明如下。

- trackbarname：进度条名称。
- winname：窗口名。
- retval：获取的进度条位置。

本案例简单实现了一个颜色混合器（或称调色板），在图像窗口中创建了 3 个进度条，进度条的值为 0~255，3 个进度条的值对应图像的 RGB 3 个通道的值，调节进度条可以查看颜色混合后的结果。案例代码如下：

```python
import cv2
import numpy as np

#定义进度条回调函数
def changeColor(x):
    #获取进度条中 R 通道颜色值
    color_r = cv2.getTrackbarPos('color_R', 'image')
    #获取进度条中 G 通道颜色值
    color_g = cv2.getTrackbarPos('color_G', 'image')
    #获取进度条中 B 通道颜色值
    color_b = cv2.getTrackbarPos('color_B', 'image')
    #给图像 img 的 BGR 3 个通道颜色赋值
    img[:] = [color_b, color_g, color_r]
    cv2.imshow("image", img)

#创建图像
```

```python
img = np.zeros([512, 512, 3], np.uint8)
#创建窗口 image
cv2.namedWindow("image")
#创建 3 个进度条，对应 3 个通道颜色
cv2.createTrackbar("color_R", 'image', 0, 255, changeColor)
cv2.createTrackbar("color_G", 'image', 0, 255, changeColor)
cv2.createTrackbar("color_B", 'image', 0, 255, changeColor)
#无限循环等待键盘事件
while(True):
    cv2.imshow("image", img)
    #如果按键为 Esc 则退出
    key = cv2.waitKey(0)
    if key == 27:
        break
cv2.destroyAllWindows()
```

案例执行结果如图 6.6 所示。

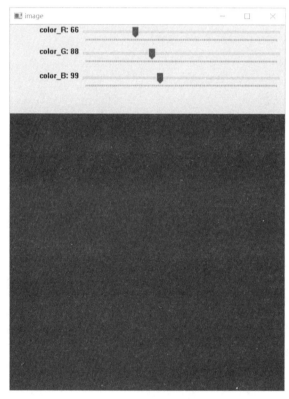

图 6.6

6.7 进阶必备：在 Qt 中使用 OpenCV

highgui 模块提供的可视化使用比较方便，但是因为 UI（用户接口）特性不是很完整，所以无法实现较为精细的可视化功能。Qt 是一个跨平台的工具包，提供丰富的 UI 功能，在客户端开发中被广泛使用，使用基于 Qt 的函数的开发弥补了 highgui 模块的不足，提升了开发效率。

本节介绍在 Qt 中使用 OpenCV 的配置方法，包括以下 4 步。

第一步，安装 Qt，安装完成的 Qt Creator 如图 6.7 所示。

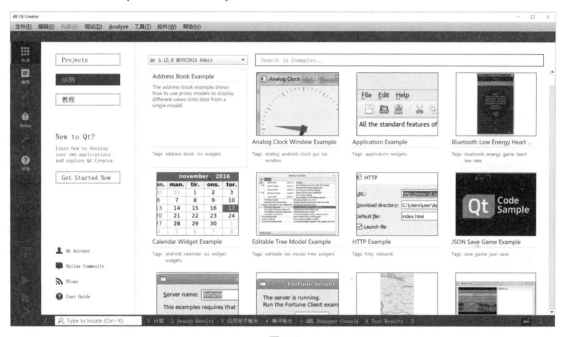

图 6.7

Qt 安装后，注意将 MinGW 加入环境变量并重启计算机（本节使用的编译器为 MinGW 编译器，如果读者使用的是 MSVC 编译器，则不需要此步骤），否则项目生成会出错，如图 6.8 所示。

图 6.8

第二步，安装 CMake 软件，1.4.1 节中对此软件有介绍，读者可以参考。

第三步，OpenCV 编译。

使用 CMake 软件生成解决方案的配置如图 6.9 所示，注意勾选"Advanced"复选框。

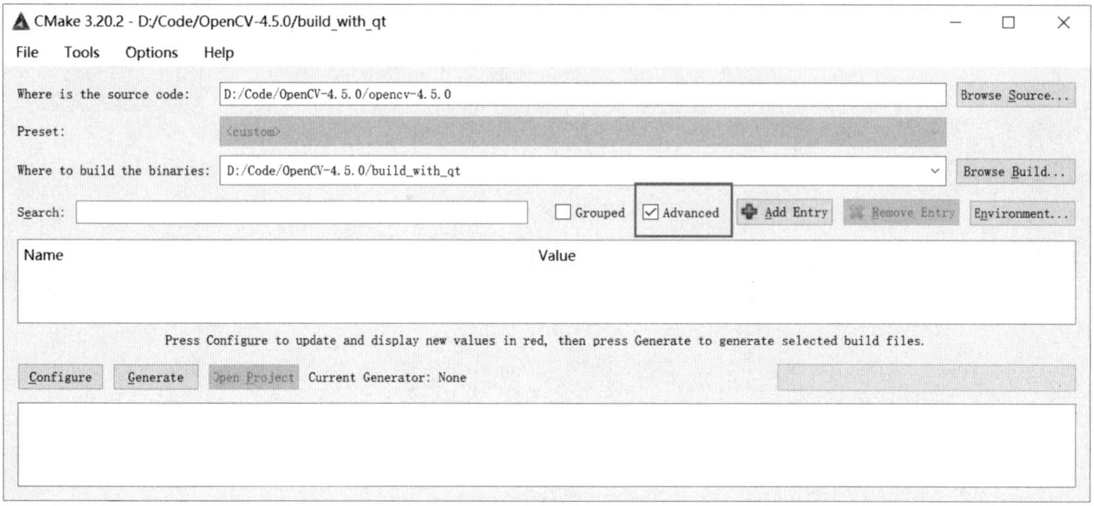

图 6.9

单击"Configure"按钮进行配置，自定义编译器，选择"MinGW Makefiles"，如图 6.10 所示。

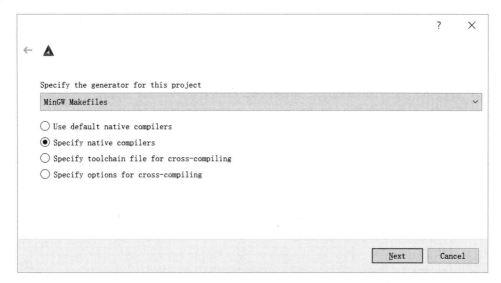

图 6.10

单击"Next"按钮，配置 C 和 C++语言的编译器，如图 6.11 所示。

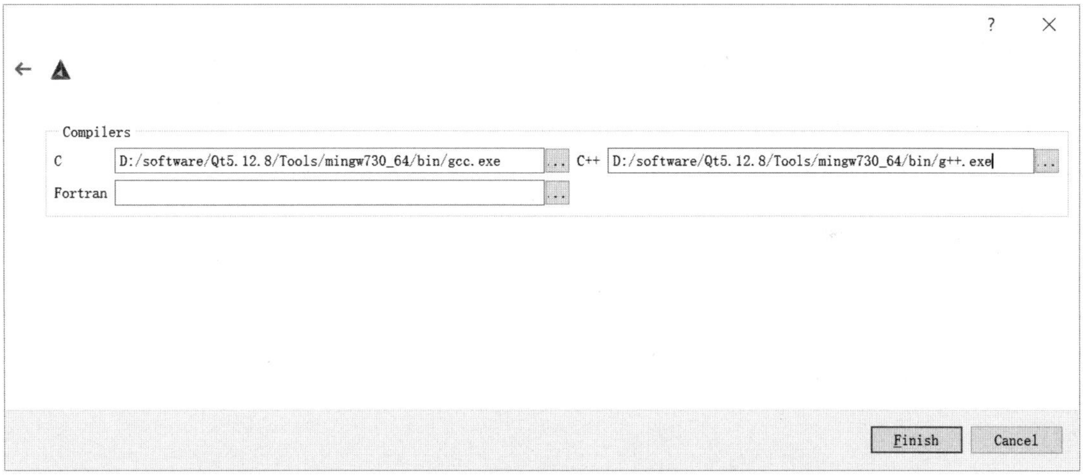

图 6.11

gcc 和 g++编译器在 Qt 安装目录的工具包中，如图 6.12 所示。

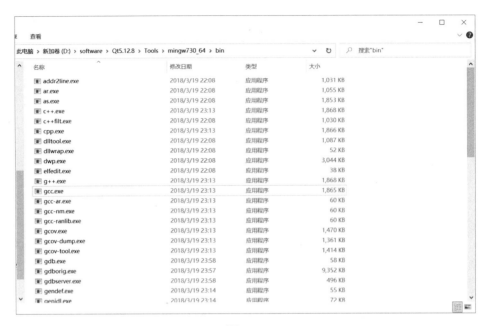

图 6.12

单击图 6.11 中的 "Finish" 按钮，完成配置，在编译选项中勾选 "WITH_OPENGL" 和 "WITH_QT" 复选框，如图 6.13 所示。

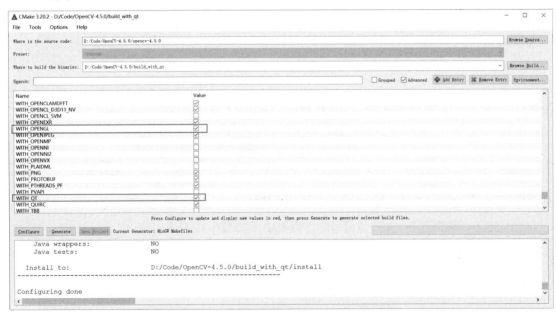

图 6.13

单击"Configure"按钮完成配置，单击"Generate"按钮生成项目工程，进入工程目录，使用 mingw32-make 命令进行项目编译，如图 6.14 所示。

```
C:\Users\lxiao217>cd D:\Code\OpenCV-4.5.0\build_with_qt
C:\Users\lxiao217>d:
D:\Code\OpenCV-4.5.0\build_with_qt>mingw32-make -j 8
```

图 6.14

查看 mingw32-make 命令的帮助信息，可以了解其中的编译参数的含义，如图 6.15 所示。在图 6.14 中，-j 参数用于指定多个任务同时进行。

```
C:\Users\lxiao217>mingw32-make --help
Usage: mingw32-make [options] [target] ...
Options:
  -b, -m                      Ignored for compatibility.
  -B, --always-make           Unconditionally make all targets.
  -C DIRECTORY, --directory=DIRECTORY
                              Change to DIRECTORY before doing anything.
  -d                          Print lots of debugging information.
  --debug[=FLAGS]             Print various types of debugging information.
  -e, --environment-overrides
                              Environment variables override makefiles.
  --eval=STRING               Evaluate STRING as a makefile statement.
  -f FILE, --file=FILE, --makefile=FILE
                              Read FILE as a makefile.
  -h, --help                  Print this message and exit.
  -i, --ignore-errors         Ignore errors from recipes.
  -I DIRECTORY, --include-dir=DIRECTORY
                              Search DIRECTORY for included makefiles.
  -j [N], --jobs[=N]          Allow N jobs at once; infinite jobs with no arg.
  -k, --keep-going            Keep going when some targets can't be made.
  -l [N], --load-average[=N], --max-load[=N]
                              Don't start multiple jobs unless load is below N.
  -L, --check-symlink-times   Use the latest mtime between symlinks and target.
  -n, --just-print, --dry-run, --recon
                              Don't actually run any recipe; just print them.
  -o FILE, --old-file=FILE, --assume-old=FILE
                              Consider FILE to be very old and don't remake it.
  -O[TYPE], --output-sync[=TYPE]
                              Synchronize output of parallel jobs by TYPE.
  -p, --print-data-base       Print make's internal database.
  -q, --question              Run no recipe; exit status says if up to date.
  -r, --no-builtin-rules      Disable the built-in implicit rules.
  -R, --no-builtin-variables  Disable the built-in variable settings.
  -s, --silent, --quiet       Don't echo recipes.
  -S, --no-keep-going, --stop
                              Turns off -k.
  -t, --touch                 Touch targets instead of remaking them.
  --trace                     Print tracing information.
  -v, --version               Print the version number of make and exit.
  -w, --print-directory       Print the current directory.
  --no-print-directory        Turn off -w, even if it was turned on implicitly.
  -W FILE, --what-if=FILE, --new-file=FILE, --assume-new=FILE
                              Consider FILE to be infinitely new.
  --warn-undefined-variables  Warn when an undefined variable is referenced.

This program built for x86_64-w64-mingw32
Report bugs to <bug-make@gnu.org>
```

图 6.15

编译完成后，执行 mingw32-make install 命令生成 OpenCV 安装包，如图 6.16 所示，可以在配置时勾选"BUILD_opencv_world"复选框生成 opencv_world 库文件。

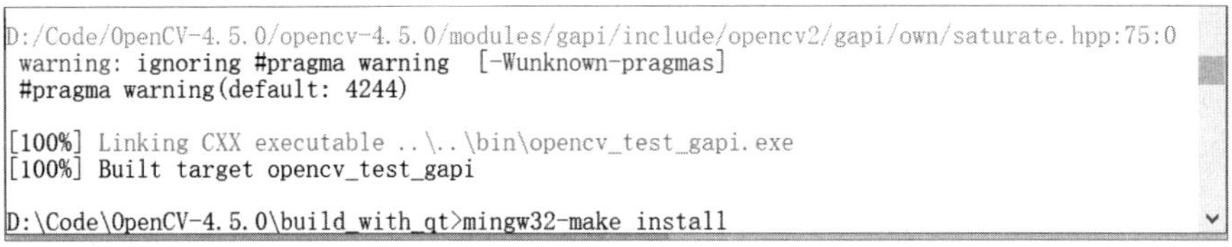

```
D:/Code/OpenCV-4.5.0/opencv-4.5.0/modules/gapi/include/opencv2/gapi/own/saturate.hpp:75:0
warning: ignoring #pragma warning [-Wunknown-pragmas]
 #pragma warning(default: 4244)

[100%] Linking CXX executable ..\..\bin\opencv_test_gapi.exe
[100%] Built target opencv_test_gapi

D:\Code\OpenCV-4.5.0\build_with_qt>mingw32-make install
```

图 6.16

完成后，在路径下生成了 OpenCV 的安装包，如图 6.17 所示。

图 6.17

第四步，编译结果测试。

创建 Qt 项目，注意配置时选用 MinGW 编译器，否则无法使用 MinGW 编译的 OpenCV，如图 6.18 所示。

图 6.18

项目创建完成后，需要在工程中增加对 OpenCV 的引用，如图 6.19 所示。

图 6.19

在测试项目过程中，在主窗口中增加了一个按钮，单击该按钮显示图像，如图 6.20 所示。

图 6.20

与 OpenCV 图像显示窗口不同，基于 Qt 的图像显示窗口包括工具栏和状态栏，工具栏上提供了图像缩放、图像保存、图像拷贝到剪贴板等功能，状态栏上提供了鼠标所在位置的坐标信息、鼠标所处当前点的 RGB 值。Qt 函数提供了创建按钮的功能，提供了更加美观的进度条与文字等元素，提供了与 OpenGL 交互的函数，还有其他特性，读者可以自行研究使用。

 提示：OpenCV 编译的详细过程在 1.4 节有介绍，此处只列出编译的关键步骤。因为在 1.4 节中的介绍使用的是 MSVC 编译器，所以本节使用 MinGW 编译器来编译。

第 7 章
视频处理模块 videoio

在视频处理中，常见的操作包括将视频写入磁盘和从磁盘读取视频，视频读/写后将其播放显示，videoio 模块提供了视频读取和保存的功能。

7.1 模块导读

videoio 模块提供了使用 OpenCV 读取和写入视频，或者读/写图像序列的功能，需要通过包含头文件"opencv2/videoio.hpp"来引入，模块说明如下：

```
#ifndef OPENCV_VIDEOIO_HPP
#define OPENCV_VIDEOIO_HPP

#include "opencv2/core.hpp"

typedef struct CvCapture CvCapture;
typedef struct CvVideoWriter CvVideoWriter;

namespace cv
{
```

如下为 VideoCapture API 后端标识符，可以在 VideoCapture::VideoCapture()类构造或 VideoCapture::open()函数中选择后端。

注意：这些可用的后端在调用时已经在 OpenCV 的库中做了编译。

```
enum VideoCaptureAPIs {
    CAP_ANY          = 0,              //!< 自动检测
```

```
    CAP_VFW              = 200,                //!< Windows 视频
    CAP_V4L              = 200,                //!< V4L/V4L2 支持
    CAP_V4L2             = CAP_V4L,            //!< 同 CAP_V4L
    CAP_FIREWIRE         = 300,                //!< IEEE 1394 驱动
    CAP_FIREWARE         = CAP_FIREWIRE,       //!< 同 CAP_FIREWIRE
    CAP_IEEE1394         = CAP_FIREWIRE,       //!< 同 CAP_FIREWIRE
    CAP_DC1394           = CAP_FIREWIRE,       //!< 同 CAP_FIREWIRE
    CAP_CMU1394          = CAP_FIREWIRE,       //!< 同 CAP_FIREWIRE
    CAP_QT               = 500,                //!< QT
    CAP_UNICAP           = 600,                //!< UniCAP 驱动
    CAP_DSHOW            = 700,                //!< DirectShow 工具
    CAP_PVAPI            = 800,                //!< PvAPI 工具
    CAP_OPENNI           = 900,                //!< Kinect 中应用的 OpenNI 工具
    CAP_OPENNI_ASUS      = 910,                //!< Asus Xtion 中应用的 OpenNI 工具
    CAP_ANDROID          = 1000,               //!< Android 系统
    CAP_XIAPI            = 1100,               //!< XIMEA 相机 API
    CAP_AVFOUNDATION     = 1200,               //!< AVFoundation 框架
    CAP_GIGANETIX        = 1300,               //!< Smartek Giganetix GigEVisionSDK 视频模块
    CAP_MSMF             = 1400,               //!< 微软 Media Foundation 视频模块
    CAP_WINRT            = 1410,               //!< 微软 Windows Runtime 视频模块
    CAP_INTELPERC        = 1500,               //!< RealSense 工具包
    CAP_REALSENSE        = 1500,               //!< 同 CAP_INTELPERC
    CAP_OPENNI2          = 1600,               //!< Kinect 应用的 OpenNI2 工具
    CAP_OPENNI2_ASUS     = 1610,               //!< Asus Xtion 等应用的 OpenNI2 工具
    CAP_GPHOTO2          = 1700,               //!< gPhoto2 连接
    CAP_GSTREAMER        = 1800,               //!< GStreamer
    CAP_FFMPEG           = 1900,               //!< 使用 FFMPEG 打开或保存视频文件或数据流
    CAP_IMAGES           = 2000,               //!< OpenCV 图像序列
    CAP_ARAVIS           = 2100,               //!< Aravis SDK 工具
    CAP_OPENCV_MJPEG     = 2200,               //!< OpenCV 内置的 MotionJPEG 编解码
    CAP_INTEL_MFX        = 2300,               //!< Intel MediaSDK 工具
    CAP_XINE             = 2400,               //!< XINE 引擎
};

/** VideoCapture 通用属性标识符
读/写属性涉及许多层，选择某种读/写方式可能会出现一些意想不到的结果，行为是否可选取决于设备硬件、驱动程序和 API 后端*/
enum VideoCaptureProperties {
    CAP_PROP_POS_MSEC          =0,    //!< 视频文件当前位置
    CAP_PROP_POS_FRAMES        =1,    //!< 下一个要解码/捕获的帧索引（从 0 开始计数）
    CAP_PROP_POS_AVI_RATIO     =2,    //!< 视频文件的相对位置：0 表示胶片开始，1 表示结束
    CAP_PROP_FRAME_WIDTH       =3,    //!< 视频流中帧的宽度
    CAP_PROP_FRAME_HEIGHT      =4,    //!< 视频流中帧的高度
```

```
CAP_PROP_FPS              =5,   //!< 帧频
CAP_PROP_FOURCC           =6,   //!< 编解码器的 4 字符代码
CAP_PROP_FRAME_COUNT      =7,   //!< 视频文件中的帧数
CAP_PROP_FORMAT           =8,   //!< Mat 格式对象
CAP_PROP_MODE             =9,   //!< 后端特定值，指示当前捕获模式
CAP_PROP_BRIGHTNESS       =10,  //!< 图像的亮度（仅适用于支持的相机）
CAP_PROP_CONTRAST         =11,  //!< 图像对比度（仅适用于相机）
CAP_PROP_SATURATION       =12,  //!< 图像的饱和度（仅适用于相机）
CAP_PROP_HUE              =13,  //!< 图像的色调（仅适用于相机）
CAP_PROP_GAIN             =14,  //!< 图像增益（仅适用于支持的相机）
CAP_PROP_EXPOSURE         =15,  //!< 曝光（仅适用于支持的相机）
CAP_PROP_CONVERT_RGB      =16,  //!< 该标志指示是否将图像转换为 RGB 格式
CAP_PROP_WHITE_BALANCE_BLUE_U  =17, //!< 暂不支持
CAP_PROP_RECTIFICATION=18, //!< 立体相机校正标志
CAP_PROP_MONOCHROME       =19,
CAP_PROP_SHARPNESS        =20,
CAP_PROP_AUTO_EXPOSURE=21, //!< DC1394：曝光控制由相机完成，读者可以使用此功能调整参考水平
CAP_PROP_GAMMA            =22,
CAP_PROP_TEMPERATURE      =23,
CAP_PROP_TRIGGER          =24,
CAP_PROP_TRIGGER_DELAY=25,
CAP_PROP_WHITE_BALANCE_RED_V   =26,
CAP_PROP_ZOOM             =27,
CAP_PROP_FOCUS            =28,
CAP_PROP_GUID             =29,
CAP_PROP_ISO_SPEED        =30,
CAP_PROP_BACKLIGHT        =32,
CAP_PROP_PAN              =33,
CAP_PROP_TILT             =34,
CAP_PROP_ROLL             =35,
CAP_PROP_IRIS             =36,
CAP_PROP_SETTINGS         =37,  //!< 弹出视频/相机过滤器对话框
CAP_PROP_BUFFERSIZE       =38,
CAP_PROP_AUTOFOCUS        =39,
CAP_PROP_SAR_NUM          =40,  //!< 样本纵横比：num/den(num)
CAP_PROP_SAR_DEN          =41,  //!< 样本纵横比：num/den (den)
CAP_PROP_BACKEND          =42,  //!< 当前后端
CAP_PROP_CHANNEL          =43,  //!< 视频输出或通道数
CAP_PROP_AUTO_WB          =44,  //!< 开启/关闭自动白平衡功能
CAP_PROP_WB_TEMPERATURE     =45, //!< 白平衡色温
CAP_PROP_CODEC_PIXEL_FORMAT =46, //!< 编解码器像素格式
CAP_PROP_BITRATE          =47,  //!< 视频比特率（kbit/s）
CAP_PROP_ORIENTATION_META =48,  //!< 由 stream meta 定义的帧旋转（仅适用于 FFMPEG 后端）
CAP_PROP_ORIENTATION_AUTO =49,  //!< 若为 true，则根据视频元数据旋转输出帧
```

```
#ifndef CV_DOXYGEN
      CV__CAP_PROP_LATEST
#endif
    };
```

如下为与视频读/写相关的标志定义：

```
/** VideoWriter 通用属性标识符*/
enum VideoWriterProperties {
  //!< 编码视频流的当前质量（0%～100%），可以在某些编解码器中动态调整
  VIDEOWRITER_PROP_QUALITY        = 1,
  //!< 编码的视频帧的大小。注意：编码顺序可能不同于表示顺序
  VIDEOWRITER_PROP_FRAMEBYTES     = 2,
  //!< 并行编码的条纹数，若将其设置为-1，则用于自动检测
  VIDEOWRITER_PROP_NSTRIPES       = 3,
  //!< 如果不为零，则编码器将期望并编码彩色帧；否则将处理灰度帧
  VIDEOWRITER_PROP_IS_COLOR       = 4
};

/** IEEE 1394 控制寄存器的模式*/
enum { CAP_PROP_DC1394_OFF              = -4, //!< 关闭该功能
       CAP_PROP_DC1394_MODE_MANUAL      = -3, //!< 当读者设置特征值时自动设置
       CAP_PROP_DC1394_MODE_AUTO        = -2,
       CAP_PROP_DC1394_MODE_ONE_PUSH_AUTO = -1,
       CAP_PROP_DC1394_MAX              = 31
     };

//! OpenNI map 生成器
enum { CAP_OPENNI_DEPTH_GENERATOR= 1 << 31,
       CAP_OPENNI_IMAGE_GENERATOR = 1 << 30,
       CAP_OPENNI_IR_GENERATOR    = 1 << 29,
       CAP_OPENNI_GENERATORS_MASK = CAP_OPENNI_DEPTH_GENERATOR + CAP_OPENNI_IMAGE_GENERATOR + CAP_OPENNI_IR_GENERATOR
     };

//! 通过 OpenNI 后端提供的摄像头属性
enum { CAP_PROP_OPENNI_OUTPUT_MODE         = 100,
       CAP_PROP_OPENNI_FRAME_MAX_DEPTH     = 101, //!< 帧最大深度，单位为 mm
       CAP_PROP_OPENNI_BASELINE            = 102, //!< 基准，单位为 mm
       CAP_PROP_OPENNI_FOCAL_LENGTH        = 103, //!< 焦距，单位为像素
       CAP_PROP_OPENNI_REGISTRATION        = 104, //!< 将重映射深度 map 与图像 map 同步的标志
       CAP_PROP_OPENNI_REGISTRATION_ON     = CAP_PROP_OPENNI_REGISTRATION,
       CAP_PROP_OPENNI_APPROX_FRAME_SYNC   = 105,
       CAP_PROP_OPENNI_MAX_BUFFER_SIZE     = 106,
       CAP_PROP_OPENNI_CIRCLE_BUFFER       = 107,
       CAP_PROP_OPENNI_MAX_TIME_DURATION   = 108,
       CAP_PROP_OPENNI_GENERATOR_PRESENT   = 109,
```

```
        CAP_PROP_OPENNI2_SYNC                   = 110,
        CAP_PROP_OPENNI2_MIRROR                  = 111
    };

//! OpenNI 快捷键
enum { CAP_OPENNI_IMAGE_GENERATOR_PRESENT         = CAP_OPENNI_IMAGE_GENERATOR +
CAP_PROP_OPENNI_GENERATOR_PRESENT,
       CAP_OPENNI_IMAGE_GENERATOR_OUTPUT_MODE     = CAP_OPENNI_IMAGE_GENERATOR +
CAP_PROP_OPENNI_OUTPUT_MODE,
       CAP_OPENNI_DEPTH_GENERATOR_PRESENT         = CAP_OPENNI_DEPTH_GENERATOR +
CAP_PROP_OPENNI_GENERATOR_PRESENT,
       CAP_OPENNI_DEPTH_GENERATOR_BASELINE        = CAP_OPENNI_DEPTH_GENERATOR +
CAP_PROP_OPENNI_BASELINE,
       CAP_OPENNI_DEPTH_GENERATOR_FOCAL_LENGTH    = CAP_OPENNI_DEPTH_GENERATOR +
CAP_PROP_OPENNI_FOCAL_LENGTH,
       CAP_OPENNI_DEPTH_GENERATOR_REGISTRATION    = CAP_OPENNI_DEPTH_GENERATOR +
CAP_PROP_OPENNI_REGISTRATION,
       CAP_OPENNI_DEPTH_GENERATOR_REGISTRATION_ON =
CAP_OPENNI_DEPTH_GENERATOR_REGISTRATION,
       CAP_OPENNI_IR_GENERATOR_PRESENT            = CAP_OPENNI_IR_GENERATOR +
CAP_PROP_OPENNI_GENERATOR_PRESENT,
    };

//! 深度发生器提供的 OpenNI 数据
enum { CAP_OPENNI_DEPTH_MAP           = 0, //!< 深度数据，单位为 mm（CV_16UC1 格式）
       CAP_OPENNI_POINT_CLOUD_MAP     = 1, //!< XYZ 数据，单位为 m（CV_32FC3 格式）
       CAP_OPENNI_DISPARITY_MAP       = 2, //!< 差异值，单位为像素（CV_8UC1 格式）
       CAP_OPENNI_DISPARITY_MAP_32F   = 3, //!< 差异值，单位为像素（CV_32FC1 格式）
       CAP_OPENNI_VALID_DEPTH_MASK    = 4, //!< CV_8UC1

       CAP_OPENNI_BGR_IMAGE           = 5, //!< RGB 图像生成器提供的数据
       CAP_OPENNI_GRAY_IMAGE          = 6, //!< 灰度图像生成器提供的数据

       CAP_OPENNI_IR_IMAGE            = 7  //!< IR 图像生成器提供的数据
    };

//! OpenNI 图像生成器支持的输出模式
enum { CAP_OPENNI_VGA_30HZ   = 0,
       CAP_OPENNI_SXGA_15HZ  = 1,
       CAP_OPENNI_SXGA_30HZ  = 2,
       CAP_OPENNI_QVGA_30HZ  = 3,
       CAP_OPENNI_QVGA_60HZ  = 4
    };

enum { CAP_PROP_GSTREAMER_QUEUE_LENGTH    = 200 //!< 默认值为 1
    };
```

```cpp
//! PVAPI
enum { CAP_PROP_PVAPI_MULTICASTIP= 300, //!< 启用多播主模式的 IP，0 为禁用多播
    CAP_PROP_PVAPI_FRAMESTARTTRIGGERMODE= 301, //!< FrameStartTriggerMode: 确定如何启动帧
    CAP_PROP_PVAPI_DECIMATIONHORIZONTAL  = 302, //!< 水平下采样图像
    CAP_PROP_PVAPI_DECIMATIONVERTICAL    = 303, //!< 垂直下采样图像
    CAP_PROP_PVAPI_BINNINGX              = 304, //!< 水平 Binning 因子
    CAP_PROP_PVAPI_BINNINGY              = 305, //!< 垂直 Binning 因子
    CAP_PROP_PVAPI_PIXELFORMAT           = 306  //!< Pixel 格式
};

//! PVAPI: FrameStartTriggerMode
enum { CAP_PVAPI_FSTRIGMODE_FREERUN   = 0,    //!< Freerun 模式
    CAP_PVAPI_FSTRIGMODE_SYNCIN1      = 1,    //!< SyncIn1 模式
    CAP_PVAPI_FSTRIGMODE_SYNCIN2      = 2,    //!< SyncIn2 模式
    CAP_PVAPI_FSTRIGMODE_FIXEDRATE    = 3,    //!< FixedRate 模式
    CAP_PVAPI_FSTRIGMODE_SOFTWARE     = 4     //!< Software 模式
};

//! PVAPI: DecimationHorizontal, DecimationVertical
enum { CAP_PVAPI_DECIMATION_OFF       = 1,    //!< 关闭该功能
    CAP_PVAPI_DECIMATION_2OUTOF4      = 2,    //!< 四选二
    CAP_PVAPI_DECIMATION_2OUTOF8      = 4,    //!< 八选二
    CAP_PVAPI_DECIMATION_2OUTOF16     = 8     //!< 十六选二
};

//! PVAPI: PixelFormat
enum { CAP_PVAPI_PIXELFORMAT_MONO8    = 1,    //!< Mono8 模式
    CAP_PVAPI_PIXELFORMAT_MONO16      = 2,    //!< Mono16 模式
    CAP_PVAPI_PIXELFORMAT_BAYER8      = 3,    //!< Bayer8 模式
    CAP_PVAPI_PIXELFORMAT_BAYER16     = 4,    //!< Bayer16 模式
    CAP_PVAPI_PIXELFORMAT_RGB24       = 5,    //!< Rgb24 模式
    CAP_PVAPI_PIXELFORMAT_BGR24       = 6,    //!< Bgr24 模式
    CAP_PVAPI_PIXELFORMAT_RGBA32      = 7,    //!< Rgba32 模式
    CAP_PVAPI_PIXELFORMAT_BGRA32      = 8,    //!< Bgra32 模式
};

//! XIMEA SDK 后端提供的摄像头属性
enum { CAP_PROP_XI_DOWNSAMPLING       = 400, //!< 通过合并或跳过更改图像分辨率
    CAP_PROP_XI_DATA_FORMAT           = 401, //!< 输出数据格式
    CAP_PROP_XI_OFFSET_X              = 402, //!< 原点到感兴趣区域的水平偏移（单位：像素）
    CAP_PROP_XI_OFFSET_Y              = 403, //!< 原点到感兴趣区域的垂直偏移（单位：像素）
    CAP_PROP_XI_TRG_SOURCE            = 404, //!< 定义触发器的源
    CAP_PROP_XI_TRG_SOFTWARE          = 405, //!< 生成内部触发器
```

属性	值	说明
CAP_PROP_XI_GPI_SELECTOR	= 406,	//!< 选择通用输入
CAP_PROP_XI_GPI_MODE	= 407,	//!< 设置通用输入模式
CAP_PROP_XI_GPI_LEVEL	= 408,	//!< 获取通用输入级别
CAP_PROP_XI_GPO_SELECTOR	= 409,	//!< 选择通用输出
CAP_PROP_XI_GPO_MODE	= 410,	//!< 设置通用输出模式
CAP_PROP_XI_LED_SELECTOR	= 411,	//!< 选择摄像头信号 LED
CAP_PROP_XI_LED_MODE	= 412,	//!< 定义摄像头信号 LED 功能
CAP_PROP_XI_MANUAL_WB	= 413,	//!< 计算白平衡
CAP_PROP_XI_AUTO_WB	= 414,	//!< 自动白平衡
CAP_PROP_XI_AEAG	= 415,	//!< 自动曝光/增益
CAP_PROP_XI_EXP_PRIORITY	= 416,	//!< 曝光优先级
CAP_PROP_XI_AE_MAX_LIMIT	= 417,	//!< AEAG 程序中的最大限制
CAP_PROP_XI_AG_MAX_LIMIT	= 418,	//!< AEAG 程序中的最大增益限制
CAP_PROP_XI_AEAG_LEVEL	= 419,	//!< 输出信号 AEAG 的平均强度应达到的水平（单位为%）
CAP_PROP_XI_TIMEOUT	= 420,	//!< 图像捕获超时（单位：ms）
CAP_PROP_XI_EXPOSURE	= 421,	//!< 曝光时间（单位：ms）
CAP_PROP_XI_EXPOSURE_BURST_COUNT	= 422,	//!< 设置一帧中的曝光次数
CAP_PROP_XI_GAIN_SELECTOR	= 423,	//!< 参数增益选择器
CAP_PROP_XI_GAIN	= 424,	//!< 增益（单位：dB）
CAP_PROP_XI_DOWNSAMPLING_TYPE	= 426,	//!< 改变图像下采样类型
CAP_PROP_XI_BINNING_SELECTOR	= 427,	//!< Binning 引擎选择器
CAP_PROP_XI_BINNING_VERTICAL	= 428,	//!< 垂直 Binning
CAP_PROP_XI_BINNING_HORIZONTAL	= 429,	//!< 水平 Binning
CAP_PROP_XI_BINNING_PATTERN	= 430,	//!< Binning 模式类型
CAP_PROP_XI_DECIMATION_SELECTOR	= 431,	//!< 采样引擎选择器
CAP_PROP_XI_DECIMATION_VERTICAL	= 432,	//!< 水平采样
CAP_PROP_XI_DECIMATION_HORIZONTAL	= 433,	//!< 垂直采样
CAP_PROP_XI_DECIMATION_PATTERN	= 434,	//!< 采样模式类型

//!< 选择由 TestPattern 特性控制的测试模式生成器

属性	值	说明
CAP_PROP_XI_TEST_PATTERN_GENERATOR_SELECTOR	= 587,	
CAP_PROP_XI_TEST_PATTERN	= 588,	//!< 选择由所选生成器生成的测试模式类型
CAP_PROP_XI_IMAGE_DATA_FORMAT	= 435,	//!< 输出数据格式
CAP_PROP_XI_SHUTTER_TYPE	= 436,	//!< 更改传感器快门类型（CMOS 传感器）
CAP_PROP_XI_SENSOR_TAPS	= 437,	//!< taps 数量
CAP_PROP_XI_AEAG_ROI_OFFSET_X	= 439,	//!< 自动曝光/增益 ROI 补偿 X
CAP_PROP_XI_AEAG_ROI_OFFSET_Y	= 440,	//!< 自动曝光/增益 ROI 补偿 Y
CAP_PROP_XI_AEAG_ROI_WIDTH	= 441,	//!< 自动曝光/增益 ROI 宽度
CAP_PROP_XI_AEAG_ROI_HEIGHT	= 442,	//!< 自动曝光/增益 ROI 高度
CAP_PROP_XI_BPC	= 445,	//!< 坏像素修正
CAP_PROP_XI_WB_KR	= 448,	//!< 白平衡红系数
CAP_PROP_XI_WB_KG	= 449,	//!< 白平衡绿系数
CAP_PROP_XI_WB_KB	= 450,	//!< 白平衡蓝系数
CAP_PROP_XI_WIDTH	= 451,	//!< 设备提供的图像宽度

```
CAP_PROP_XI_HEIGHT                    = 452, //!< 设备提供的图像高度
CAP_PROP_XI_REGION_SELECTOR           = 589, //!< 按宽度、高度、区域模式等参数选择多 ROI 中的区域
CAP_PROP_XI_REGION_MODE               = 595, //!< 激活/停用区域选择器选择的区域
CAP_PROP_XI_LIMIT_BANDWIDTH           = 459, //!< 设置/获取 bandwidth
CAP_PROP_XI_SENSOR_DATA_BIT_DEPTH     = 460, //!< 传感器输出数据位深度
CAP_PROP_XI_OUTPUT_DATA_BIT_DEPTH     = 461, //!< 设备输出数据位深度
CAP_PROP_XI_IMAGE_DATA_BIT_DEPTH      = 462, //!< xiGetImage 函数返回的位深度
CAP_PROP_XI_OUTPUT_DATA_PACKING       = 463, //!< 设备输出数据打包（或分组）已启用
CAP_PROP_XI_OUTPUT_DATA_PACKING_TYPE  = 464, //!< 数据打包类型
CAP_PROP_XI_IS_COOLED                 = 465, //!< 对于支持冷却的相机返回 1
CAP_PROP_XI_COOLING                   = 466, //!< 开启相机冷却
CAP_PROP_XI_TARGET_TEMP               = 467, //!< 设置冷却的传感器目标温度
CAP_PROP_XI_CHIP_TEMP                 = 468, //!< 相机传感器温度
CAP_PROP_XI_HOUS_TEMP                 = 469, //!< 相机外壳温度
CAP_PROP_XI_HOUS_BACK_SIDE_TEMP       = 590, //!< 摄像头外壳背面温度
CAP_PROP_XI_SENSOR_BOARD_TEMP         = 596, //!< 摄像头传感器板温度
CAP_PROP_XI_CMS                       = 470, //!< 颜色管理系统模式
CAP_PROP_XI_APPLY_CMS                 = 471, //!< 启用对 xiGetImage 应用 CMS 配置文件
CAP_PROP_XI_IMAGE_IS_COLOR            = 474, //!< 彩色相机返回 1
CAP_PROP_XI_COLOR_FILTER_ARRAY        = 475, //!< 返回原始数据颜色过滤器数组类型
CAP_PROP_XI_GAMMAY                    = 476, //!< 亮度伽马
CAP_PROP_XI_GAMMAC                    = 477, //!< 色度伽马
CAP_PROP_XI_SHARPNESS                 = 478, //!< 锐度强度
CAP_PROP_XI_CC_MATRIX_00              = 479, //!< 矩阵元素[0][0]修正
CAP_PROP_XI_CC_MATRIX_01              = 480, //!< 矩阵元素[0][1]修正
CAP_PROP_XI_CC_MATRIX_02              = 481, //!< 矩阵元素[0][2]修正
CAP_PROP_XI_CC_MATRIX_03              = 482, //!< 矩阵元素[0][3]修正
CAP_PROP_XI_CC_MATRIX_10              = 483, //!< 矩阵元素[1][0]修正
CAP_PROP_XI_CC_MATRIX_11              = 484, //!< 矩阵元素[1][1]修正
CAP_PROP_XI_CC_MATRIX_12              = 485, //!< 矩阵元素[1][2]修正
CAP_PROP_XI_CC_MATRIX_13              = 486, //!< 矩阵元素[1][3]修正
CAP_PROP_XI_CC_MATRIX_20              = 487, //!< 矩阵元素[2][0]修正
CAP_PROP_XI_CC_MATRIX_21              = 488, //!< 矩阵元素[2][1]修正
CAP_PROP_XI_CC_MATRIX_22              = 489, //!< 矩阵元素[2][2]修正
CAP_PROP_XI_CC_MATRIX_23              = 490, //!< 矩阵元素[2][3]修正
CAP_PROP_XI_CC_MATRIX_30              = 491, //!< 矩阵元素[3][0]修正
CAP_PROP_XI_CC_MATRIX_31              = 492, //!< 矩阵元素[3][1]修正
CAP_PROP_XI_CC_MATRIX_32              = 493, //!< 矩阵元素[3][2]修正
CAP_PROP_XI_CC_MATRIX_33              = 494, //!< 矩阵元素[3][3]修正
CAP_PROP_XI_DEFAULT_CC_MATRIX         = 495, //!< 设置修正矩阵默认颜色
CAP_PROP_XI_TRG_SELECTOR              = 498, //!< 选择触发器类型
CAP_PROP_XI_ACQ_FRAME_BURST_COUNT     = 499, //!< 设置突发获取的帧数
```

```
    CAP_PROP_XI_DEBOUNCE_EN                    = 507, //!< 启用/禁用对选定 GPI 的去抖动功能
    CAP_PROP_XI_DEBOUNCE_T0                    = 508, //!< T0 去抖动事件(x * 10us)
    CAP_PROP_XI_DEBOUNCE_T1                    = 509, //!< T1 去抖动事件(x * 10us)
    CAP_PROP_XI_DEBOUNCE_POL                   = 510, //!< 去抖动极性
    CAP_PROP_XI_LENS_MODE                      = 511, //!< 镜头控制函数的状态
    CAP_PROP_XI_LENS_APERTURE_VALUE            = 512, //!< 镜头光圈值
    CAP_PROP_XI_LENS_FOCUS_MOVEMENT_VALUE      = 513, //!< 镜头当前焦点移动值
    CAP_PROP_XI_LENS_FOCUS_MOVE                = 514,
//!< 根据 XI_PRM_LENS_FOCUS_MOVEMENT_VALUE 设置的步长移动镜头聚焦控制电机
    CAP_PROP_XI_LENS_FOCUS_DISTANCE            = 515, //!< 镜头焦距（单位：cm）
    CAP_PROP_XI_LENS_FOCAL_LENGTH              = 516, //!< 镜头焦距（单位：mm）
    //!< 选择 XI_PRM_LENS_FEATURE 授权的可访问的当前功能
    CAP_PROP_XI_LENS_FEATURE_SELECTOR          = 517,
    CAP_PROP_XI_LENS_FEATURE                   = 518,
//!< 允许访问由 XI_PRM_LENS_FEATURE_SELECTOR 选择器选择的镜头值
    CAP_PROP_XI_DEVICE_MODEL_ID                = 521, //!< 返回设备模型 ID
    CAP_PROP_XI_DEVICE_SN                      = 522, //!< 返回设备序列号
    CAP_PROP_XI_IMAGE_DATA_FORMAT_RGB32_ALPHA  = 529, //!< RGB32 输出图像的 alpha 通道
    //!< 缓冲区大小（字节），足以容纳 xiGetImage 返回的输出图像
    CAP_PROP_XI_IMAGE_PAYLOAD_SIZE= 530,
    CAP_PROP_XI_TRANSPORT_PIXEL_FORMAT         = 531, //!< 传输层上像素的当前格式
    CAP_PROP_XI_SENSOR_CLOCK_FREQ_HZ           = 532, //!< 传感器时钟频率（Hz）
    CAP_PROP_XI_SENSOR_CLOCK_FREQ_INDEX        = 533, //!< 传感器时钟频率指数
    CAP_PROP_XI_SENSOR_OUTPUT_CHANNEL_COUNT    = 534, //!< 用于数据传输的传感器输出通道数
    CAP_PROP_XI_FRAMERATE                      = 535, //!< 以 Hz 为单位定义帧速率
    CAP_PROP_XI_COUNTER_SELECTOR               = 536, //!< 选择计数器
    CAP_PROP_XI_COUNTER_VALUE                  = 537, //!< 计数器状态
    CAP_PROP_XI_ACQ_TIMING_MODE                = 538, //!< 传感器帧的类型
    CAP_PROP_XI_AVAILABLE_BANDWIDTH            = 539, //!< 计算并返回可用的函数带宽
    CAP_PROP_XI_BUFFER_POLICY                  = 540, //!< 数据移动策略
    CAP_PROP_XI_LUT_EN                         = 541, //!< 激活 LUT
    CAP_PROP_XI_LUT_INDEX                      = 542, //!< 控制要在 LUT 中访问的系数的索引（偏移）
    CAP_PROP_XI_LUT_VALUE                      = 543, //!< LUT 的输入 LUTIndex 处的值
    //!< 标识接收到指定触发信号后，在激活它之前应用的延迟（单位：μs）
    CAP_PROP_XI_TRG_DELAY                      = 544,
    CAP_PROP_XI_TS_RST_MODE                    = 545, //!< 定义时间戳重置引擎的待命方式
    CAP_PROP_XI_TS_RST_SOURCE                  = 546, //!< 定义将用于时间戳重置的源
    CAP_PROP_XI_IS_DEVICE_EXIST                = 547, //!< 相机已连接且工作正常返回 1
    CAP_PROP_XI_ACQ_BUFFER_SIZE                = 548, //!< 采集缓冲区大小
    CAP_PROP_XI_ACQ_BUFFER_SIZE_UNIT           = 549, //!< 采集缓冲区大小单位（字节）
    CAP_PROP_XI_ACQ_TRANSPORT_BUFFER_SIZE      = 550, //!< 采集传输缓冲区大小
    CAP_PROP_XI_BUFFERS_QUEUE_SIZE             = 551, //!< 帧缓冲区队列
```

```
    CAP_PROP_XI_ACQ_TRANSPORT_BUFFER_COMMIT  = 552,  //!< 提交到低级别缓冲区数
    CAP_PROP_XI_RECENT_FRAME                 = 553,  //!< GetImage 返回最近帧
    CAP_PROP_XI_DEVICE_RESET                 = 554,  //!< 相机复位为默认状态
    CAP_PROP_XI_COLUMN_FPN_CORRECTION        = 555,  //!< 校正 FPN 列
    CAP_PROP_XI_ROW_FPN_CORRECTION           = 591,  //!< 校正 FPN 行
    CAP_PROP_XI_SENSOR_MODE                  = 558,  //!< 当前传感器模式
    CAP_PROP_XI_HDR                          = 559,  //!< 启用高动态范围功能
    CAP_PROP_XI_HDR_KNEEPOINT_COUNT          = 560,  //!< PWLR 中的 kneepoints 数
    CAP_PROP_XI_HDR_T1                       = 561,  //!< 第一个 kneepoint 位置
    CAP_PROP_XI_HDR_T2                       = 562,  //!< 第二个 kneepoint 位置
    CAP_PROP_XI_KNEEPOINT1                   = 563,  //!< 第一个 kneepoint 值
    CAP_PROP_XI_KNEEPOINT2                   = 564,  //!< 第二个 kneepoint 值
    CAP_PROP_XI_IMAGE_BLACK_LEVEL            = 565,  //!< 最后一个图像黑色级数
    CAP_PROP_XI_HW_REVISION                  = 571,  //!< 返回硬件版本号
    CAP_PROP_XI_DEBUG_LEVEL                  = 572,  //!< 设置调试级别
    CAP_PROP_XI_AUTO_BANDWIDTH_CALCULATION   = 573,  //!< 自动带宽计算
    CAP_PROP_XI_FFS_FILE_ID                  = 594,  //!< 文件数
    CAP_PROP_XI_FFS_FILE_SIZE                = 580,  //!< 文件尺寸
    CAP_PROP_XI_FREE_FFS_SIZE                = 581,  //!< 空闲相机 FFS 的大小
    CAP_PROP_XI_USED_FFS_SIZE                = 582,  //!< 在用相机 FFS 的大小
    CAP_PROP_XI_FFS_ACCESS_KEY               = 583,  //!< 设置键可在某些相机上启用文件操作
    CAP_PROP_XI_SENSOR_FEATURE_SELECTOR      = 585,
//!< 选择由 XI_PRM_SENSOR_FEATURE_VALUE 授权的可访问的当前功能
    CAP_PROP_XI_SENSOR_FEATURE_VALUE         = 586,
//!< 允许访问 XI_PRM_SENSOR_FEATURE_SELECTOR 选择的当前传感器特征值
};

//! 通过 ARAVIS 后端提供的摄像头属性，如果相机配置了软件触发开关，则自动触发帧捕获
enum { CAP_PROP_ARAVIS_AUTOTRIGGER          = 600
};

//! 通过 AVFOUNDATION 后端提供的相机属性
enum { CAP_PROP_IOS_DEVICE_FOCUS            = 9001,
    CAP_PROP_IOS_DEVICE_EXPOSURE            = 9002,
    CAP_PROP_IOS_DEVICE_FLASH               = 9003,
    CAP_PROP_IOS_DEVICE_WHITEBALANCE        = 9004,
    CAP_PROP_IOS_DEVICE_TORCH               = 9005
};

//! 通过 Smartek Giganitex 以太网视觉后端提供的摄像头属性
enum { CAP_PROP_GIGA_FRAME_OFFSET_X         = 10001,
    CAP_PROP_GIGA_FRAME_OFFSET_Y            = 10002,
    CAP_PROP_GIGA_FRAME_WIDTH_MAX           = 10003,
    CAP_PROP_GIGA_FRAME_HEIGH_MAX           = 10004,
```

```
        CAP_PROP_GIGA_FRAME_SENS_WIDTH            = 10005,
        CAP_PROP_GIGA_FRAME_SENS_HEIGH            = 10006
    };

//! Intel 感知计算 SDK
enum { CAP_PROP_INTELPERC_PROFILE_COUNT               = 11001,
       CAP_PROP_INTELPERC_PROFILE_IDX                 = 11002,
       CAP_PROP_INTELPERC_DEPTH_LOW_CONFIDENCE_VALUE  = 11003,
       CAP_PROP_INTELPERC_DEPTH_SATURATION_VALUE      = 11004,
       CAP_PROP_INTELPERC_DEPTH_CONFIDENCE_THRESHOLD  = 11005,
       CAP_PROP_INTELPERC_DEPTH_FOCAL_LENGTH_HORZ     = 11006,
       CAP_PROP_INTELPERC_DEPTH_FOCAL_LENGTH_VERT     = 11007
    };

//! Intel 感知流
enum { CAP_INTELPERC_DEPTH_GENERATOR  = 1 << 29,
       CAP_INTELPERC_IMAGE_GENERATOR  = 1 << 28,
       CAP_INTELPERC_IR_GENERATOR     = 1 << 27,
       CAP_INTELPERC_GENERATORS_MASK  = CAP_INTELPERC_DEPTH_GENERATOR +
CAP_INTELPERC_IMAGE_GENERATOR + CAP_INTELPERC_IR_GENERATOR
    };

//!< 每个像素是一个 16 位整数, 该值表示从对象到相机 XY 平面的距离或笛卡儿深度
enum { CAP_INTELPERC_DEPTH_MAP    = 0,
       //!< 每个像素包含两个 0~1 的 32 位浮点值, 表示深度坐标到颜色坐标的映射
       CAP_INTELPERC_UVDEPTH_MAP  = 1,
       CAP_INTELPERC_IR_MAP  = 2, //!< 每个像素是一个 16 位整数, 表示反射激光束的强度
       CAP_INTELPERC_IMAGE        = 3
    };

/** gPhoto2 属性*/
enum { CAP_PROP_GPHOTO2_PREVIEW            = 17001, //!< 仅从 liveview 模式捕获预览
       CAP_PROP_GPHOTO2_WIDGET_ENUMERATE   = 17002, //!< 只读, 返回(const char *)
       CAP_PROP_GPHOTO2_RELOAD_CONFIG      = 17003, //!< 触发器设置
       CAP_PROP_GPHOTO2_RELOAD_ON_CHANGE   = 17004, //!< 重载所有设置
       CAP_PROP_GPHOTO2_COLLECT_MSGS       = 17005, //!< 收集信息详情
       CAP_PROP_GPHOTO2_FLUSH_MSGS         = 17006, //!< 只读, 返回(const char *)
       CAP_PROP_SPEED                      = 17007, //!< 曝光速度, 可只读, 取决于摄像头程序
       CAP_PROP_APERTURE                   = 17008, //!< 光圈, 可只读, 取决于摄像头程序
       CAP_PROP_EXPOSUREPROGRAM            = 17009, //!< 相机曝光程序
       CAP_PROP_VIEWFINDER                 = 17010  //!< 输入 liveview 模式
    };

enum { CAP_PROP_IMAGES_BASE = 18000,
       CAP_PROP_IMAGES_LAST = 19000 // 排除
```

```
    };

class IVideoCapture;

namespace internal { class VideoCapturePrivateAccessor; }
```

如下为 VideoCapture 类的定义，该类用于从视频文件、图像序列或相机中捕获视频：

```
class CV_EXPORTS_W VideoCapture
{
public:
    /** 构造函数*/
    CV_WRAP VideoCapture();
    CV_WRAP explicit VideoCapture(const String& filename, int apiPreference = CAP_ANY);
    CV_WRAP explicit VideoCapture(int index, int apiPreference= CAP_ANY);

    /** 析构函数*/
    virtual ~VideoCapture();

    /** 打开用于视频捕获的视频文件、捕获设备或 IP 视频流*/
    CV_WRAP virtual bool open(const String& filename, int apiPreference = CAP_ANY);

    /** 打开相机进行视频捕获*/
    CV_WRAP virtual bool open(int index, int apiPreference    = CAP_ANY);

    /** 如果已初始化视频捕获，则返回 true*/
    CV_WRAP virtual bool isOpened() const;

    /** 关闭视频文件或捕获设备*/
    CV_WRAP virtual void release();

    /** 从视频文件或捕获设备中抓取下一帧*/
    CV_WRAP virtual bool grab();

    /** 解码并返回抓取的视频帧*/
    CV_WRAP virtual bool retrieve(OutputArray image, int flag = 0);

    /** Stream 操作，读取下一个视频帧*/
    virtual VideoCapture& operator >> (CV_OUT Mat& image);
    virtual VideoCapture& operator >> (CV_OUT UMat& image);

    /** 获取、解码并返回下一个视频帧*/
    CV_WRAP virtual bool read(OutputArray image);

    /** 在 VideoCapture 中设置属性*/
    CV_WRAP virtual bool set(int propId, double value);
```

```cpp
    /** 返回指定的 VideoCapture 属性*/
    CV_WRAP virtual double get(int propId) const;

    /** 获取使用的后端名称*/
    CV_WRAP String getBackendName() const;

    /** 切换异常模式*/
    CV_WRAP void setExceptionMode(bool enable) { throwOnFail = enable; }

    /// 查询异常模式是否激活
    CV_WRAP bool getExceptionMode() { return throwOnFail; }

    /** 等待 VideoCapture 的就绪帧*/
    static /*CV_WRAP*/
    bool waitAny(
            const std::vector<VideoCapture>& streams,
            CV_OUT std::vector<int>& readyIndex,
            int64 timeoutNs = 0);

protected:
    Ptr<CvCapture> cap;
    Ptr<IVideoCapture> icap;
    bool throwOnFail;

    friend class internal::VideoCapturePrivateAccessor;
};

class IVideoWriter;
```

如下为 VideoWriter 的定义，用于视频写入存储（保存）：

```cpp
class CV_EXPORTS_W VideoWriter
{
public:
    /** 构造函数*/
    CV_WRAP VideoWriter();
    CV_WRAP VideoWriter(const String& filename, int fourcc, double fps,
                       Size frameSize, bool isColor = true);
    CV_WRAP VideoWriter(const String& filename, int apiPreference,
                       int fourcc, double fps, Size frameSize, bool isColor = true);
    CV_WRAP VideoWriter(const String& filename, int fourcc, double fps,
                       const Size& frameSize, const std::vector<int>& params);
    CV_WRAP VideoWriter(const String& filename, int apiPreference, int fourcc,
                       double fps, const Size& frameSize, const std::vector<int>& params);
```

```cpp
/** 析构函数*/
virtual ~VideoWriter();

/** 初始化或重新初始化视频写入器*/
CV_WRAP virtual bool open(const String& filename, int fourcc, double fps,
            Size frameSize, bool isColor    = true);
CV_WRAP bool open(const String& filename, int apiPreference, int fourcc,
            double fps, Size frameSize, bool isColor    = true);
CV_WRAP bool open(const String& filename, int fourcc, double fps, const Size& frameSize,
            const std::vector<int>& params);
CV_WRAP bool open(const String& filename, int apiPreference, int fourcc, double fps,
            const Size& frameSize, const std::vector<int>& params);

/** 如果视频写入器初始化成功，则返回 true*/
CV_WRAP virtual bool isOpened() const;

/** 关闭视频写入器*/
CV_WRAP virtual void release();

/** Stream 操作以写入下一个视频帧*/
virtual VideoWriter& operator << (const Mat& image);
virtual VideoWriter& operator << (const UMat& image);

/** 写下一个视频帧*/
CV_WRAP virtual void write(InputArray image);

/** 在 VideoWriter 中设置属性*/
CV_WRAP virtual bool set(int propId, double value);

/** 返回指定的 VideoWriter 属性*/
CV_WRAP virtual double get(int propId) const;

/** 将 4 个字符连接到 fourcc 代码*/
CV_WRAP static int fourcc(char c1, char c2, char c3, char c4);

/** 返回使用的后端名称*/
CV_WRAP String getBackendName() const;

protected:
    Ptr<CvVideoWriter> writer;
    Ptr<IVideoWriter> iwriter;

    static Ptr<IVideoWriter> create(const String& filename, int fourcc, double fps,
                                    Size frameSize, bool isColor = true);
};
```

```
template<> struct DefaultDeleter<CvCapture>{ CV_EXPORTS void operator ()(CvCapture* obj) 
const; };
template<> struct DefaultDeleter<CvVideoWriter>{ CV_EXPORTS void operator ()(CvVideoWriter* obj) 
const; };
} // cv

#endif //OPENCV_VIDEOIO_HPP
```

7.2 视频读取

OpenCV 中提供了用于执行视频读取操作的类 VideoCapture（类定义见 7.1 节），可以使用视频文件初始化该类（从文件读取视频），也可以使用相机等设备的索引初始化该类（从设备读取视频）。

VideoCapture 类提供了初始化、打开视频文件或设备、视频帧捕获、视频文件或设备关闭（或释放）、属性设置或获取等功能。

7.2.1 案例 84：从文件读取视频

VideoCapture 类的成员函数 isOpened 的作用是检查视频是否能够成功打开，读者在做视频处理时最好使用该函数做一下判断，因为视频打开失败的因素很多，如视频文件不存在、视频编码时的编码器未知、没有打开权限、视频损坏等。该函数定义如下：

```
retval = isOpened(self)
```

参数说明如下。

- retval：视频是否成功打开的标识，打开成功返回 true，否则返回 false。

VideoCapture 类的成员函数 get 用于获取视频的一些参数或属性，其定义如下：

```
retval = get(self, propId)
```

参数说明如下。

- propId：由 VideoCaptureProperties（见 7.1 节）定义的属性 ID。
- retval：帧率等返回结果（返回值）。

VideoCapture 类的成员函数 read 用于捕获、解码并返回下一帧的视频图像，其定义如下：

```
retval, image = read(self, image=None)
```

参数说明如下。

- image：返回的视频帧图像（返回值）。
- retval：视频帧捕获是否成功的标识，当没有视频帧时，返回 false（返回值）。

本案例使用的视频资源来源于 OpenCV 示例资源中的视频文件 Megamind.avi，案例中读取该视频文件，并打印该视频的帧频、宽高，读取视频的数据帧，并将前两帧数据显示出来，读者也可以读取视频帧数据，将其保存为图像文件。案例代码如下：

```
import cv2
#打开视频文件
video=cv2.VideoCapture('./Megamind.avi')

#判断视频文件是否打开成功
if video.isOpened():
    print('Video Opened Success!')
    #返回视频的帧率
    fps = video.get(cv2.CAP_PROP_FPS)
    #返回视频的宽度
    width = video.get(cv2.CAP_PROP_FRAME_WIDTH)
    #返回视频的高度
    height = video.get(cv2.CAP_PROP_FRAME_HEIGHT)
    print('fps:', fps,'\nwidth:',width,'\nheight:',height)
    i=0
    while 1:
        if i==2:
            break
        else:
            i=i+1
            #读取一帧视频
            ret, frame = video.read()
            #显示图像帧
            cv2.imshow("video_frame"+str(i), frame)
            cv2.waitKey(0)
else:
    print('Video Open failed!')
#销毁图像窗口
cv2.destroyAllWindows()
```

打印信息如下：

```
Video Opened Success!
fps: 23.976
width: 720.0
height: 528.0
```

由打印信息可知，该视频的帧频为 23.976，宽度为 720.0 像素，高度为 528.0 像素，显示的第二帧的图像如图 7.1 所示。

图 7.1

7.2.2 案例 85：从设备读取视频

相机（或摄像机）是生活中视频采集的常用设备，VideoCapture 类可以传入相机的 ID，0 表示打开第一个摄像头，1 表示打开第二个摄像头，依次类推，−1 表示随机打开一个摄像头。

从相机设备读取视频的案例代码如下：

```python
import cv2

#用相机设备初始化视频捕获类
device = cv2.VideoCapture(0)#传入参数 0，表示打开第一个摄像头
if device.isOpened():    #判断设备是否打开
    print("Camera device opened success!")
    while (True):
        #读取视频帧
        ret, frame = device.read()
        #显示视频帧
        cv2.imshow("device_frame", frame)
        cv2.waitKey(1)
    #释放资源并关闭显示窗口
    device.release()
    cv2.destroyAllWindows()
else:
    print("Camera device opened failed!")
```

从设备捕获的视频结果如图 7.2 所示。

7.3 视频保存

OpenCV 中提供了用于视频保存的类 VideoWriter（定义见 7.1 节），该类可以将图像

图 7.2

文件写入视频文件中。

VideoWriter 中的第一个参数 filename 表示目标存储的视频文件名或全路径；第二个参数为 fourcc，由函数 cv2.VideoWriter_fourcc 返回，其定义如下：

```
retval = VideoWriter_fourcc(c1, c2, c3, c4)
```

参数说明如下。

- c1、c2、c3、c4：4 字符编码，表示的是视频编码格式，常用的 4 字符及对应的编码类型如表 7.1 所示。
- retval：返回 fourcc 编码。

表 7.1

4 字符参数	编 码 类 型
'I', '4', '2', '0'	YUV 编码类型，文件名后缀为.avi
'P', 'I', 'M', 'I'	MPEG-1 编码类型，文件名后缀为.avi
'X', 'V', 'I', 'D'	MPEG-4 编码类型，文件名后缀为.avi
'T', 'H', 'E', 'O'	Ogg Vorbis，文件名后缀为.ogv
'F', 'L', 'V', '1'	Flash 视频，文件名后缀为.flv

如下代码即表示获取 YUV 编码类型的参数 fourcc：

```
fourcc = cv2.VideoWriter_fourcc('I', '4', '2', '0')
```

也可以使用如下写法：

```
fourcc = cv2.VideoWriter_fourcc(*"I420")
```

更多 fourcc 编码设置可以参考官方网站。

7.3.1 案例 86：从图片文件创建视频

本案例选取了 OpenCV 示例资源中的一组图像，图像大小均为 640×480（单位为像素），将这组图像保存为视频，保存的帧频为每秒钟 2 帧图像，案例代码如下：

```python
import os
import cv2

#遍历路径下所有的图像文件
path = './images/'
filelist = os.listdir(path)
#创建视频编解码器
fourcc = cv2.VideoWriter_fourcc('I', '4', '2', '0')
#设置帧频，每秒钟 2 帧图像
fps = 2
#设置待转换图像的尺寸
size = (640, 480)
#创建视频写入对象
```

```
video = cv2.VideoWriter('./images/img2video.avi', fourcc, fps, size)
if video.isOpened():
    #遍历图像并将其写入视频文件
    for item in filelist:
        #将后缀为.jpg的图像文件写入视频
        if item.endswith('.jpg'):
            item = path + item
            img = cv2.imread(item)
            video.write(img)
    video.release()
else:
    print("VideoWriter open failed!")
```

保存后的视频文件及图像文件存于 images 文件夹下，如图 7.3 所示。

图 7.3

7.3.2　案例 87：保存相机采集的视频

本案例由相机采集视频帧数据，并将每帧的图像保存到视频文件中，案例代码如下：

```
import cv2

#用相机设备初始化视频捕获类
device = cv2.VideoCapture(0)
#创建视频编解码器
fourcc = cv2.VideoWriter_fourcc('I', '4', '2', '0')
#获取设备帧频
fps = device.get(cv2.CAP_PROP_FPS)
#获取设备帧的尺寸
width = device.get(cv2.CAP_PROP_FRAME_WIDTH)
height = device.get(cv2.CAP_PROP_FRAME_HEIGHT)
size = (int(width), int(height))
#创建视频写入对象
i = 0
video = cv2.VideoWriter('./video_capture.avi', fourcc, int(fps), size)
if video.isOpened():
    if device.isOpened():
        print("Camera device opened success!")
        while (True):
```

```
            #读取视频帧
            i = i+1
            ret, frame = device.read()
            video.write(frame)
            #显示视频帧
            cv2.imshow("device_frame", frame)
            #等待1ms
            cv2.waitKey(1)
            #只采集1000帧就结束，读者可以自行设置
            if i > 1000:
                break
        #释放资源并关闭显示窗口
        device.release()
        cv2.destroyAllWindows()
    else:
        print("Camera device opened failed!")
else:
    print("VideoWriter create failed!")
```

本案例中的操作会打开第一个摄像头，将摄像头采集到的帧图像保存到视频文件中，对于没有额外加装摄像头设备的笔记本电脑，会打开摄像头，并将采集到的视频（案例中为 1000 帧图像）保存到本地文件 video_capture.avi 中。

7.4　进阶必备：视频编/解码工具 FFMPEG

视频编/解码中最常用的工具为 FFMPEG（Fast Forward MPEG），这是一个跨平台的多媒体框架，提供了编码、解码、转码、音频混合等丰富的功能，被很多常用视频播放器采用。

FFMPEG 名称中的 MPEG（Moving Picture Experts Group，动态图像专家组）是专门针对运动图像和语音压缩标准的制定而成立的国际组织，该组织制定了 MPEG-1、MPEG-2、MPEG-4、MPEG-7 及 MPEG-21 共 5 个标准。

FFMPEG 是一款开源软件，可以在 Linux、macOS、Windows 等各种系统环境中编译运行，包含了 libavcodec、libavutil、libavformat、libavfilter、libavdevice、libswscale 和 libswresample 等库文件，还有 ffmpeg、ffplay、ffprobe 和 ffserver 等工具用于转码与播放。

其中各库文件和工具的功能如下。

libavcodec 提供了更广泛编/解码器的实现。

libavformat 实现流协议、容器格式和基本 I/O 访问。

libavutil 包括哈希器、解压缩器和其他实用程序函数。

libavfilter 提供了一种通过过滤器链改变解码音频和视频的方法。

libavdevice 提供了访问捕获和回放设备的抽象。

libswresample 实现音频混合和重采样例程。

libswscale 实现颜色转换和缩放例程。

ffmpeg 是一个用于操作、转换和流式传输多媒体内容的命令行工具箱。

ffplay 是一款简约的多媒体播放器。

ffprobe 是一个检查多媒体内容的简单分析工具。

ffserver 是一个用于直播的多媒体流服务器。

另外，还有一些其他的小型工具，如 aviocat、ismindex 和 qt-faststart。

OpenCV 中也使用 FFMPEG 进行视频编/解码，在编译的时候，可以通过开关 WITH_FFMPEG 控制是否需要 FFMPEG：

```
OCV_OPTION(WITH_FFMPEG "Include FFMPEG support" (NOT ANDROID)
  VISIBLE_IF NOT IOS AND NOT WINRT
  VERIFY HAVE_FFMPEG)
```

如果需要 FFMPEG，则在编译时会下载需要的链接库，编译前会判断是否有相应的工具包：

```
if(WITH_FFMPEG OR HAVE_FFMPEG)
  if(OPENCV_FFMPEG_USE_FIND_PACKAGE)
    status("FFMPEG:"    HAVE_FFMPEG   THEN "YES (find_package)"    ELSE "NO (find_package)")
  elseif(WIN32)
    status("FFMPEG:"    HAVE_FFMPEG   THEN "YES (prebuilt binaries)" ELSE NO)
  else()
    status("FFMPEG:"    HAVE_FFMPEG   THEN YES ELSE NO)
  endif()
  status("  avcodec:"    FFMPEG_libavcodec_VERSION   THEN "YES (${FFMPEG_libavcodec_VERSION})" ELSE NO)
  status("  avformat:"   FFMPEG_libavformat_VERSION  THEN "YES (${FFMPEG_libavformat_VERSION})" ELSE NO)
  status("  avutil:"  FFMPEG_libavutil_VERSION    THEN "YES (${FFMPEG_libavutil_VERSION})" ELSE NO)
  status("  swscale:"   FFMPEG_libswscale_VERSION   THEN "YES (${FFMPEG_libswscale_VERSION})" ELSE NO)
  status("  avresample:" FFMPEG_libavresample_VERSION THEN "YES (${FFMPEG_libavresample_VERSION})" ELSE NO)
endif()
```

读者在编译 OpenCV 时，可以根据需要选择是否编译 FFMPEG，若编译，则设置 WITH_FFMPEG 开关为 ON；否则设置为 OFF。

第 8 章
视频分析模块 video

图像和视频是视觉信息的两种重要来源,视频由一帧帧的图像构成,因此,视频处理过程也是图像处理过程。OpenCV 中的 video 模块提供视频分析功能,主要封装了运动分析和目标跟踪两种算法。

8.1 运动分析

8.1.1 模块导读

在 C++语言编程中,video 模块的引入需要通过包含头文件"opencv2/video.hpp"来实现。该头文件的内容如下:

```
#ifndef OPENCV_VIDEO_HPP
#define OPENCV_VIDEO_HPP

#include "opencv2/video/tracking.hpp"
#include "opencv2/video/background_segm.hpp"

#endif //OPENCV_VIDEO_HPP
```

该头文件中引入了另外两个内部头文件"opencv2/video/tracking.hpp"和"opencv2/video/background_segm.hpp",其中,background_segm.hpp 定义了运动分析的相关算法,如 MOG2、KNN(K 最近邻)等背景/前景分割算法,这些算法在运动分析中有重要作用。background_segm.hpp 头文件的内容如下:

```
#ifndef OPENCV_BACKGROUND_SEGM_HPP
```

```cpp
#define OPENCV_BACKGROUND_SEGM_HPP

#include "opencv2/core.hpp"

namespace cv
{
```

如下为 BackgroundSubtractor 的定义，该类为前景/背景分割基类：

```cpp
class CV_EXPORTS_W BackgroundSubtractor : public Algorithm
{
public:
    /** 计算前景 mask*/
    CV_WRAP virtual void apply(InputArray image,
                               OutputArray fgmask, double learningRate=-1) = 0;

    /** 计算背景图像，注意：有时背景图像可能非常模糊，因为它包含平均背景统计信息*/
    CV_WRAP virtual void getBackgroundImage(OutputArray backgroundImage) const = 0;
};
```

如下为 BackgroundSubtractorMOG2 的定义，该类为基于高斯混合的背景/前景分割算法的实现：

```cpp
class CV_EXPORTS_W BackgroundSubtractorMOG2 : public BackgroundSubtractor
{
public:
    /** 返回影响背景模型的最后帧数*/
    CV_WRAP virtual int getHistory() const = 0;
    /** 设置影响背景模型的最后帧数*/
    CV_WRAP virtual void setHistory(int history) = 0;

    /** 返回背景模型中高斯分量的数目*/
    CV_WRAP virtual int getNMixtures() const = 0;
    /** 设置背景模型中高斯分量的数目，模型需要重新初始化以保留存储*/
    CV_WRAP virtual void setNMixtures(int nmixtures) = 0;

    /** 返回算法中的背景比率参数*/
    CV_WRAP virtual double getBackgroundRatio() const = 0;
    /** 设置算法中的背景比率参数*/
    CV_WRAP virtual void setBackgroundRatio(double ratio) = 0;

    /** 返回像素模型匹配的方差阈值*/
    CV_WRAP virtual double getVarThreshold() const = 0;
    /** 设置像素模型匹配的方差阈值*/
    CV_WRAP virtual void setVarThreshold(double varThreshold) = 0;

    /** 返回用于生成新混合组件的像素模型匹配的方差阈值*/
```

```cpp
    CV_WRAP virtual double getVarThresholdGen() const = 0;
    /** 设置用于生成新混合组件的像素模型匹配的方差阈值*/
    CV_WRAP virtual void setVarThresholdGen(double varThresholdGen) = 0;

    /** 返回每个高斯分量的初始方差*/
    CV_WRAP virtual double getVarInit() const = 0;
    /** 设置每个高斯分量的初始方差*/
    CV_WRAP virtual void setVarInit(double varInit) = 0;

    CV_WRAP virtual double getVarMin() const = 0;
    CV_WRAP virtual void setVarMin(double varMin) = 0;

    CV_WRAP virtual double getVarMax() const = 0;
    CV_WRAP virtual void setVarMax(double varMax) = 0;

    /** 返回复杂度降低阈值*/
    CV_WRAP virtual double getComplexityReductionThreshold() const = 0;
    /** 设置复杂度降低阈值*/
    CV_WRAP virtual void setComplexityReductionThreshold(double ct) = 0;

    /** 返回阴影检测标志*/
    CV_WRAP virtual bool getDetectShadows() const = 0;
    /** 允许或禁止阴影检测*/
    CV_WRAP virtual void setDetectShadows(bool detectShadows) = 0;

    /** 返回阴影值*/
    CV_WRAP virtual int getShadowValue() const = 0;
    /** 设置阴影值*/
    CV_WRAP virtual void setShadowValue(int value) = 0;

    /** 返回阴影阈值*/
    CV_WRAP virtual double getShadowThreshold() const = 0;
    /** 设置阴影阈值*/
    CV_WRAP virtual void setShadowThreshold(double threshold) = 0;

    /** 计算前景 mask*/
    CV_WRAP virtual void apply(InputArray image, OutputArray fgmask, double learningRate=-1)
CV_OVERRIDE = 0;
};

/** 创建 MOG2 背景提取器*/
CV_EXPORTS_W Ptr<BackgroundSubtractorMOG2>
    createBackgroundSubtractorMOG2(int history=500, double varThreshold=16,
                                   bool detectShadows=true);
```

如下为 BackgroundSubtractorKNN 类的定义，该类为基于 KNN（K 最近邻）算法的背景/前景分割算法的实现：

```
class CV_EXPORTS_W BackgroundSubtractorKNN : public BackgroundSubtractor
{
public:
    /** 返回影响背景模型的最后帧数*/
    CV_WRAP virtual int getHistory() const = 0;
    /** 设置影响背景模型的最后帧数*/
    CV_WRAP virtual void setHistory(int history) = 0;

    /** 返回背景模型中的数据样本数*/
    CV_WRAP virtual int getNSamples() const = 0;
    /** 设置背景模型中的数据样本数*/
    CV_WRAP virtual void setNSamples(int _nN) = 0;//needs reinitialization!

    /** 返回像素和样本之间的平方距离的阈值*/
    CV_WRAP virtual double getDist2Threshold() const = 0;
    /** 设置像素和样本之间的平方距离的阈值*/
    CV_WRAP virtual void setDist2Threshold(double _dist2Threshold) = 0;

    /** 返回邻居数，即 KNN 中的 k*/
    CV_WRAP virtual int getkNNSamples() const = 0;
    /** 设置 KNN 中的 k*/
    CV_WRAP virtual void setkNNSamples(int _nkNN) = 0;

    /** 返回阴影检测标志*/
    CV_WRAP virtual bool getDetectShadows() const = 0;
    /** 允许或禁止阴影检测*/
    CV_WRAP virtual void setDetectShadows(bool detectShadows) = 0;

    /** 返回阴影值*/
    CV_WRAP virtual int getShadowValue() const = 0;
    /** 设置阴影值*/
    CV_WRAP virtual void setShadowValue(int value) = 0;

    /** 返回阴影阈值*/
    CV_WRAP virtual double getShadowThreshold() const = 0;
    /** 设置阴影阈值*/
    CV_WRAP virtual void setShadowThreshold(double threshold) = 0;
};

/** 创建 KNN 背景提取器*/
CV_EXPORTS_W Ptr<BackgroundSubtractorKNN>
    createBackgroundSubtractorKNN(int history=500, double dist2Threshold=400.0,
```

```
                              bool detectShadows=true);
} // cv

#endif
```

8.1.2 案例 88:基于 MOG2 与 KNN 算法的运动分析

本节介绍基于 MOG2 与 KNN 算法的运动分析的案例,本案例中使用的视频文件为 OpenCV 提供的资源文件 vtest.avi。案例中读取视频中的每一帧,获取前景掩模并显示。

案例中使用的 MOG2 算法背景提取器对象由 cv2.createBackgroundSubtractorMOG2() 创建(BackgroundSubtractorMOG2 的定义见 8.1.1 节),KNN 算法背景提取器对象由 cv2.createBackgroundSubtractorKNN() 创建(BackgroundSubtractorKNN 的定义见 8.1.1 节)。

案例代码如下:

```python
import cv2

#选择使用的背景提取算法
algo = 'MOG2'
#定义输入文件
input_file = "vtest.avi"
#创建背景提取器对象
if algo == 'MOG2':
    backSub = cv2.createBackgroundSubtractorMOG2()
else:
    backSub = cv2.createBackgroundSubtractorKNN()

##视频捕获
capture = cv2.VideoCapture(input_file)
if not capture.isOpened:
    print('Unable to open: ' + input_file)
    exit(0)

while True:
    #读取视频中的帧
    ret, frame = capture.read()
    if frame is None:
        break

    #更新背景模型
    fgMask = backSub.apply(frame)

    #获取帧索引并将其写入当前帧
```

```
cv2.rectangle(frame, (10, 2), (100,20), (255,255,255), -1)
cv2.putText(frame, str(capture.get(cv2.CAP_PROP_POS_FRAMES)), (15, 15),
        cv2.FONT_HERSHEY_SIMPLEX, 0.5 , (0,0,0))

#显示当前帧图像与前景掩模
cv2.imshow('Frame', frame)
cv2.imshow('FG Mask', fgMask)

#键盘操作
keyboard = cv2.waitKey(30)
if keyboard == 'q' or keyboard == 27:
    break
```

基于 MOG2 算法提取的前景 mask 及对应的图像帧显示结果如图 8.1 所示。

图 8.1

基于 KNN 算法提取的前景 mask 及对应的图像帧显示结果如图 8.2 所示。

图 8.2

8.2 目标跟踪

8.2.1 模块导读

video 模块中的另外一个重要算法就是目标跟踪，该算法定义在头文件 tracking.hpp 中，其内容如下：

```cpp
#ifndef OPENCV_TRACKING_HPP
#define OPENCV_TRACKING_HPP

#include "opencv2/core.hpp"
#include "opencv2/imgproc.hpp"

namespace cv
{
```

如下为与光流算法相关的标志定义与函数定义：

```cpp
enum { OPTFLOW_USE_INITIAL_FLOW     = 4,
       OPTFLOW_LK_GET_MIN_EIGENVALS = 8,
       OPTFLOW_FARNEBACK_GAUSSIAN   = 256
     };

/** 查找对象中心、大小和方向*/
CV_EXPORTS_W RotatedRect CamShift( InputArray probImage, CV_IN_OUT Rect& window,
                                   TermCriteria criteria );

/** 在反投影图像上查找对象*/
CV_EXPORTS_W int meanShift( InputArray probImage, CV_IN_OUT Rect& window, TermCriteria criteria );

/** 构造可以传递给 calcOpticalFlowPyrLK 的图像金字塔*/
CV_EXPORTS_W int buildOpticalFlowPyramid( InputArray img, OutputArrayOfArrays pyramid,
                                          Size winSize, int maxLevel, bool withDerivatives = true,
                                          int pyrBorder = BORDER_REFLECT_101,
                                          int derivBorder = BORDER_CONSTANT,
                                          bool tryReuseInputImage = true );

/** 使用带金字塔的迭代 Lucas-Kanade 方法计算稀疏特征集的光流*/
CV_EXPORTS_W void calcOpticalFlowPyrLK( InputArray prevImg, InputArray nextImg,
                                        InputArray prevPts, InputOutputArray nextPts,
                                        OutputArray status, OutputArray err,
                                        Size winSize = Size(21,21), int maxLevel = 3,
                                        TermCriteria criteria =
                                    TermCriteria(TermCriteria::COUNT+TermCriteria::EPS, 30, 0.01),
                                        int flags = 0, double minEigThreshold = 1e-4 );
```

```
/** 用 Gunnar-Farneback 算法计算稠密光流*/
CV_EXPORTS_W void calcOpticalFlowFarneback( InputArray prev, InputArray next,
                                            InputOutputArray flow,
                                            double pyr_scale, int levels, int winsize,
                                            int iterations, int poly_n, double poly_sigma,
                                            int flags );

/** 计算两个二维点集之间的最佳仿射变换*/
CV_DEPRECATED CV_EXPORTS Mat estimateRigidTransform( InputArray src,
                                            InputArray dst, bool fullAffine );

enum
{
    MOTION_TRANSLATION = 0,
    MOTION_EUCLIDEAN   = 1,
    MOTION_AFFINE      = 2,
    MOTION_HOMOGRAPHY  = 3
};

/** 计算两幅图像之间的增强相关系数数值*/
CV_EXPORTS_W double computeECC(InputArray templateImage, InputArray inputImage,
                               InputArray inputMask = noArray());

/** 根据 ECC 准则查找两幅图像之间的几何变换（warp）*/
CV_EXPORTS_W double findTransformECC( InputArray templateImage, InputArray inputImage,
                                      InputOutputArray warpMatrix, int motionType,
                                      TermCriteria criteria,
                                      InputArray inputMask, int gaussFiltSize);

CV_EXPORTS double findTransformECC(InputArray templateImage, InputArray inputImage,
    InputOutputArray warpMatrix, int motionType = MOTION_AFFINE,
    TermCriteria criteria = TermCriteria(TermCriteria::COUNT+TermCriteria::EPS, 50, 0.001),
    InputArray inputMask = noArray());
```

如下为 Kalman 滤波器定义的类：

```
class CV_EXPORTS_W KalmanFilter
{
public:
    CV_WRAP KalmanFilter();
    CV_WRAP KalmanFilter( int dynamParams, int measureParams,
                          int controlParams = 0, int type = CV_32F );

    /** Kalman 滤波器初始化*/
    void init( int dynamParams, int measureParams, int controlParams = 0, int type = CV_32F );
    /** 计算预测状态*/
```

```cpp
    CV_WRAP const Mat& predict( const Mat& control = Mat() );
    /** 更新测量的预测状态*/
    CV_WRAP const Mat& correct( const Mat& measurement );

    CV_PROP_RW Mat statePre;
    CV_PROP_RW Mat statePost;
    CV_PROP_RW Mat transitionMatrix;
    CV_PROP_RW Mat controlMatrix;
    CV_PROP_RW Mat measurementMatrix;
    CV_PROP_RW Mat processNoiseCov;
    CV_PROP_RW Mat measurementNoiseCov;
    CV_PROP_RW Mat errorCovPre;
    CV_PROP_RW Mat gain;
    CV_PROP_RW Mat errorCovPost;

    // 临时矩阵
    Mat temp1;
    Mat temp2;
    Mat temp3;
    Mat temp4;
    Mat temp5;
};

/** 读取.flo文件*/
CV_EXPORTS_W Mat readOpticalFlow( const String& path );
/** 保存.flo文件*/
CV_EXPORTS_W bool writeOpticalFlow( const String& path, InputArray flow );
```

如下为DenseOpticalFlow类的定义，该类为稠密光流算法基类：

```cpp
class CV_EXPORTS_W DenseOpticalFlow : public Algorithm
{
public:
    /** 计算光流*/
    CV_WRAP virtual void calc( InputArray I0, InputArray I1, InputOutputArray flow ) = 0;
    /** 释放所有内部缓存*/
    CV_WRAP virtual void collectGarbage() = 0;
};

/** 稀疏光流算法的基本函数*/
class CV_EXPORTS_W SparseOpticalFlow : public Algorithm
{
public:
    /** 计算稀疏光流*/
    CV_WRAP virtual void calc(InputArray prevImg, InputArray nextImg,
                  InputArray prevPts, InputOutputArray nextPts,
```

```
                OutputArray status,
                OutputArray err = cv::noArray()) = 0;
};
```

如下为 DenseOpticalFlow 类的定义，该类为使用 Gunnar-Farneback 算法计算稠密光流的类：

```
class CV_EXPORTS_W FarnebackOpticalFlow : public DenseOpticalFlow
{
public:
    CV_WRAP virtual int getNumLevels() const = 0;
    CV_WRAP virtual void setNumLevels(int numLevels) = 0;

    CV_WRAP virtual double getPyrScale() const = 0;
    CV_WRAP virtual void setPyrScale(double pyrScale) = 0;

    CV_WRAP virtual bool getFastPyramids() const = 0;
    CV_WRAP virtual void setFastPyramids(bool fastPyramids) = 0;

    CV_WRAP virtual int getWinSize() const = 0;
    CV_WRAP virtual void setWinSize(int winSize) = 0;

    CV_WRAP virtual int getNumIters() const = 0;
    CV_WRAP virtual void setNumIters(int numIters) = 0;

    CV_WRAP virtual int getPolyN() const = 0;
    CV_WRAP virtual void setPolyN(int polyN) = 0;

    CV_WRAP virtual double getPolySigma() const = 0;
    CV_WRAP virtual void setPolySigma(double polySigma) = 0;

    CV_WRAP virtual int getFlags() const = 0;
    CV_WRAP virtual void setFlags(int flags) = 0;

    CV_WRAP static Ptr<FarnebackOpticalFlow> create(
            int numLevels = 5,
            double pyrScale = 0.5,
            bool fastPyramids = false,
            int winSize = 13,
            int numIters = 10,
            int polyN = 5,
            double polySigma = 1.1,
            int flags = 0);
};
```

如下为 VariationalRefinement 类的定义，该类用于变分光流优化：

```cpp
class CV_EXPORTS_W VariationalRefinement : public DenseOpticalFlow
{
public:
    /** 处理单独的水平（u）和垂直（v）流分量（避免额外的拆分/合并）*/
    CV_WRAP virtual void calcUV(InputArray I0, InputArray I1,
                                InputOutputArray flow_u, InputOutputArray flow_v) = 0;

    /** 获取最小化过程中的外部（定点）迭代次数*/
    CV_WRAP virtual int getFixedPointIterations() const = 0;
    /** 设置最小化过程中的外部（定点）迭代次数*/
    CV_WRAP virtual void setFixedPointIterations(int val) = 0;

    /** 求解相关线性系统的极小化过程中的内部连续过松弛（SOR）迭代次数*/
    CV_WRAP virtual int getSorIterations() const = 0;
    CV_WRAP virtual void setSorIterations(int val) = 0;

    /** SOR 中的松弛因子*/
    CV_WRAP virtual float getOmega() const = 0;
    CV_WRAP virtual void setOmega(float val) = 0;

    /** 平滑项的权重*/
    CV_WRAP virtual float getAlpha() const = 0;
    CV_WRAP virtual void setAlpha(float val) = 0;

    /** 颜色不变项的权重*/
    CV_WRAP virtual float getDelta() const = 0;
    CV_WRAP virtual void setDelta(float val) = 0;

    /** 梯度不变项的权重*/
    CV_WRAP virtual float getGamma() const = 0;
    CV_WRAP virtual void setGamma(float val) = 0;

    /** 创建 VariationalRefinement 实例*/
    CV_WRAP static Ptr<VariationalRefinement> create();
};
```

如下为 DISOpticalFlow 类的定义，该类为 DIS 光流算法类：

```cpp
class CV_EXPORTS_W DISOpticalFlow : public DenseOpticalFlow
{
public:
    enum
    {
        PRESET_ULTRAFAST = 0,
        PRESET_FAST = 1,
        PRESET_MEDIUM = 2
```

```cpp
};

    /** 计算光流的高斯金字塔的最优级别*/
    CV_WRAP virtual int getFinestScale() const = 0;
    CV_WRAP virtual void setFinestScale(int val) = 0;

    /** 获取匹配的图像块的大小（单位：像素）*/
    CV_WRAP virtual int getPatchSize() const = 0;
    CV_WRAP virtual void setPatchSize(int val) = 0;

    /** 获取相邻 patch 之间的步长*/
    CV_WRAP virtual int getPatchStride() const = 0;
    CV_WRAP virtual void setPatchStride(int val) = 0;

    /** 在 patch 逆搜索中，最大次数的梯度下降迭代次数*/
    CV_WRAP virtual int getGradientDescentIterations() const = 0;
    CV_WRAP virtual void setGradientDescentIterations(int val) = 0;

    /** 每个 scale 中变分优化固定点迭代次数*/
    CV_WRAP virtual int getVariationalRefinementIterations() const = 0;
    CV_WRAP virtual void setVariationalRefinementIterations(int val) = 0;

    /** 平滑项的权重*/
    CV_WRAP virtual float getVariationalRefinementAlpha() const = 0;
    CV_WRAP virtual void setVariationalRefinementAlpha(float val) = 0;

    /** 颜色不变项的权重*/
    CV_WRAP virtual float getVariationalRefinementDelta() const = 0;
    CV_WRAP virtual void setVariationalRefinementDelta(float val) = 0;

    /** 梯度不变项的权重*/
    CV_WRAP virtual float getVariationalRefinementGamma() const = 0;
    CV_WRAP virtual void setVariationalRefinementGamma(float val) = 0;

    /** 计算 patch 距离时是否使用平均的 patch 标准化*/
    CV_WRAP virtual bool getUseMeanNormalization() const = 0;
    CV_WRAP virtual void setUseMeanNormalization(bool val) = 0;

    /** 是否使用良好光流矢量的空间传播*/
    CV_WRAP virtual bool getUseSpatialPropagation() const = 0;
    CV_WRAP virtual void setUseSpatialPropagation(bool val) = 0;

    /** 创建 DISOpticalFlow 实例*/
    CV_WRAP static Ptr<DISOpticalFlow> create(int preset = DISOpticalFlow::PRESET_FAST);
};
```

如下为 SparsePyrLKOpticalFlow 类的定义，是用于计算稀疏光流的类：

```
class CV_EXPORTS_W SparsePyrLKOpticalFlow : public SparseOpticalFlow
{
public:
    CV_WRAP virtual Size getWinSize() const = 0;
    CV_WRAP virtual void setWinSize(Size winSize) = 0;

    CV_WRAP virtual int getMaxLevel() const = 0;
    CV_WRAP virtual void setMaxLevel(int maxLevel) = 0;

    CV_WRAP virtual TermCriteria getTermCriteria() const = 0;
    CV_WRAP virtual void setTermCriteria(TermCriteria& crit) = 0;

    CV_WRAP virtual int getFlags() const = 0;
    CV_WRAP virtual void setFlags(int flags) = 0;

    CV_WRAP virtual double getMinEigThreshold() const = 0;
    CV_WRAP virtual void setMinEigThreshold(double minEigThreshold) = 0;

    CV_WRAP static Ptr<SparsePyrLKOpticalFlow> create(
            Size winSize = Size(21, 21),
            int maxLevel = 3, TermCriteria crit =
            TermCriteria(TermCriteria::COUNT+TermCriteria::EPS, 30, 0.01),
            int flags = 0,
            double minEigThreshold = 1e-4);
};
} // cv

#endif
```

8.2.2 案例 89：基于 CamShift 算法的目标跟踪

CamShift 算法为基于颜色分布的连续自适应均值漂移算法，可以根据目标在图像中发生变化（如由于距离变化或发生形变）时自适应调节窗口的大小，有效解决了跟踪过程中的目标变形问题，不过该算法需要在简单背景且目标与环境颜色特征差别较明显的场景才能有较好的跟踪效果。

OpenCV 中的 CamShift 目标跟踪算法由 CamShift 函数实现，其定义如下：

```
retval, window = CamShift(probImage, window, criteria)
```

参数说明如下。

- probImage：目标直方图的反投影（Back projection）结果。
- window：跟踪矩形窗口（返回值）。
- criteria：停止迭代条件。

- retval:对象中心、大小和方向数据的组合(返回值)。

本案例中使用的视频文件为 OpenCV 提供的资源文件 vtest.avi,案例代码如下:

```python
import cv2
import numpy as np

#定义输入文件
input_file = "vtest.avi"
cap = cv2.VideoCapture(input_file)

#设置窗口初始化位置
x, y, w, h = 640, 240, 50, 91
track_window = (x, y, w, h)

#获取视频第一帧
ret,frame = cap.read()
print(frame.shape)
cv2.rectangle(frame,track_window, color=(0,0,255), thickness=2)
cv2.imshow("frame", frame)
cv2.waitKey(0)

#设置追踪的感兴趣区域
roi = frame[y:y+h, x:x+w]
hsv_roi =  cv2.cvtColor(roi, cv2.COLOR_BGR2HSV)
mask = cv2.inRange(hsv_roi, np.array((0., 60.,32.)), np.array((180.,255.,255.)))
roi_hist = cv2.calcHist([hsv_roi],[0],mask,[180],[0,180])
cv2.normalize(roi_hist,roi_hist,0,255,cv2.NORM_MINMAX)

#设置终止条件,10 次迭代或至少移动 1pt
term_crit = ( cv2.TERM_CRITERIA_EPS | cv2.TERM_CRITERIA_COUNT, 10, 1 )

while(1):
    ret, frame = cap.read()

    if ret == True:
        hsv = cv2.cvtColor(frame, cv2.COLOR_BGR2HSV)
        dst = cv2.calcBackProject([hsv],[0],roi_hist,[0,180],1)

        #应用 CamShift 算法获取新的位置
        ret, track_window = cv2.CamShift(dst, track_window, term_crit)

        #在图像上绘制追踪位置
        pts = cv2.boxPoints(ret)
        pts = np.int0(pts)
        img2 = cv2.polylines(frame,[pts],True, 255,2)
        cv2.imshow('img2',img2)
```

```
        k = cv2.waitKey(3000) & 0xff
        if k == 27:
            break
    else:
        break
```

本案例读取第一帧图像，在其中标注了一个待追踪的目标，然后硬编码了该目标的位置，读者可以通过图像分割的办法选取自己的追踪目标。对选取的追踪目标使用红色矩形框标注，如图 8.3 所示。

执行跟踪的结果如图 8.4 所示。

图 8.3

图 8.4

由于目标较小且干扰较多，所以后期的追踪效果并不是很理想，如图 8.5 所示。

图 8.5

8.2.3　案例 90：基于 meanShift 算法的目标跟踪

OpenCV 中的 meanShift 目标跟踪算法由 meanShift 函数实现，其定义如下：

```
retval, window = meanShift(probImage, window, criteria)
```

参数说明如下。
- probImage：目标直方图的反投影（Back projection）结果。
- window：跟踪矩形窗口（返回值）。
- criteria：停止迭代条件。
- retval：对象中心、大小和方向数据的组合（返回值）。

本案例中使用的视频文件为 OpenCV 提供的资源文件 vtest.avi，案例代码如下：

```python
import numpy as np
import cv2

#定义输入文件
input_file = "vtest.avi"
cap = cv2.VideoCapture(input_file)

#设置窗口初始化位置
x, y, w, h = 640, 240, 50, 91
track_window = (x, y, w, h)

#获取视频第一帧
ret,frame = cap.read()

#设置追踪的感兴趣区域
roi = frame[y:y+h, x:x+w]
hsv_roi = cv2.cvtColor(roi, cv2.COLOR_BGR2HSV)
mask = cv2.inRange(hsv_roi, np.array((0., 60.,32.)), np.array((180.,255.,255.)))
roi_hist = cv2.calcHist([hsv_roi],[0],mask,[180],[0,180])
cv2.normalize(roi_hist,roi_hist,0,255,cv2.NORM_MINMAX)

#设置终止条件，10 次迭代或至少移动 1pt
term_crit = ( cv2.TERM_CRITERIA_EPS | cv2.TERM_CRITERIA_COUNT, 10, 1 )

while(1):
    ret, frame = cap.read()

    if ret == True:
        hsv = cv2.cvtColor(frame, cv2.COLOR_BGR2HSV)
        dst = cv2.calcBackProject([hsv],[0],roi_hist,[0,180],1)

        #应用 meanShift 算法获取新的位置
        ret, track_window = cv2.meanShift(dst, track_window, term_crit)

        #绘制追踪结果
        x,y,w,h = track_window
        img2 = cv2.rectangle(frame, (x,y), (x+w,y+h), 255,2)
```

```
        cv2.imshow('img2',img2)

        k = cv2.waitKey(3000) & 0xff
        if k == 27:
            break
    else:
        break
```

本案例设置的初始跟踪窗口的方法和位置与案例 89 的相同，目标跟踪的结果如图 8.6 所示。

图 8.6

8.2.4　案例 91：稀疏光流法运动目标跟踪

OpenCV 中提供了稀疏光流法运动目标跟踪算法函数 calcOpticalFlowPyrLK，该算法使用带金字塔的迭代 Lucas-Kanade 方法计算稀疏特征集的光流进行运动跟踪，函数定义如下：
nextPts, status, err = calcOpticalFlowPyrLK(prevImg, nextImg, prevPts, nextPts, status=None, err=None, winSize=None, maxLevel=None, criteria=None, flags=None, minEigThreshold=None)

参数说明如下。

- prevImg：8 位输入图像或图像金字塔。
- nextImg：与 prevImg 同大小同类型的 8 位输入图像或图像金字塔。
- prevPts：待寻找的光流点的二维点坐标。
- nextPts：输出的二维点坐标（返回值）。
- status：输出状态向量（返回值）。
- err：输出错误的向量（返回值）。
- winSize：每个等级金字塔的搜索窗口尺寸。
- maxLevel：基于 0 的最大金字塔层数，如果将其设置为 0，则不使用金字塔（只有单个级别）；如果设置为 1，则使用两个级别，依次类推。

- criteria：指定迭代搜索算法的终止条件。
- flags：操作标志，可以设置的标志定义见 8.2.1 节，可用参数值含义如下。

OPTFLOW_USE_INITIAL_FLOW：使用存储在 nextPts 中的初始估计，如果未设置该标志，则将 prevPts 复制到 nextPts 中并将其视为初始估计。

OPTFLOW_LK_GET_MIN_EIGENVALS：使用最小特征值作为误差度量，如果未设置该标志，则使用原始点和移动点周围的 patches 之间的 L1 距离除以窗口中的像素数作为误差度量。

- minEigThreshold：算法计算光流方程的 2×2 标准矩阵的最小特征值除以窗口中的像素数，如果计算的结果小于 minEigThreshold，那么相应的特征就会被过滤掉，并且该特征值对应的光流不会被处理，因此它允许删除坏点并获得性能提升。

案例代码如下：

```python
import numpy as np
import cv2

#定义输入文件
input_file = "vtest.avi"
cap = cv2.VideoCapture(input_file)

#定义 Shi-Tomasi 角点检测参数
feature_params = dict( maxCorners = 100,
                       qualityLevel = 0.3,
                       minDistance = 7,
                       blockSize = 7 )

#定义 Lucas-Kanade 光流算法参数
lk_params = dict( winSize  = (15,15),
                  maxLevel = 2,
                  criteria = (cv2.TERM_CRITERIA_EPS | cv2.TERM_CRITERIA_COUNT, 10, 0.03))

#创建随机颜色
color = np.random.randint(0,255,(100,3))

#读取第一帧并进行角点检测
ret, old_frame = cap.read()
old_gray = cv2.cvtColor(old_frame, cv2.COLOR_BGR2GRAY)
p0 = cv2.goodFeaturesToTrack(old_gray, mask = None, **feature_params)

#创建掩模图像以备图像绘制之用
mask = np.zeros_like(old_frame)

while(1):
```

```
ret,frame = cap.read()
frame_gray = cv2.cvtColor(frame, cv2.COLOR_BGR2GRAY)

#光流计算
p1, st, err = cv2.calcOpticalFlowPyrLK(old_gray, frame_gray, p0, None, **lk_params)

#选择 st==1 标识的运动了的角点
good_new = p1[st==1]
good_old = p0[st==1]

#绘制轨迹
for i,(new,old) in enumerate(zip(good_new, good_old)):
    a,b = new.ravel()
    c,d = old.ravel()
    mask = cv2.line(mask, (a,b),(c,d), color[i].tolist(), 2)
    frame = cv2.circle(frame,(a,b),5,color[i].tolist(),-1)
img = cv2.add(frame,mask)

cv2.imshow('frame',img)
k = cv2.waitKey(30) & 0xff
if k == 27:
    break

#刷新前一帧图像和角点位置
old_gray = frame_gray.copy()
p0 = good_new.reshape(-1,1,2)
```

运动目标跟踪的结果如图 8.7 所示。

图 8.7

8.2.5　案例 92：稠密光流法运动目标跟踪

OpenCV 中提供了稠密光流法运动目标跟踪算法函数 calcOpticalFlowFarneback，该函数封装了 Gunnar-Farneback 算法，该算法是基于前后两帧所有像素点的移动估算目标运动的算法，其跟踪效果要比稀疏光流算法的跟踪效果好。calcOpticalFlowFarneback 函数的定义如下：

```
flow= calcOpticalFlowFarneback(prev, next, flow, pyr_scale, levels, winsize, iterations, poly_n, poly_sigma, flags)
```

参数说明如下。

- prev：第一个 8 位单通道输入图像。
- next：与 prev 同大小同类型的输入图像。
- flow：计算的光流图像（返回值）。
- pyr_scale：金字塔上下两层之间的尺度关系。
- levels：金字塔级数。
- winsize：均值窗口大小。
- iterations：迭代次数。
- poly_n：像素邻域大小，典型值为 5 和 7。
- poly_sigma：高斯标准差。
- flags：计算标志，其可用参数值如下（定义见 8.2.1 节）。

OPTFLOW_USE_INITIAL_FLOW：使用输入流作为初始流近似值。

OPTFLOW_FARNEBACK_GAUSSIAN：使用高斯滤波代替方框滤波。

使用稠密光流算法进行运动目标跟踪的案例代码如下：

```python
import numpy as np
import cv2

cap = cv2.VideoCapture("vtest.avi")
#获取第一帧图像
ret, frame1 = cap.read()
prvs = cv2.cvtColor(frame1,cv2.COLOR_BGR2GRAY)
hsv = np.zeros_like(frame1)

#遍历每一行的第一列
hsv[...,1] = 255
while(1):
    ret, frame2 = cap.read()
    next = cv2.cvtColor(frame2,cv2.COLOR_BGR2GRAY)
    #返回一个两通道的光流向量，即每个点的像素位移
    flow = cv2.calcOpticalFlowFarneback(prvs,next, None, 0.5, 3, 15, 3, 5, 1.2, 0)
    #将笛卡儿坐标转换为极坐标，获得极轴和极角
    mag, ang = cv2.cartToPolar(flow[...,0], flow[...,1])
    hsv[...,0] = ang*180/np.pi/2
```

```
hsv[...,2] = cv2.normalize(mag,None,0,255,cv2.NORM_MINMAX)
bgr = cv2.cvtColor(hsv,cv2.COLOR_HSV2BGR)
#光流结果显示
cv2.imshow('frame2',bgr)
k = cv2.waitKey(30) & 0xff
if k == 27:
    break
elif k == ord('s'):
    cv2.imshow('opticalfb',frame2)
    cv2.imshow('opticalhsv',bgr)
    cv2.waitKey(10000)
prvs = next
```

本案例的执行结果如图 8.8 所示。

图 8.8

8.3 进阶必备：深度学习光流算法

深度学习的方法在计算机视觉领域大放异彩，因此很多传统算法的研究课题都有深度学习的实现方法，FlowNet 是第一个采用深度学习 CNN 方法预测光流的算法，该算法发表于 2015 年，算法总体架构如图 8.9 所示。

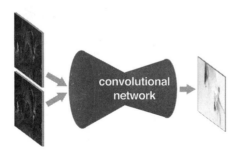

图 8.9

FlowNet 架构采用和 U-Net 架构类似的编码器与解码器，其中，编码器（下采样网络收缩）用于从两个连续图像中提取特征，而解码器（上采样网络放大）则用于放大编码器特征图并获得最终的光流预测。

算法作者设计了两种编码器特征提取网络：FlowNetSimple 和 FlowNetCorr，如图 8.10 所示。

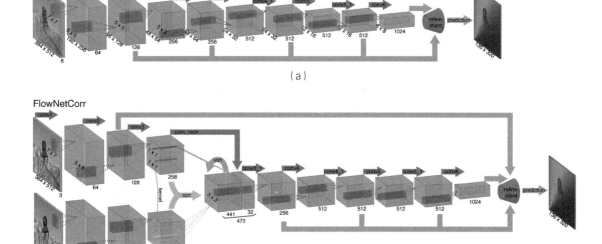

图 8.10

如图 8.10（a）所示，该网络将两个相邻帧的输入图像串联在一起，让它们通过一个普通的 9 层卷积网络，由这个网络从这一组图像中提取出光流信息特征。

如图 8.10（b）所示，该网络先独立地提取两幅图像的特征，然后在高层次中把这两种特征混合在一起。

解码器的架构如图 8.11 所示。

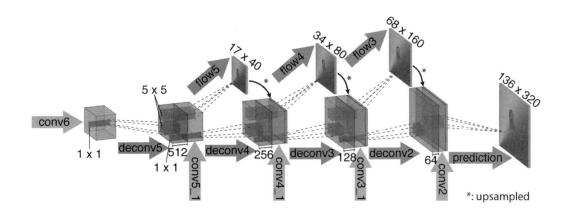

图 8.11

正如 U-Net 那样，解码器使用可训练的上卷积网络放大编码器光流输出，每个解码器层将上一层的放大结果和编码器相应层的特征图连接起来，这种使用来自编码器的数据有助于细节预测。

对 FlowNet 算法的深入研究可以参考论文 *FlowNet: Learning Optical Flow with Convolutional Networks*，还有很多别的深度学习光流算法，如非常优秀的 Raft 算法，对该算法的研究可以参考论文 *RAFT: Recurrent All-Pairs Field Transforms for Optical Flow*。

第 9 章
照片处理模块 photo

随着智能手机的拍照功能越来越强大,照片美颜的需求也在不断增加,各大拍照软件在照片美化上也是不遗余力。OpenCV 中的 photo 模块封装了照片处理算法(如照片修复、去噪、HDR 成像)和高动态范围成像算法(如色调映射等)。

9.1 模块导读

OpenCV 中引入 photo 模块需要包含头文件"opencv2/photo.hpp",通过该头文件,读者可以了解模块封装的算法。该头文件的定义如下:

```
#ifndef OPENCV_PHOTO_HPP
#define OPENCV_PHOTO_HPP

#include "opencv2/core.hpp"
#include "opencv2/imgproc.hpp"

namespace cv
{
```
如下为与图像修复、图像去噪相关的函数定义(见案例 95):
```
enum
{
    INPAINT_NS    = 0, //!<使用基于 Navier-Stokes 的方法
    INPAINT_TELEA = 1 //!< 使用 Alexandru Telea 算法
};
```

```cpp
/** 使用区域邻域恢复图像中的选定区域*/
CV_EXPORTS_W void inpaint( InputArray src, InputArray inpaintMask,
        OutputArray dst, double inpaintRadius, int flags );

/** 使用非局部均值去噪算法进行图像去噪，并进行若干计算优化，其噪声应为高斯白噪声*/
CV_EXPORTS_W void fastNlMeansDenoising( InputArray src, OutputArray dst, float h = 3,
                                    int templateWindowSize = 7, int searchWindowSize = 21);

CV_EXPORTS_W void fastNlMeansDenoising( InputArray src, OutputArray dst,
          const std::vector<float>& h, int templateWindowSize = 7,
          int searchWindowSize = 21, int normType = NORM_L2);

/** 彩色图像的 fastNlMeansDenoising 函数修正*/
CV_EXPORTS_W void fastNlMeansDenoisingColored( InputArray src, OutputArray dst,
            float h = 3, float hColor = 3, int templateWindowSize = 7, int searchWindowSize = 21);

/** 图像序列的 fastNlMeansDenoising 函数修正（用于短时间内连续拍摄的图像）*/
CV_EXPORTS_W void fastNlMeansDenoisingMulti( InputArrayOfArrays srcImgs, OutputArray dst,
              int imgToDenoiseIndex, int temporalWindowSize, float h = 3,
              int templateWindowSize = 7, int searchWindowSize = 21);

CV_EXPORTS_W void fastNlMeansDenoisingMulti( InputArrayOfArrays srcImgs, OutputArray dst,
              int imgToDenoiseIndex, int temporalWindowSize, const std::vector<float>& h,
              int templateWindowSize = 7, int searchWindowSize = 21, int normType = NORM_L2);

/** 该函数将图像转换为 CIELAB 颜色空间，然后调用 FastNLMeansDenoise Multi 函数，使用给定的 h 参数分别
对 L 和 AB 分量进行去噪*/
CV_EXPORTS_W void fastNlMeansDenoisingColoredMulti( InputArrayOfArrays srcImgs,
              OutputArray dst, int imgToDenoiseIndex, int temporalWindowSize, float h = 3,
              float hColor = 3, int templateWindowSize = 7, int searchWindowSize = 21);

/**原始-对偶算法是一种求解特殊类型变分问题的算法，图像去噪可以看作一个变分问题，因此可以使用原始-对偶
算法进行去噪*/
CV_EXPORTS_W void denoise_TVL1(const std::vector<Mat>& observations,Mat& result,
              double lambda=1.0, int niters=30);

enum { LDR_SIZE = 256 };
```

如下为与色调映射相关的类的定义：

```cpp
/** 色调映射算法的基类，用于将 HDR 图像映射到 8 位范围的工具*/
class CV_EXPORTS_W Tonemap : public Algorithm
{
public:
    CV_WRAP virtual void process(InputArray src, OutputArray dst) = 0;
    CV_WRAP virtual float getGamma() const = 0;
    CV_WRAP virtual void setGamma(float gamma) = 0;
```

```
};

/** 创建带有 gamma 校正的简单线性映射器*/
CV_EXPORTS_W Ptr<Tonemap> createTonemap(float gamma = 1.0f);

/** 自适应对数映射是一种快速的全局色调映射算法，它在对数域内对图像进行缩放。因为它是一个全局运算符，所
以相同的函数应用于所有像素，由 bias 参数控制 */
class CV_EXPORTS_W TonemapDrago : public Tonemap
{
public:
    CV_WRAP virtual float getSaturation() const = 0;
    CV_WRAP virtual void setSaturation(float saturation) = 0;
    CV_WRAP virtual float getBias() const = 0;
    CV_WRAP virtual void setBias(float bias) = 0;
};

/** 创建 TonemapDrago 对象*/
CV_EXPORTS_W Ptr<TonemapDrago> createTonemapDrago(float gamma = 1.0f,
                          float saturation = 1.0f, float bias = 0.85f);

/** 这是一个模拟人类视觉系统的全局色调映射操作符，映射函数由自适应参数控制，即通过光自适应和颜色自适应
来计算*/
class CV_EXPORTS_W TonemapReinhard : public Tonemap
{
public:
    CV_WRAP virtual float getIntensity() const = 0;
    CV_WRAP virtual void setIntensity(float intensity) = 0;

    CV_WRAP virtual float getLightAdaptation() const = 0;
    CV_WRAP virtual void setLightAdaptation(float light_adapt) = 0;

    CV_WRAP virtual float getColorAdaptation() const = 0;
    CV_WRAP virtual void setColorAdaptation(float color_adapt) = 0;
};

/** 创建 TonemapReinhard 对象*/
CV_EXPORTS_W Ptr<TonemapReinhard>
createTonemapReinhard(float gamma = 1.0f, float intensity = 0.0f, float light_adapt = 1.0f,
                      float color_adapt = 0.0f);

/** 该算法利用高斯金字塔的各个层次上的梯度将图像转化为对比度值，将对比度值转化为 HVS 响应，并对响应进行
缩放，然后根据新的对比度值重建图像*/
class CV_EXPORTS_W TonemapMantiuk : public Tonemap
{
public:
    CV_WRAP virtual float getScale() const = 0;
```

```cpp
    CV_WRAP virtual void setScale(float scale) = 0;

    CV_WRAP virtual float getSaturation() const = 0;
    CV_WRAP virtual void setSaturation(float saturation) = 0;
};

/** 创建 TonemapMantiuk 对象*/
CV_EXPORTS_W Ptr<TonemapMantiuk>
createTonemapMantiuk(float gamma = 1.0f, float scale = 0.7f, float saturation = 1.0f);
```

如下为 AlignExposures 类的定义，该类为用于将同一场景的图像与不同曝光对齐的算法的基类：

```cpp
class CV_EXPORTS_W AlignExposures : public Algorithm
{
public:
    CV_WRAP virtual void process(InputArrayOfArrays src, std::vector<Mat>& dst,
                                 InputArray times, InputArray response) = 0;
};
```

如下为 AlignMTB 类的定义，该类用于将图像转换为中值阈值位图（亮度大于中值的像素为 1，否则为 0），然后使用位操作对齐生成的位图：

```cpp
class CV_EXPORTS_W AlignMTB : public AlignExposures
{
public:
    CV_WRAP virtual void process(InputArrayOfArrays src, std::vector<Mat>& dst,
                                 InputArray times, InputArray response) CV_OVERRIDE = 0;
    CV_WRAP virtual void process(InputArrayOfArrays src, std::vector<Mat>& dst) = 0;
    CV_WRAP virtual Point calculateShift(InputArray img0, InputArray img1) = 0;
    CV_WRAP virtual void shiftMat(InputArray src, OutputArray dst, const Point shift) = 0;
    CV_WRAP virtual void computeBitmaps(InputArray img, OutputArray tb, OutputArray eb) = 0;
    CV_WRAP virtual int getMaxBits() const = 0;
    CV_WRAP virtual void setMaxBits(int max_bits) = 0;
    CV_WRAP virtual int getExcludeRange() const = 0;
    CV_WRAP virtual void setExcludeRange(int exclude_range) = 0;
    CV_WRAP virtual bool getCut() const = 0;
    CV_WRAP virtual void setCut(bool value) = 0;
};

/** 创建 AlignMTB 对象*/
CV_EXPORTS_W Ptr<AlignMTB> createAlignMTB(int max_bits = 6, int exclude_range = 4, bool cut = true);

/** 相机响应标定算法的基类*/
class CV_EXPORTS_W CalibrateCRF : public Algorithm
{
```

```
public:
    CV_WRAP virtual void process(InputArrayOfArrays src, OutputArray dst, InputArray times) = 0;
};
```

如下为 CalibrateDebevec 类的定义，该类通过将目标函数最小化为线性系统，为每个亮度值提取逆相机响应函数；利用所有图像中同一位置的像素值构造目标函数，并加入额外的项使结果更加平滑：

```
class CV_EXPORTS_W CalibrateDebevec : public CalibrateCRF
{
public:
    CV_WRAP virtual float getLambda() const = 0;
    CV_WRAP virtual void setLambda(float lambda) = 0;

    CV_WRAP virtual int getSamples() const = 0;
    CV_WRAP virtual void setSamples(int samples) = 0;

    CV_WRAP virtual bool getRandom() const = 0;
    CV_WRAP virtual void setRandom(bool random) = 0;
};

/** 创建 CalibrateDebevec 对象*/
CV_EXPORTS_W Ptr<CalibrateDebevec> createCalibrateDebevec(int samples = 70,
                                        float lambda = 10.0f, bool random = false);

/** 通过将目标函数最小化为线性系统为每个亮度值提取逆相机响应函数, 此算法使用所有图像像素*/
class CV_EXPORTS_W CalibrateRobertson : public CalibrateCRF
{
public:
    CV_WRAP virtual int getMaxIter() const = 0;
    CV_WRAP virtual void setMaxIter(int max_iter) = 0;

    CV_WRAP virtual float getThreshold() const = 0;
    CV_WRAP virtual void setThreshold(float threshold) = 0;

    CV_WRAP virtual Mat getRadiance() const = 0;
};

/** 创建 CalibrateRobertson 对象*/
CV_EXPORTS_W Ptr<CalibrateRobertson> createCalibrateRobertson(int max_iter = 30,
                                        float threshold = 0.01f);

/** 将曝光序列合并到单个图像的基类算法*/
class CV_EXPORTS_W MergeExposures : public Algorithm
{
public:
```

```
    CV_WRAP virtual void process(InputArrayOfArrays src, OutputArray dst,
                                 InputArray times, InputArray response) = 0;
};
```

如下为 MergeDebevec 类的定义，该类将得到的 HDR 图像计算为考虑曝光值和相机响应的曝光的加权平均值：

```
class CV_EXPORTS_W MergeDebevec : public MergeExposures
{
public:
    CV_WRAP virtual void process(InputArrayOfArrays src, OutputArray dst,
                                 InputArray times, InputArray response) CV_OVERRIDE = 0;
    CV_WRAP virtual void process(InputArrayOfArrays src, OutputArray dst, InputArray times) = 0;
};

/** 创建 MergeDebevec 对象*/
CV_EXPORTS_W Ptr<MergeDebevec> createMergeDebevec();

/** 使用对比度、饱和度和曝光度对像素进行加权，使用拉普拉斯金字塔对图像进行组合。得到的图像权重被构造为对比度、饱和度和曝光度的加权平均值。生成的图像不需要色调映射，可以通过乘以 255 转换为 8 位图像，但建议应用 gamma 校正和/或线性色调映射*/
class CV_EXPORTS_W MergeMertens : public MergeExposures
{
public:
    CV_WRAP virtual void process(InputArrayOfArrays src, OutputArray dst,
                                 InputArray times, InputArray response) CV_OVERRIDE = 0;
    CV_WRAP virtual void process(InputArrayOfArrays src, OutputArray dst) = 0;

    CV_WRAP virtual float getContrastWeight() const = 0;
    CV_WRAP virtual void setContrastWeight(float contrast_weiht) = 0;

    CV_WRAP virtual float getSaturationWeight() const = 0;
    CV_WRAP virtual void setSaturationWeight(float saturation_weight) = 0;

    CV_WRAP virtual float getExposureWeight() const = 0;
    CV_WRAP virtual void setExposureWeight(float exposure_weight) = 0;
};

/** 创建 MergeMertens 对象*/
CV_EXPORTS_W Ptr<MergeMertens>
createMergeMertens(float contrast_weight = 1.0f, float saturation_weight = 1.0f,
                   float exposure_weight = 0.0f);

/** 将得到的 HDR 图像计算为考虑曝光值和相机响应的曝光的加权平均值*/
class CV_EXPORTS_W MergeRobertson : public MergeExposures
{
```

```
public:
    CV_WRAP virtual void process(InputArrayOfArrays src, OutputArray dst,
                                 InputArray times, InputArray response) CV_OVERRIDE = 0;
    CV_WRAP virtual void process(InputArrayOfArrays src, OutputArray dst, InputArray times) = 0;
};

/** 创建 MergeRobertson 对象*/
CV_EXPORTS_W Ptr<MergeRobertson> createMergeRobertson();
```

如下为对比度保留脱色函数的定义，该函数用于将彩色图像转换为灰度图像，是数字印刷、风格化黑白照片渲染及许多单通道图像处理应用中的基本工具（见案例94）：

```
CV_EXPORTS_W void decolor( InputArray src, OutputArray grayscale, OutputArray color_boost);
```

如下为与无缝克隆算法相关的定义（见案例93）：

```
//! 无缝克隆算法标志
enum
{
    /** 标准克隆方法*/
    NORMAL_CLONE = 1,
    /** 混合无缝克隆*/
    MIXED_CLONE  = 2,
    /** Monochrome 迁移*/
    MONOCHROME_TRANSFER = 3
};

/** 无缝克隆算法*/
CV_EXPORTS_W void seamlessClone(InputArray src, InputArray dst, InputArray mask,
                                Point p, OutputArray blend, int flags);

/** 给定一个原始彩色图像，两个不同颜色的图像可以无缝融合*/
CV_EXPORTS_W void colorChange(InputArray src, InputArray mask, OutputArray dst,
                float red_mul = 1.0f, float green_mul = 1.0f, float blue_mul = 1.0f);

/** 照度修改*/
CV_EXPORTS_W void illuminationChange(InputArray src, InputArray mask, OutputArray dst,
                float alpha = 0.2f, float beta = 0.4f);

/** 纹理展平*/
CV_EXPORTS_W void textureFlattening(InputArray src, InputArray mask, OutputArray dst,
                float low_threshold = 30, float high_threshold = 45, int kernel_size = 3);
```

如下为边缘保留滤波相关操作的定义（见案例97）：

```
enum
{
    RECURS_FILTER = 1, //!< 递归滤波
```

```
    NORMCONV_FILTER= 2    //!< 归一化卷积滤波
};

/** 边缘保留平滑滤波*/
CV_EXPORTS_W void edgePreservingFilter(InputArray src, OutputArray dst, int flags = 1,
                                       float sigma_s = 60, float sigma_r = 0.4f);
```

如下为用于图像细节增强的函数的定义（见案例 98）：

```
CV_EXPORTS_W void detailEnhance(InputArray src, OutputArray dst, float sigma_s = 10,
                                float sigma_r = 0.15f);
```

如下为用于铅笔样式非真实感线绘制的函数的定义（见案例 99）：

```
CV_EXPORTS_W void pencilSketch(InputArray src, OutputArray dst1, OutputArray dst2,
                               float sigma_s = 60, float sigma_r = 0.07f, float shade_factor = 0.02f);
```

如下为风格化函数的定义（见案例 100）：

```
CV_EXPORTS_W void stylization(InputArray src, OutputArray dst, float sigma_s = 60,
                              float sigma_r = 0.45f);
} // cv

#endif
```

9.2 案例 93：基于 OpenCV 的无缝克隆

"身不能至，心向往之"是很多人在看到如画美景时的感叹，如今先进的计算机技术可以帮我们实现这种"入画"的梦想，这种技术就是"PS"，即使用 Photoshop 软件进行处理，但是这个软件的使用还是有一定的难度的，而 OpenCV 提供的 PS 技术可以帮助读者轻松实现自己的想法。图 9.1 是一幅美丽的风景图像。

另外有一幅飞机图像，如图 9.2 所示。

图 9.1

图 9.2

图 9.1 所示的风景图像中并没有出现如图 9.2 所示的飞机，如果想要这架飞机出现在风景

图像中,则需要利用 PS 技术把飞机"P"到风景图像中。

如上所述,"把飞机 P 到风景图像中",意思是将飞机图像融合到风景图像中去,Photoshop 是实现这种融合的强大工具。然而,Photoshop 工具比较复杂,想用好这项技术有一定的难度,图 9.3 所示的直接图像叠加看起来是很糟糕的。

真正希望得到的结果如图 9.4 所示,浑然天成,毫无违和感。

图 9.3

图 9.4

OpenCV 中提供了用于无缝克隆的函数 seamlessClone,其定义如下:
blend = seamlessClone(src, dst, mask, p, flags, blend=None)

参数说明如下。

- src:输入图像,需要输入 8 位 3 通道图像。
- dst:输入图像,需要输入 8 位 3 通道图像。
- mask:掩模,需要输入 8 位 1 或 3 通道图像。
- p:src 图像置于 dst 图像中的位置。
- flags:克隆方法,如 cv::NORMAL_CLONE 等,定义见 9.1 节。
- blend:输出图像(返回值)。

案例代码如下:

```
import cv2
import numpy as np

#图像读取
src = cv2.imread("airplane.jpg")
dst = cv2.imread("sky.jpg")

#在飞机周围创建一个强mask(选取飞机所在的区域)
src_mask = np.zeros(src.shape, src.dtype)
poly = np.array([ [4,80], [30,54], [151,63], [254,37], [298,90], [272,134], [43,122] ], np.int32)
cv2.fillPoly(src_mask, [poly], (255, 255, 255))

#定义飞机将要放置位置的中心点坐标
```

```
center = (800,100)

#无缝克隆
output = cv2.seamlessClone(src, dst, src_mask, center, cv2.NORMAL_CLONE)

#结果显示
cv2.imshow("seamless-cloning", output)
cv2.waitKey(0)
cv2.destroyAllWindows()
```

设置的飞机 mask 如图 9.5 所示，图像融合结果如图 9.6 所示。

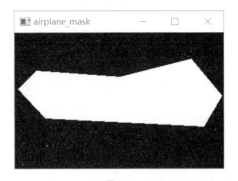
图 9.5

图 9.6

提示：本节案例及后续章节部分案例来源于 OpenCV 官方推荐案例，有兴趣的读者可以参考 GitHub 仓库 learnopencv。

9.3 案例 94：基于 OpenCV 的图像对比度保留脱色

图像脱色即将彩色图像转换为灰度图像的过程，相关应用在数字印刷、风格化黑白摄影和许多单通道图像处理应用程序中比较常见。图像对比度保留脱色是指最大限度地保留原始颜色对比度。

OpenCV 中提供了用于图像对比度保留脱色的函数 decolor，其定义如下：

grayscale, color_boost = decolor(src, grayscale=None, color_boost=None)

参数说明如下。

- src：输入图像。
- grayscale：对比度保留脱色结果（返回值）。
- color_boost：颜色增强结果（返回值）。

案例代码如下：

```python
import cv2

img = cv2.imread("flower.png")
#应用对比度保留脱色
grayscale, color_boost = cv2.decolor(img)
#为了对比，直接转为灰度图像
cvt_gray = cv2.cvtColor(img, cv2.COLOR_BGR2GRAY)
#结果展示
cv2.imshow("color_boost", color_boost)      #显示原始图像
cv2.imshow("color_boost", color_boost)      #显示颜色增强结果图像
cv2.imshow("grayscale", grayscale)          #显示对比度保留脱色的结果
cv2.imshow("cvt_gray", cvt_gray)            #显示直接转为灰度图像的结果
cv2.waitKey(0)
cv2.destroyAllWindows()
```

本案例使用的输入图像如图 9.7 所示。

图 9.7

原始图像与颜色增强的结果对比如图 9.8 所示。

图 9.8

对比度保留脱色与转为灰度图像的结果对比如图 9.9 所示。

图 9.9

9.4 案例 95：基于 OpenCV 的图像修复

OpenCV 中提供了用于图像修复的函数 inpaint，其定义如下：

```
dst = inpaint(src, inpaintMask, inpaintRadius, flags, dst=None)
```

参数说明如下。

- src：输入图像。
- inpaintMask：图像修复 mask。
- inpaintRadius：算法考虑的每个修复点的圆形邻域半径。
- flags：修复方法，可用值为 cv::INPAINT_NS 和 cv::INPAINT_TELEA（定义见 9.1 节）。
- dst：输出图像（返回值）。

案例代码如下：

```python
import numpy as np
import cv2

#封装一个实用的鼠标操作类
class Sketcher:
    def __init__(self, windowname, dests, colors_func):
        self.prev_pt = None
        self.windowname = windowname
        self.dests = dests
        self.colors_func = colors_func
        self.dirty = False
        self.show()
        cv2.setMouseCallback(self.windowname, self.on_mouse)
```

```python
    def show(self):
        cv2.imshow(self.windowname, self.dests[0])
        cv2.imshow(self.windowname + ": mask", self.dests[1])

    #鼠标回调函数
    def on_mouse(self, event, x, y, flags, param):
        pt = (x, y)
        if event == cv2.EVENT_LBUTTONDOWN:
            self.prev_pt = pt
        elif event == cv2.EVENT_LBUTTONUP:
            self.prev_pt = None

        if self.prev_pt and flags & cv2.EVENT_FLAG_LBUTTON:
            for dst, color in zip(self.dests, self.colors_func()):
                cv2.line(dst, self.prev_pt, pt, color, 5)
            self.dirty = True
            self.prev_pt = pt
            self.show()


def main():
    #图像读取
    img = cv2.imread("sample.jpeg", cv2.IMREAD_COLOR)

    #图像读取失败处理
    if img is None:
        print('Failed to load image file!')
        return

    #创建原始图像备份
    img_mask = img.copy()
    #创建原始图像大小的 mask
    inpaintMask = np.zeros(img.shape[:2], np.uint8)
    #创建 Sketcher 类对象
    sketch = Sketcher('image', [img_mask, inpaintMask], lambda : ((255, 255, 255), 255))

    while True:
        ch = cv2.waitKey()
        if ch == 27:
            break
        if ch == ord('t'):
            #使用 Alexendra Telea 提出的快速行进算法（Fast Marching Method）
            res = cv2.inpaint(src=img_mask, inpaintMask=inpaintMask, inpaintRadius=3,
                              flags=cv2.INPAINT_TELEA)
            cv2.imshow('Inpaint Output using FMM', res)
```

```
        if ch == ord('n'):
            #使用 Bertalmio 等提出的算法 Navier-Stokes
            res = cv2.inpaint(src=img_mask, inpaintMask=inpaintMask, inpaintRadius=3,
                              flags=cv2.INPAINT_NS)
            cv2.imshow('Inpaint Output using NS Technique', res)
        if ch == ord('r'):
            #图像复原
            img_mask[:] = img
            inpaintMask[:] = 0
            sketch.show()
if __name__ == '__main__':
    main()
    cv2.destroyAllWindows()
```

本案例使用的输入图像如图 9.7 所示，按下 T 键，选择 cv2.INPAINT_TELEA 修复方法，修复结果对比如图 9.10 所示。

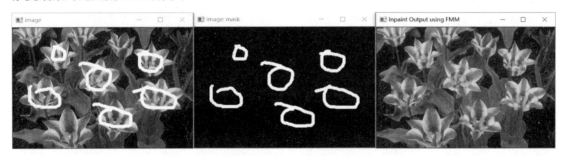

图 9.10

在图 9.10 中，左图为鼠标在原始图像上选择待修复区域的 mask，中图为 mask，右图为修复结果。

按下 N 键，选择 cv2.INPAINT_NS 修复方法，修复结果对比如图 9.11 所示。

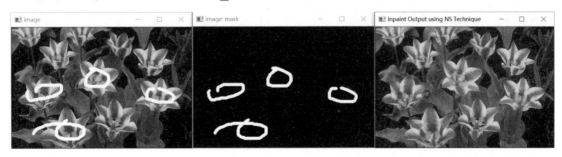

图 9.11

9.5 案例 96：基于 OpenCV 的 HDR 成像

大多数数码相机和显示器都以 24 位矩阵的形式捕获或显示彩色图像。每个颜色空间有 8 位，因此每个通道的像素值为 0～255。换句话说，普通相机或显示器的动态范围有限。

然而，我们周围的世界具有非常大的动态范围，OpenCV 封装了通过不同曝光的图像生成高动态范围（HDR）图像的算法，步骤如下。

- 第 1 步，拍摄多幅不同曝光度的图像。

本案例中使用的曝光图像分别如图 9.12～图 9.15 所示。

曝光时间为 0.033s 的图像如图 9.12 所示，曝光时间为 0.25s 的图像如图 9.13 所示。

图 9.12

图 9.13

曝光时间为 2.5s 的图像如图 9.14 所示，曝光时间为 15s 的图像如图 9.15 所示。

图 9.14

图 9.15

- 第 2 步，对齐图像。

图像对齐在 HDR 图像生成中非常重要，如果不对齐，则会出现严重的伪影，如图 9.16 所示，其中左侧为未对齐图像，右侧为对齐图像。

图 9.16

- 第 3 步，恢复相机响应功能。
- 第 4 步，合并图像。
- 第 5 步，色调映射。

案例代码如下：

```python
import cv2
import numpy as np

#获取图像和曝光时间
def readImagesAndTimes():
    times = np.array([ 1/30.0, 0.25, 2.5, 15.0 ], dtype=np.float32)
    filenames = ["img_0.033.jpg", "img_0.25.jpg", "img_2.5.jpg", "img_15.jpg"]
    images = []
    for filename in filenames:
        im = cv2.imread(filename)
        images.append(im)
    return images, times

if __name__ == '__main__':
    #读取图像和曝光时间
    images, times = readImagesAndTimes()
    
    #图像对齐
    alignMTB = cv2.createAlignMTB()
    alignMTB.process(images, images)
    
    #获取相机响应函数（CRF）
    calibrateDebevec = cv2.createCalibrateDebevec()
    responseDebevec = calibrateDebevec.process(images, times)
```

```
#合并对齐后的图像为HDR线性图像
mergeDebevec = cv2.createMergeDebevec()
hdrDebevec = mergeDebevec.process(images, times, responseDebevec)
#图像尺寸较大故直接保存后展示
cv2.imwrite("hdrDebevec.jpg", hdrDebevec)

#使用Drago方法的色调映射以获取24位彩色图像
tonemapDrago = cv2.createTonemapDrago(1.0, 0.7)
ldrDrago = tonemapDrago.process(hdrDebevec)
ldrDrago = 3 * ldrDrago
cv2.imwrite("ldr-Drago.jpg", ldrDrago * 255)

#使用Reinhard方法的色调映射以获取24位彩色图像
tonemapReinhard = cv2.createTonemapReinhard(1.5, 0,0,0)
ldrReinhard = tonemapReinhard.process(hdrDebevec)
cv2.imwrite("ldr-Reinhard.jpg", ldrReinhard * 255)

#使用Mantiuk方法的色调映射以获取24位彩色图像
tonemapMantiuk = cv2.createTonemapMantiuk(2.2,0.85, 1.2)
ldrMantiuk = tonemapMantiuk.process(hdrDebevec)
ldrMantiuk = 3 * ldrMantiuk
cv2.imwrite("ldr-Mantiuk.jpg", ldrMantiuk * 255)
```

合并后的图像如图9.17所示。

使用Drago方法的色调映射获取的24位彩色图像如图9.18所示。

图9.17　　　　　　　　　　　　　　　图9.18

使用Reinhard方法的色调映射获取的24位彩色图像如图9.19所示。

使用 Mantiuk 方法的色调映射获取的 24 位彩色图像如图 9.20 所示。

图 9.19

图 9.20

读者可以将合并后的图像保存为.hdr 文件，保存的 HDR 图像可以在 Photoshop 中加载并进行色调映射：

```
cv2.imwrite("hdrDebevec.hdr", hdrDebevec)
```

9.6 图像非真实感渲染

OpenCV 中封装了多种用于非真实感渲染的算法，如边缘保留滤波、细节增强、铅笔素描和风格化。

9.6.1 案例 97：边缘保留滤波

OpenCV 中提供的边缘保留滤波函数为 edgePreservingFilter，其定义如下：

```
dst = edgePreservingFilter(src, dst=None, flags=None, sigma_s=None, sigma_r=None)
```

参数说明如下。
- src：输入图像。
- dst：输出图像（返回值）。
- flags：边缘保留滤波标志，可用值为 cv::RECURS_FILTER 和 cv::NORMCONV_FILTER，定义见 9.1 节。
- sigma_s：取值在 0 到 200 之间。
- sigma_r：取值在 0 到 1 之间。

边缘保留滤波的案例代码如下：

```
import cv2
```

```
#输入图像读取
img = cv2.imread("cow.jpg")
cv2.imshow("input", img)

#以两种不同标志进行边缘保留滤波
imout = cv2.edgePreservingFilter(img, flags=cv2.RECURS_FILTER)
cv2.imshow("edge-preserving-recursive-filter", imout)

imout = cv2.edgePreservingFilter(img, flags=cv2.NORMCONV_FILTER)
cv2.imshow("edge-preserving-normalized-convolution-filter", imout)

cv2.waitKey(0)
cv2.destroyAllWindows()
```

使用标志 cv2.RECURS_FILTER（递归滤波）的边缘保留滤波结果与原始图像对比如图 9.21 所示。

图 9.21

使用标志 cv2.NORMCONV_FILTER（归一化卷积滤波）的边缘保留滤波结果与原始图像对比如图 9.22 所示。

图 9.22

9.6.2 案例98：图像细节增强

OpenCV 中提供的用于细节增强的函数为 detailEnhance，其定义如下：

```
dst = detailEnhance(src, dst=None, sigma_s=None, sigma_r=None)
```

参数说明如下。
- src：输入图像。
- dst：输出图像（返回值）。
- sigma_s：取值在 0 到 200 之间。
- sigma_r：取值在 0 到 1 之间。

案例代码如下：

```
import cv2

#输入图像读取
img = cv2.imread("cow.jpg")
cv2.imshow("input", img)

#细节增强滤波
imout = cv2.detailEnhance(img)
cv2.imshow("detail-enhance", imout)

cv2.waitKey(0)
cv2.destroyAllWindows()
```

细节增强结果与原始图像对比如图 9.23 所示。

图 9.23

9.6.3 案例99：铅笔素描

OpenCV 中提供的铅笔素描函数为 pencilSketch，其定义如下：

```
dst1, dst2 = pencilSketch(src, dst1=None, dst2=None, sigma_s=None, sigma_r=None,
shade_factor=None)
```

参数说明如下。

- src：输入图像。
- dst1：输出图像，8 位单通道图像（返回值）。
- dst2：输出图像，与 src 同大小同类型（返回值）。
- sigma_s：取值在 0 到 200 之间。
- sigma_r：取值在 0 到 1 之间。
- shade_factor：取值在 0 到 0.1 之间。

案例代码如下：

```
import cv2

#输入图像读取
img = cv2.imread("cow.jpg")
cv2.imshow("input", img)

#铅笔素描滤波
imout_gray, imout = cv2.pencilSketch(img, sigma_s=60, sigma_r=0.07, shade_factor=0.05)
cv2.imshow("pencil-sketch", imout_gray)
cv2.imshow("pencil-sketch-color", imout)
```

铅笔素描结果与原始图像对比如图 9.24 所示。

图 9.24

9.6.4　案例 100：风格化图像

OpenCV 中提供的风格化函数为 stylization，其定义如下：

```
dst = stylization(src, dst=None, sigma_s=None, sigma_r=None)
```

参数说明如下。

- src：输入图像。
- dst：输出图像（返回值）。

- sigma_s：取值在 0 到 200 之间。
- sigma_r：取值在 0 到 1 之间。

案例代码如下：

```python
import cv2

#输入图像读取
img = cv2.imread("cow.jpg")
cv2.imshow("input", img)

#风格化滤波
cv2.stylization(img,imout)
cv2.imshow("stylization", imout)
cv2.waitKey(0)
cv2.destroyAllWindows()
```

风格化结果与原始图像对比如图 9.25 所示。

图 9.25

9.7 进阶必备：照片处理算法概述

OpenCV 中的 photo 模块提供了很多照片处理函数，本节介绍部分函数使用的算法。

1. 图像去噪算法

OpenCV 中提供了 fastNlMeansDenoising 函数用于图像去噪，该函数使用的是非局部平均 NL-Means（Non-Local Means）。

常见的去噪方法如高斯滤波、中值滤波等都有一个滤波核，去除噪声使用核覆盖范围内的像素计算锚点处的像素值（如计算高斯平均值或核范围内的中值），这些方法在噪声较小时效果较好，这些方法对噪声的去除是在局部邻域内完成的。

与高斯滤波等常用的利用图像局部信息去噪的算法不同，NL-Means 利用整幅图像的信息去噪，即以图像块为单位在图像中寻找相似区域，再对这些区域求平均，可以比较好地去掉图像中存在的高斯噪声。该算法的去噪效果较传统方法的去噪效果好，但是耗时更久。

2. 对比度保留脱色算法

将彩色图像转为灰度图像的算法较多，如将 RGB 彩色图像转为灰度图像的著名计算方法如下：

```
Gray = R*0.299 + G*0.587 + B*0.114
```

对于如图 9.26 所示图像，使用上述算法得到的灰度图像如图 9.27（a）所示。

图 9.26

图 9.26 为原图，图 9.27（a）～（e）为常用算法的处理结果，这 6 幅图像中的晚霞和阳光在水中的倒影在脱色之后就几乎无法分辨出来了，这就是脱色后失去了对比度。

图 9.27（f）即对比度保留脱色算法的处理结果，脱色后的图像最大限度地保留原始颜色对比度。

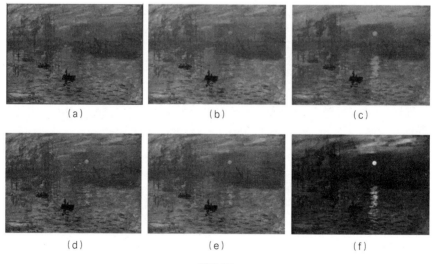

图 9.27

读者可以参考 Cewu Lu 发表的两篇论文 *Contrast Preserving Decolorization* 和 *Contrast Preserving Decolorization with Perception-Based Quality Metrics* 来深入了解对比度保留脱色算法。

3. 无缝克隆

无缝克隆技术使用的是泊松图像编辑算法，该算法可以将图像很好地融合到原始图像中，无须额外的图像抠取操作，处理结果无缝融合，毫无断层感。

读者可以参考论文 *Poisson Image Editing* 来深入了解无缝克隆像素插入的细节。

4. 图像非真实感渲染

图像非真实感渲染可以参考论文 *Domain Transform for Edge-Aware Image and Video Processing*，论文中展示了在边缘保留滤波、细节增强、风格化、图像重新着色、铅笔画、景深几个方向上的研究与应用。

第 10 章
2D 特征模块 features2d

features2d 模块封装了特征检测与描述、描述符匹配、特征点绘制和匹配绘制、对象分类等算法。

10.1 模块导读

在 OpenCV 中引入 features2d 模块需要包含头文件"opencv2/features2d.hpp"，通过该头文件，读者可以了解模块封装的算法。该头文件的定义如下：

```
#ifndef OPENCV_FEATURES_2D_HPP
#define OPENCV_FEATURES_2D_HPP

#include "opencv2/opencv_modules.hpp"
#include "opencv2/core.hpp"

#ifdef HAVE_OPENCV_FLANN
#include "opencv2/flann/miniflann.hpp"
#endif

namespace cv
{
```

如下为 KeyPointsFilter 类的定义，该类用于特征点向量的过滤：

```
class CV_EXPORTS KeyPointsFilter
{
public:
    KeyPointsFilter(){}
```

```
    static void runByImageBorder( std::vector<KeyPoint>& keypoints,
                                  Size imageSize, int borderSize );
    static void runByKeypointSize( std::vector<KeyPoint>& keypoints,
                                   float minSize, float maxSize=FLT_MAX );
    static void runByPixelsMask( std::vector<KeyPoint>& keypoints, const Mat& mask );
    static void removeDuplicated( std::vector<KeyPoint>& keypoints );
    static void removeDuplicatedSorted( std::vector<KeyPoint>& keypoints );
    static void retainBest( std::vector<KeyPoint>& keypoints, int npoints );
};
```

如下为 Feature2D 类的定义，该类为二维图像特征检测器和描述符提取器的抽象基类：

```
#ifdef __EMSCRIPTEN__
class CV_EXPORTS_W Feature2D : public Algorithm
#else
class CV_EXPORTS_W Feature2D : public virtual Algorithm
#endif
{
public:
    virtual ~Feature2D();

    /** 特征点检测*/
    CV_WRAP virtual void detect( InputArray image, CV_OUT std::vector<KeyPoint>& keypoints,
                                 InputArray mask=noArray() );

    CV_WRAP virtual void detect( InputArrayOfArrays images,
                                 CV_OUT std::vector<std::vector<KeyPoint> >& keypoints,
                                 InputArrayOfArrays masks=noArray() );

    /** 计算图像中检测到的一组特征点的描述符*/
    CV_WRAP virtual void compute( InputArray image,
                                  CV_OUT CV_IN_OUT std::vector<KeyPoint>& keypoints,
                                  OutputArray descriptors );

    CV_WRAP virtual void compute( InputArrayOfArrays images,
                                  CV_OUT CV_IN_OUT std::vector<std::vector<KeyPoint> >& keypoints,
                                  OutputArrayOfArrays descriptors );

    /** 检测特征点并计算描述符*/
    CV_WRAP virtual void detectAndCompute( InputArray image, InputArray mask,
                     CV_OUT std::vector<KeyPoint>& keypoints,
                     OutputArray descriptors, bool useProvidedKeypoints=false );

    CV_WRAP virtual int descriptorSize() const;
    CV_WRAP virtual int descriptorType() const;
    CV_WRAP virtual int defaultNorm() const;
```

```
    CV_WRAP void write( const String& fileName ) const;

    CV_WRAP void read( const String& fileName );

    virtual void write( FileStorage&) const CV_OVERRIDE;

    CV_WRAP virtual void read( const FileNode&) CV_OVERRIDE;

    //! 判断检测器对象是否为空
    CV_WRAP virtual bool empty() const CV_OVERRIDE;
    CV_WRAP virtual String getDefaultName() const CV_OVERRIDE;

    CV_WRAP inline void write(const Ptr<FileStorage>& fs, const String& name = String()) const
    { Algorithm::write(fs, name); }
};

/** 特征检测器*/
typedef Feature2D FeatureDetector;

/** 特征点描述符提取器*/
typedef Feature2D DescriptorExtractor;
```

如下为 AffineFeature 类的定义，该类为包装器 wrapper 实现类，该包装器使检测器和提取器具有仿射不变性：

```
class CV_EXPORTS_W AffineFeature : public Feature2D
{
public:
    //对象创建函数
    CV_WRAP static Ptr<AffineFeature> create(const Ptr<Feature2D>& backend,
                        int maxTilt = 5, int minTilt = 0,
                        float tiltStep = 1.4142135623730951f, float rotateStepBase = 72);

    CV_WRAP virtual void setViewParams(const std::vector<float>& tilts,
                        const std::vector<float>& rolls) = 0;
    CV_WRAP virtual void getViewParams(std::vector<float>& tilts,
                        std::vector<float>& rolls) const = 0;
    CV_WRAP virtual String getDefaultName() const CV_OVERRIDE;
};

typedef AffineFeature AffineFeatureDetector;
typedef AffineFeature AffineDescriptorExtractor;
```

如下为 SIFT 特征点检测和描述符提取算法实现类（见案例 101）：

```
class CV_EXPORTS_W SIFT : public Feature2D
{
```

```cpp
public:
    /** SIFT 对象创建类*/
    CV_WRAP static Ptr<SIFT> create(int nfeatures = 0, int nOctaveLayers = 3,
                    double contrastThreshold = 0.04, double edgeThreshold = 10,
                    double sigma = 1.6);

    CV_WRAP static Ptr<SIFT> create(int nfeatures, int nOctaveLayers,
                    double contrastThreshold, double edgeThreshold,
                    double sigma, int descriptorType);

    CV_WRAP virtual String getDefaultName() const CV_OVERRIDE;
};

typedef SIFT SiftFeatureDetector;
typedef SIFT SiftDescriptorExtractor;
```

如下为 BRISK 特征点检测和描述符提取算法类（见案例 103）：

```cpp
class CV_EXPORTS_W BRISK : public Feature2D
{
public:
    /** BRISK 对象创建*/
    CV_WRAP static Ptr<BRISK> create(int thresh=30, int octaves=3, float patternScale=1.0f);
    CV_WRAP static Ptr<BRISK> create(const std::vector<float> &radiusList,
                    const std::vector<int> &numberList, float dMax=5.85f,
                    float dMin=8.2f, const std::vector<int>& indexChange=std::vector<int>());

    CV_WRAP static Ptr<BRISK> create(int thresh, int octaves,
                    const std::vector<float> &radiusList, const std::vector<int> &numberList,
                    float dMax=5.85f, float dMin=8.2f,
                    const std::vector<int>& indexChange=std::vector<int>());
    CV_WRAP virtual String getDefaultName() const CV_OVERRIDE;

    /** 设置检测阈值*/
    CV_WRAP virtual void setThreshold(int threshold) { CV_UNUSED(threshold); return; }
    CV_WRAP virtual int getThreshold() const { return -1; }

    CV_WRAP virtual void setOctaves(int octaves) { CV_UNUSED(octaves); return; }
    CV_WRAP virtual int getOctaves() const { return -1; }
};
```

如下为 ORB 特征点检测和描述符提取算法类（见案例 104）：

```cpp
class CV_EXPORTS_W ORB : public Feature2D
{
public:
    enum ScoreType { HARRIS_SCORE=0, FAST_SCORE=1 };
```

```cpp
    static const int kBytes = 32;

    /** ORB 对象创建*/
    CV_WRAP static Ptr<ORB> create(int nfeatures=500, float scaleFactor=1.2f, int nlevels=8,
                int edgeThreshold=31, int firstLevel=0, int WTA_K=2,
                ORB::ScoreType scoreType=ORB::HARRIS_SCORE,
                int patchSize=31, int fastThreshold=20);

    CV_WRAP virtual void setMaxFeatures(int maxFeatures) = 0;
    CV_WRAP virtual int getMaxFeatures() const = 0;

    CV_WRAP virtual void setScaleFactor(double scaleFactor) = 0;
    CV_WRAP virtual double getScaleFactor() const = 0;

    CV_WRAP virtual void setNLevels(int nlevels) = 0;
    CV_WRAP virtual int getNLevels() const = 0;

    CV_WRAP virtual void setEdgeThreshold(int edgeThreshold) = 0;
    CV_WRAP virtual int getEdgeThreshold() const = 0;

    CV_WRAP virtual void setFirstLevel(int firstLevel) = 0;
    CV_WRAP virtual int getFirstLevel() const = 0;

    CV_WRAP virtual void setWTA_K(int wta_k) = 0;
    CV_WRAP virtual int getWTA_K() const = 0;

    CV_WRAP virtual void setScoreType(ORB::ScoreType scoreType) = 0;
    CV_WRAP virtual ORB::ScoreType getScoreType() const = 0;

    CV_WRAP virtual void setPatchSize(int patchSize) = 0;
    CV_WRAP virtual int getPatchSize() const = 0;

    CV_WRAP virtual void setFastThreshold(int fastThreshold) = 0;
    CV_WRAP virtual int getFastThreshold() const = 0;
    CV_WRAP virtual String getDefaultName() const CV_OVERRIDE;
};
```

如下为 MSER 算法实现类:

```cpp
class CV_EXPORTS_W MSER : public Feature2D
{
public:
    /** MSER 对象创建*/
    CV_WRAP static Ptr<MSER> create( int _delta=5, int _min_area=60, int _max_area=14400,
              double _max_variation=0.25, double _min_diversity=.2,
              int _max_evolution=200, double _area_threshold=1.01,
              double _min_margin=0.003, int _edge_blur_size=5 );
```

```cpp
    CV_WRAP virtual void detectRegions( InputArray image,
                        CV_OUT std::vector<std::vector<Point> >& msers,
                        CV_OUT std::vector<Rect>& bboxes ) = 0;

    CV_WRAP virtual void setDelta(int delta) = 0;
    CV_WRAP virtual int getDelta() const = 0;

    CV_WRAP virtual void setMinArea(int minArea) = 0;
    CV_WRAP virtual int getMinArea() const = 0;

    CV_WRAP virtual void setMaxArea(int maxArea) = 0;
    CV_WRAP virtual int getMaxArea() const = 0;

    CV_WRAP virtual void setPass2Only(bool f) = 0;
    CV_WRAP virtual bool getPass2Only() const = 0;
    CV_WRAP virtual String getDefaultName() const CV_OVERRIDE;
};
```

如下为 FAST 算法实现类（见案例 108）：

```cpp
class CV_EXPORTS_W FastFeatureDetector : public Feature2D
{
public:
    enum DetectorType
    {
        TYPE_5_8 = 0, TYPE_7_12 = 1, TYPE_9_16 = 2
    };
    enum
    {
        THRESHOLD = 10000, NONMAX_SUPPRESSION=10001, FAST_N=10002
    };


    CV_WRAP static Ptr<FastFeatureDetector> create( int threshold=10,
                    bool nonmaxSuppression=true,
                    FastFeatureDetector::DetectorType
type=FastFeatureDetector::TYPE_9_16 );

    CV_WRAP virtual void setThreshold(int threshold) = 0;
    CV_WRAP virtual int getThreshold() const = 0;

    CV_WRAP virtual void setNonmaxSuppression(bool f) = 0;
    CV_WRAP virtual bool getNonmaxSuppression() const = 0;

    CV_WRAP virtual void setType(FastFeatureDetector::DetectorType type) = 0;
    CV_WRAP virtual FastFeatureDetector::DetectorType getType() const = 0;
    CV_WRAP virtual String getDefaultName() const CV_OVERRIDE;
};
```

```cpp
CV_EXPORTS void FAST( InputArray image, CV_OUT std::vector<KeyPoint>& keypoints,
                      int threshold, bool nonmaxSuppression=true );


/** FAST 角点检测*/
CV_EXPORTS void FAST( InputArray image, CV_OUT std::vector<KeyPoint>& keypoints,
                      int threshold, bool nonmaxSuppression,
                      FastFeatureDetector::DetectorType type );
```

如下为 AGAST 算法实现类（见案例 107）：

```cpp
class CV_EXPORTS_W AgastFeatureDetector : public Feature2D
{
public:
    enum DetectorType
    {
        AGAST_5_8 = 0, AGAST_7_12d = 1, AGAST_7_12s = 2, OAST_9_16 = 3,
    };

    enum
    {
        THRESHOLD = 10000, NONMAX_SUPPRESSION = 10001,
    };

    CV_WRAP static Ptr<AgastFeatureDetector> create( int threshold=10,
                    bool nonmaxSuppression=true,
                    AgastFeatureDetector::DetectorType type =
AgastFeatureDetector::OAST_9_16);

    CV_WRAP virtual void setThreshold(int threshold) = 0;
    CV_WRAP virtual int getThreshold() const = 0;

    CV_WRAP virtual void setNonmaxSuppression(bool f) = 0;
    CV_WRAP virtual bool getNonmaxSuppression() const = 0;

    CV_WRAP virtual void setType(AgastFeatureDetector::DetectorType type) = 0;
    CV_WRAP virtual AgastFeatureDetector::DetectorType getType() const = 0;
    CV_WRAP virtual String getDefaultName() const CV_OVERRIDE;
};

CV_EXPORTS void AGAST( InputArray image, CV_OUT std::vector<KeyPoint>& keypoints,
                       int threshold, bool nonmaxSuppression=true );


/** AGAST 角点检测*/
CV_EXPORTS void AGAST( InputArray image, CV_OUT std::vector<KeyPoint>& keypoints,
                       int threshold, bool nonmaxSuppression,
                       AgastFeatureDetector::DetectorType type );
```

如下为使用 goodFeaturesToTrack 函数做特征点检测的类:
```cpp
class CV_EXPORTS_W GFTTDetector : public Feature2D
{
public:
    CV_WRAP static Ptr<GFTTDetector> create( int maxCorners=1000, double qualityLevel=0.01,
                                             double minDistance=1, int blockSize=3,
                                             bool useHarrisDetector=false, double k=0.04 );
    CV_WRAP static Ptr<GFTTDetector> create( int maxCorners, double qualityLevel,
                                             double minDistance, int blockSize, int gradiantSize,
                                             bool useHarrisDetector=false, double k=0.04 );
    CV_WRAP virtual void setMaxFeatures(int maxFeatures) = 0;
    CV_WRAP virtual int getMaxFeatures() const = 0;

    CV_WRAP virtual void setQualityLevel(double qlevel) = 0;
    CV_WRAP virtual double getQualityLevel() const = 0;

    CV_WRAP virtual void setMinDistance(double minDistance) = 0;
    CV_WRAP virtual double getMinDistance() const = 0;

    CV_WRAP virtual void setBlockSize(int blockSize) = 0;
    CV_WRAP virtual int getBlockSize() const = 0;

    CV_WRAP virtual void setHarrisDetector(bool val) = 0;
    CV_WRAP virtual bool getHarrisDetector() const = 0;

    CV_WRAP virtual void setK(double k) = 0;
    CV_WRAP virtual double getK() const = 0;
    CV_WRAP virtual String getDefaultName() const CV_OVERRIDE;
};
```

如下为 SimpleBlobDetector 类的定义,这个类实现了一个从图像中提取斑点的简单算法:
```cpp
class CV_EXPORTS_W SimpleBlobDetector : public Feature2D
{
public:
  struct CV_EXPORTS_W_SIMPLE Params
  {
      CV_WRAP Params();
      CV_PROP_RW float thresholdStep;
      CV_PROP_RW float minThreshold;
      CV_PROP_RW float maxThreshold;
      CV_PROP_RW size_t minRepeatability;
      CV_PROP_RW float minDistBetweenBlobs;

      CV_PROP_RW bool filterByColor;
      CV_PROP_RW uchar blobColor;
```

```cpp
    CV_PROP_RW bool filterByArea;
    CV_PROP_RW float minArea, maxArea;

    CV_PROP_RW bool filterByCircularity;
    CV_PROP_RW float minCircularity, maxCircularity;

    CV_PROP_RW bool filterByInertia;
    CV_PROP_RW float minInertiaRatio, maxInertiaRatio;

    CV_PROP_RW bool filterByConvexity;
    CV_PROP_RW float minConvexity, maxConvexity;

    void read( const FileNode& fn );
    void write( FileStorage& fs ) const;
};

CV_WRAP static Ptr<SimpleBlobDetector>
    create(const SimpleBlobDetector::Params &parameters = SimpleBlobDetector::Params());
CV_WRAP virtual String getDefaultName() const CV_OVERRIDE;
};
```

如下为 KAZE 特征点检测和特征点提取的类（见案例 105）：

```cpp
class CV_EXPORTS_W KAZE : public Feature2D
{
public:
    enum DiffusivityType
    {
        DIFF_PM_G1 = 0,
        DIFF_PM_G2 = 1,
        DIFF_WEICKERT = 2,
        DIFF_CHARBONNIER = 3
    };

    /** KAZE 对象创建*/
    CV_WRAP static Ptr<KAZE> create(bool extended=false, bool upright=false,
                    float threshold = 0.001f, int nOctaves = 4, int nOctaveLayers = 4,
                    KAZE::DiffusivityType diffusivity = KAZE::DIFF_PM_G2);

    CV_WRAP virtual void setExtended(bool extended) = 0;
    CV_WRAP virtual bool getExtended() const = 0;

    CV_WRAP virtual void setUpright(bool upright) = 0;
    CV_WRAP virtual bool getUpright() const = 0;

    CV_WRAP virtual void setThreshold(double threshold) = 0;
    CV_WRAP virtual double getThreshold() const = 0;
```

```cpp
    CV_WRAP virtual void setNOctaves(int octaves) = 0;
    CV_WRAP virtual int getNOctaves() const = 0;

    CV_WRAP virtual void setNOctaveLayers(int octaveLayers) = 0;
    CV_WRAP virtual int getNOctaveLayers() const = 0;

    CV_WRAP virtual void setDiffusivity(KAZE::DiffusivityType diff) = 0;
    CV_WRAP virtual KAZE::DiffusivityType getDiffusivity() const = 0;
    CV_WRAP virtual String getDefaultName() const CV_OVERRIDE;
};
```

如下为 AKAZE 特征点检测和特征点检测算法的类（见案例 106）：

```cpp
class CV_EXPORTS_W AKAZE : public Feature2D
{
public:
    // AKAZE 描述子类型
    enum DescriptorType
    {
        DESCRIPTOR_KAZE_UPRIGHT = 2,
        DESCRIPTOR_KAZE = 3,
        DESCRIPTOR_MLDB_UPRIGHT = 4,
        DESCRIPTOR_MLDB = 5
    };

    /** AKAZE 对象创建*/
    CV_WRAP static Ptr<AKAZE> create(
                        AKAZE::DescriptorType descriptor_type = AKAZE::DESCRIPTOR_MLDB,
                        int descriptor_size = 0, int descriptor_channels = 3,
                        float threshold = 0.001f, int nOctaves = 4,
                        int nOctaveLayers = 4,
                        KAZE::DiffusivityType diffusivity = KAZE::DIFF_PM_G2);

    CV_WRAP virtual void setDescriptorType(AKAZE::DescriptorType dtype) = 0;
    CV_WRAP virtual AKAZE::DescriptorType getDescriptorType() const = 0;

    CV_WRAP virtual void setDescriptorSize(int dsize) = 0;
    CV_WRAP virtual int getDescriptorSize() const = 0;

    CV_WRAP virtual void setDescriptorChannels(int dch) = 0;
    CV_WRAP virtual int getDescriptorChannels() const = 0;

    CV_WRAP virtual void setThreshold(double threshold) = 0;
    CV_WRAP virtual double getThreshold() const = 0;

    CV_WRAP virtual void setNOctaves(int octaves) = 0;
    CV_WRAP virtual int getNOctaves() const = 0;
```

```cpp
    CV_WRAP virtual void setNOctaveLayers(int octaveLayers) = 0;
    CV_WRAP virtual int getNOctaveLayers() const = 0;

    CV_WRAP virtual void setDiffusivity(KAZE::DiffusivityType diff) = 0;
    CV_WRAP virtual KAZE::DiffusivityType getDiffusivity() const = 0;
    CV_WRAP virtual String getDefaultName() const CV_OVERRIDE;
};
```

如下为一些结构的定义：

```cpp
template<typename T>
struct CV_EXPORTS Accumulator
{
    typedef T Type;
};

template<> struct Accumulator<unsigned char>  { typedef float Type; };
template<> struct Accumulator<unsigned short> { typedef float Type; };
template<> struct Accumulator<char>  { typedef float Type; };
template<> struct Accumulator<short> { typedef float Type; };

template<class T>
struct CV_EXPORTS SL2
{
    static const NormTypes normType = NORM_L2SQR;
    typedef T ValueType;
    typedef typename Accumulator<T>::Type ResultType;

    ResultType operator()( const T* a, const T* b, int size ) const
    {
        return normL2Sqr<ValueType, ResultType>(a, b, size);
    }
};

template<class T>
struct L2
{
    static const NormTypes normType = NORM_L2;
    typedef T ValueType;
    typedef typename Accumulator<T>::Type ResultType;

    ResultType operator()( const T* a, const T* b, int size ) const
    {
        return (ResultType)std::sqrt((double)normL2Sqr<ValueType, ResultType>(a, b, size));
    }
};

/** 曼哈顿距离 */
```

```
template<class T>
struct L1
{
    static const NormTypes normType = NORM_L1;
    typedef T ValueType;
    typedef typename Accumulator<T>::Type ResultType;

    ResultType operator()( const T* a, const T* b, int size ) const
    {
        return normL1<ValueType, ResultType>(a, b, size);
    }
};
```

如下为 DescriptorMatcher 类的定义，该类为特征点匹配抽象基类：

```
class CV_EXPORTS_W DescriptorMatcher : public Algorithm
{
public:
    enum MatcherType
    {
        FLANNBASED            = 1,
        BRUTEFORCE            = 2,
        BRUTEFORCE_L1         = 3,
        BRUTEFORCE_HAMMING    = 4,
        BRUTEFORCE_HAMMINGLUT = 5,
        BRUTEFORCE_SL2        = 6
    };

    virtual ~DescriptorMatcher();

    /** 添加描述符以训练 CPU 或 GPU 描述符集合*/
    CV_WRAP virtual void add( InputArrayOfArrays descriptors );
    CV_WRAP const std::vector<Mat>& getTrainDescriptors() const;
    CV_WRAP virtual void clear() CV_OVERRIDE;
    CV_WRAP virtual bool empty() const CV_OVERRIDE;
    CV_WRAP virtual bool isMaskSupported() const = 0;

    /** 训练描述符匹配器*/
    CV_WRAP virtual void train();

    /** 从查询集中查找每个描述符的最佳匹配*/
    CV_WRAP void match( InputArray queryDescriptors, InputArray trainDescriptors,
                CV_OUT std::vector<DMatch>& matches, InputArray mask=noArray() ) const;

    /** 从查询集中为每个描述符查找 k 个最佳匹配项*/
    CV_WRAP void knnMatch( InputArray queryDescriptors, InputArray trainDescriptors,
                CV_OUT std::vector<std::vector<DMatch> >& matches, int k,
```

```cpp
                          InputArray mask=noArray(), bool compactResult=false ) const;
/** 对于每个查询描述符，查找不超过指定距离的训练描述符*/
CV_WRAP void radiusMatch( InputArray queryDescriptors, InputArray trainDescriptors,
                          CV_OUT std::vector<std::vector<DMatch> >& matches,
                          float maxDistance, InputArray mask=noArray(),
                          bool compactResult=false ) const;

CV_WRAP void match( InputArray queryDescriptors, CV_OUT std::vector<DMatch>& matches,
                    InputArrayOfArrays masks=noArray() );

CV_WRAP void knnMatch( InputArray queryDescriptors,
                       CV_OUT std::vector<std::vector<DMatch> >& matches, int k,
                       InputArrayOfArrays masks=noArray(), bool compactResult=false );

CV_WRAP void radiusMatch( InputArray queryDescriptors,
                          CV_OUT std::vector<std::vector<DMatch> >& matches,
                          float maxDistance, InputArrayOfArrays masks=noArray(),
                          bool compactResult=false );

CV_WRAP void write( const String& fileName ) const
{
    FileStorage fs(fileName, FileStorage::WRITE);
    write(fs);
}

CV_WRAP void read( const String& fileName )
{
    FileStorage fs(fileName, FileStorage::READ);
    read(fs.root());
}

CV_WRAP virtual void read( const FileNode& ) CV_OVERRIDE;

virtual void write( FileStorage& ) const CV_OVERRIDE;

CV_WRAP virtual Ptr<DescriptorMatcher> clone( bool emptyTrainData=false ) const = 0;

/** 使用默认参数创建给定类型的描述符匹配器（使用默认构造函数）*/
CV_WRAP static Ptr<DescriptorMatcher> create( const String& descriptorMatcherType );

CV_WRAP static Ptr<DescriptorMatcher> create( const DescriptorMatcher::MatcherType&
matcherType );

CV_WRAP inline void write(const Ptr<FileStorage>& fs, const String& name = String()) const
{ Algorithm::write(fs, name); }
```

```
protected:
    class CV_EXPORTS DescriptorCollection
    {
    public:
        DescriptorCollection();
        DescriptorCollection( const DescriptorCollection& collection );
        virtual ~DescriptorCollection();
        void set( const std::vector<Mat>& descriptors );
        virtual void clear();

        const Mat& getDescriptors() const;
        const Mat getDescriptor( int imgIdx, int localDescIdx ) const;
        const Mat getDescriptor( int globalDescIdx ) const;
        void getLocalIdx(int globalDescIdx, int& imgIdx, int& localDescIdx) const;

        int size() const;

    protected:
        Mat mergedDescriptors;
        std::vector<int> startIdxs;
    };

    virtual void knnMatchImpl( InputArray queryDescriptors,
                               std::vector<std::vector<DMatch> >& matches, int k,
                               InputArrayOfArrays masks=noArray(),
                               bool compactResult=false ) = 0;
    virtual void radiusMatchImpl( InputArray queryDescriptors,
                                  std::vector<std::vector<DMatch> >& matches,
                                  float maxDistance, InputArrayOfArrays masks=noArray(),
                                  bool compactResult=false ) = 0;

    static bool isPossibleMatch( InputArray mask, int queryIdx, int trainIdx );
    static bool isMaskedOut( InputArrayOfArrays masks, int queryIdx );

    static Mat clone_op( Mat m ) { return m.clone(); }
    void checkMasks( InputArrayOfArrays masks, int queryDescriptorsCount ) const;

    std::vector<Mat> trainDescCollection;
    std::vector<UMat> utrainDescCollection;
};
```

如下为 BFMatcher 类的定义，该类为暴力描述符匹配器类（见案例 107）：

```
class CV_EXPORTS_W BFMatcher : public DescriptorMatcher
{
public:
    CV_WRAP BFMatcher( int normType=NORM_L2, bool crossCheck=false );
    virtual ~BFMatcher() {}
```

```cpp
    virtual bool isMaskSupported() const CV_OVERRIDE { return true; }
    /**创建 BFMatcher 对象*/
    CV_WRAP static Ptr<BFMatcher> create( int normType=NORM_L2, bool crossCheck=false ) ;

    virtual Ptr<DescriptorMatcher> clone( bool emptyTrainData=false ) const CV_OVERRIDE;
protected:
    virtual void knnMatchImpl( InputArray queryDescriptors,
                   std::vector<std::vector<DMatch> >& matches, int k,
                   InputArrayOfArrays masks=noArray(),
                   bool compactResult=false ) CV_OVERRIDE;
    virtual void radiusMatchImpl( InputArray queryDescriptors,
                   std::vector<std::vector<DMatch> >& matches, float maxDistance,
                   InputArrayOfArrays masks=noArray(),
                   bool compactResult=false ) CV_OVERRIDE;

    int normType;
    bool crossCheck;
};

#if defined(HAVE_OPENCV_FLANN) || defined(CV_DOXYGEN)
```

如下为 FlannBasedMatcher 类的定义,该类为基于 FLANN 的描述符匹配器类(见案例110):

```cpp
class CV_EXPORTS_W FlannBasedMatcher : public DescriptorMatcher
{
public:
    CV_WRAP FlannBasedMatcher(
            const Ptr<flann::IndexParams>& indexParams=makePtr<flann::KDTreeIndexParams>(),
            const Ptr<flann::SearchParams>& searchParams=makePtr<flann::SearchParams>() );

    virtual void add( InputArrayOfArrays descriptors ) CV_OVERRIDE;
    virtual void clear() CV_OVERRIDE;
    virtual void read( const FileNode& ) CV_OVERRIDE;
    virtual void write( FileStorage& ) const CV_OVERRIDE;
    virtual void train() CV_OVERRIDE;
    virtual bool isMaskSupported() const CV_OVERRIDE;

    CV_WRAP static Ptr<FlannBasedMatcher> create();

    virtual Ptr<DescriptorMatcher> clone( bool emptyTrainData=false ) const CV_OVERRIDE;
protected:
    static void convertToDMatches( const DescriptorCollection& descriptors,
                   const Mat& indices, const Mat& distances,
                   std::vector<std::vector<DMatch> >& matches );

    virtual void knnMatchImpl( InputArray queryDescriptors,
```

```cpp
                    std::vector<std::vector<DMatch> >& matches, int k,
                    InputArrayOfArrays masks=noArray(), bool compactResult=false )
CV_OVERRIDE;
    virtual void radiusMatchImpl( InputArray queryDescriptors,
                    std::vector<std::vector<DMatch> >& matches, float maxDistance,
                    InputArrayOfArrays masks=noArray(),
                    bool compactResult=false ) CV_OVERRIDE;

    Ptr<flann::IndexParams> indexParams;
    Ptr<flann::SearchParams> searchParams;
    Ptr<flann::Index> flannIndex;

    DescriptorCollection mergedDescriptors;
    int addedDescCount;
};

#endif
```

如下为与特征点匹配绘制相关的标志定义和函数定义：

```cpp
// 特征点匹配绘制标志
enum struct DrawMatchesFlags
{
  DEFAULT = 0,
  DRAW_OVER_OUTIMG = 1,
  NOT_DRAW_SINGLE_POINTS = 2,
  DRAW_RICH_KEYPOINTS = 4
};
CV_ENUM_FLAGS(DrawMatchesFlags)

/** 特征点绘制*/
CV_EXPORTS_W void drawKeypoints( InputArray image, const std::vector<KeyPoint>& keypoints,
                                 InputOutputArray outImage, const Scalar& color=Scalar::all(-1),
                                 DrawMatchesFlags flags=DrawMatchesFlags::DEFAULT );

/** 在两幅图像中绘制找到的特征点匹配*/
CV_EXPORTS_W void drawMatches( InputArray img1, const std::vector<KeyPoint>& keypoints1,
                InputArray img2, const std::vector<KeyPoint>& keypoints2,
                const std::vector<DMatch>& matches1to2, InputOutputArray outImg,
                const Scalar& matchColor=Scalar::all(-1),
                const Scalar& singlePointColor=Scalar::all(-1),
                const std::vector<char>& matchesMask=std::vector<char>(),
                DrawMatchesFlags flags=DrawMatchesFlags::DEFAULT );

CV_EXPORTS_AS(drawMatchesKnn) void drawMatches( InputArray img1,
                const std::vector<KeyPoint>& keypoints1, InputArray img2,
                const std::vector<KeyPoint>& keypoints2,
                const std::vector<std::vector<DMatch> >& matches1to2, InputOutputArray outImg,
```

```
                    const Scalar& matchColor=Scalar::all(-1),
                    const Scalar& singlePointColor=Scalar::all(-1),
                    const std::vector<std::vector<char> >&
matchesMask=std::vector<std::vector<char> >(),
                    DrawMatchesFlags flags=DrawMatchesFlags::DEFAULT );
```

如下为特征检测器和描述符匹配器评估函数的定义：

```
CV_EXPORTS void evaluateFeatureDetector( const Mat& img1, const Mat& img2, const Mat& H1to2,
                 std::vector<KeyPoint>* keypoints1, std::vector<KeyPoint>* keypoints2,
                 float& repeatability, int& correspCount,
                 const Ptr<FeatureDetector>& fdetector=Ptr<FeatureDetector>() );

CV_EXPORTS void computeRecallPrecisionCurve(
                 const std::vector<std::vector<DMatch> >& matches1to2,
                 const std::vector<std::vector<uchar> >& correctMatches1to2Mask,
                 std::vector<Point2f>& recallPrecisionCurve );

CV_EXPORTS float getRecall( const std::vector<Point2f>& recallPrecisionCurve,
                 float l_precision );
CV_EXPORTS int getNearestPoint( const std::vector<Point2f>& recallPrecisionCurve,
                 float l_precision );
```

如下为与词袋（BOW）训练器相关的类的定义：

```
class CV_EXPORTS_W BOWTrainer
{
public:
    BOWTrainer();
    virtual ~BOWTrainer();
    CV_WRAP void add( const Mat& descriptors );
    CV_WRAP const std::vector<Mat>& getDescriptors() const;
    CV_WRAP int descriptorsCount() const;
    CV_WRAP virtual void clear();
    CV_WRAP virtual Mat cluster() const = 0;
    CV_WRAP virtual Mat cluster( const Mat& descriptors ) const = 0;

protected:
    std::vector<Mat> descriptors;
    int size;
};

/** BOW 训练器类*/
class CV_EXPORTS_W BOWKMeansTrainer : public BOWTrainer
{
public:
    CV_WRAP BOWKMeansTrainer( int clusterCount, const TermCriteria& termcrit=TermCriteria(),
```

```cpp
                                      int attempts=3, int flags=KMEANS_PP_CENTERS );
    virtual ~BOWKMeansTrainer();

    CV_WRAP virtual Mat cluster() const CV_OVERRIDE;
    CV_WRAP virtual Mat cluster( const Mat& descriptors ) const CV_OVERRIDE;

protected:
    int clusterCount;
    TermCriteria termcrit;
    int attempts;
    int flags;
};

/** 基于 BOW 计算图像描述符类*/
class CV_EXPORTS_W BOWImgDescriptorExtractor
{
public:
    CV_WRAP BOWImgDescriptorExtractor( const Ptr<DescriptorExtractor>& dextractor,
                const Ptr<DescriptorMatcher>& dmatcher );

    BOWImgDescriptorExtractor( const Ptr<DescriptorMatcher>& dmatcher );
    virtual ~BOWImgDescriptorExtractor();

    CV_WRAP void setVocabulary( const Mat& vocabulary );
    CV_WRAP const Mat& getVocabulary() const;

    void compute( InputArray image, std::vector<KeyPoint>& keypoints,
                  OutputArray imgDescriptor,
                  std::vector<std::vector<int> >* pointIdxsOfClusters=0, Mat*
descriptors=0 );
    void compute( InputArray keypointDescriptors, OutputArray imgDescriptor,
                  std::vector<std::vector<int> >* pointIdxsOfClusters=0 );
    CV_WRAP_AS(compute) void compute2( const Mat& image, std::vector<KeyPoint>& keypoints,
                CV_OUT Mat& imgDescriptor )
    { compute(image,keypoints,imgDescriptor); }

    CV_WRAP int descriptorSize() const;
    CV_WRAP int descriptorType() const;

protected:
    Mat vocabulary;
    Ptr<DescriptorExtractor> dextractor;
    Ptr<DescriptorMatcher> dmatcher;
};
} /* namespace cv */

#endif
```

10.2 特征点检测

10.2.1 案例 101：SIFT 特征点检测

SIFT（Scale Invariant Feature Transform，尺度不变特征变换）算法在图像识别、图像检索等领域有着广泛应用，该算法申请了专利，因此，在 OpenCV 3.x 的很多版本中，需要安装库 opencv-contrib-python，并按如下方式创建 SIFT 对象：

```
sift = cv2.xfeatures2d.SIFT_create()
```

在 OpenCV 4.x 的较新版本（如 OpenCV 4.5）中采用如上方式调用会出现如下的报错：

```
cv::xfeatures2d::SIFT_create DEPRECATED: cv.xfeatures2d.SIFT_create() is deprecated due SIFT tranfer to the main repository.
```

报错显示 SIFT 算法已经被移植到 OpenCV 主仓库中，这是因为该专利过了专利保护期，OpenCV 新版本可以通过如下方法调用：

```
sift = cv2.SIFT_create()
```

SIFT_create 函数的定义如下：

```
retval = SIFT_create(nfeatures=None, nOctaveLayers=None, contrastThreshold=None, edgeThreshold=None, sigma=None)
```

参数说明如下。

- nfeatures：保留的最优特征数量。
- nOctaveLayers：每个 octave 的层数。
- contrastThreshold：对比度阈值，用于在低对比度区域滤除弱特征，阈值越大，检测的特征越少。
- edgeThreshold：用于过滤疑似边缘的特征的阈值，该阈值越大，保留的特征越多。
- sigma：高斯模糊参数 sigma。
- retval：创建的 SIFT 对象（返回值）。

SIFT 算法类继承于 Feature2D 类（见 10.1 节）。Feature2D 中定义了特征点检测函数 detect、描述符计算函数 compute，以及特征点检测与描述符计算两个过程结合的函数 detectAndCompute。

detect 函数的定义如下：

```
keypoints = detect(self, image, mask=None)
```

参数说明如下。

- image：输入图像。

- mask：指定特征点检测的区域。
- keypoints：存储检测的特征点（返回值）。

compute 函数的定义如下：

```
keypoints, descriptors = compute(image, keypoints, descriptors=None)
```

参数说明如下。
- image：输入图像。
- keypoints：检测到的特征点（返回值）。
- descriptors：计算的特征描述符（返回值）。

detectAndCompute 函数的定义如下：

```
keypoints, descriptors = detectAndCompute(self, image, mask, descriptors=None,
useProvidedKeypoints=None)
```

参数说明如下。
- image：输入图像。
- keypoints：存储检测的特征点（返回值）。
- mask：指定特征点检测的区域。
- descriptors：计算的特征描述符（返回值）。
- useProvidedKeypoints：是否使用已有的特征点。

特征点检测完成后，需要进行特征点绘制，OpenCV 提供了用于特征点绘制的函数 drawKeypoints。该函数的定义如下：

```
outImage = drawKeypoints(image, keypoints, outImage, color=None, flags=None)
```

参数说明如下。
- image：输入图像。
- keypoints：检测到的特征点。
- outImage：输出图像（返回值）。
- color：绘制特征点的颜色。
- flags：绘制方式，由 DrawMatchesFlags 定义（见 10.1 节）。

使用 SIFT 算法进行特征点检测的案例代码如下：

```python
import cv2

#输入图像读取
src = cv2.imread("src.jpg")
if src is None:
    print('Could not open or find the images!')
    exit(0)
gray = cv2.cvtColor(src, cv2.COLOR_BGR2GRAY)
```

```
#SIFT 算法特征点检测
sift = cv2.SIFT_create()
key_point = sift.detect(gray, None)

#绘制特征点
draw_keypoints = cv2.drawKeypoints(src, key_point, src, color=(0, 0, 255))
cv2.imshow("SIFT Keypoints", draw_keypoints)
cv2.waitKey(0)
cv2.destroyAllWindows()
```

本案例使用的输入图像如图 4.9 所示，特征点检测与绘制结果如图 10.1 所示。

图 10.1

10.2.2　案例 102：SURF 特征点检测

SURF（Speeded Up Robust Feature，加速稳健特征）算法也是一种尺度不变特征检测算法，是 SIFT 算法的加速版本，该算法还在专利保护期内，因此该算法的使用需要安装库 opencv-contrib-python。

opencv-contrib-python 库安装后，SURF 算法的引用需要通过 cv2.xfeatures2d_SURF.create()函数创建 SURF 对象。create 函数的定义如下：

```
retval = create(hessianThreshold=None, nOctaves=None, nOctaveLayers=None, extended=None, upright=None)
```

参数说明如下。

- hessianThreshold：hessian 特征点检测器使用的阈值。
- nOctaves：金字塔 octave 的数量。
- nOctaveLayers：每个 octave 的层数。
- extended：是否增强描述子。
- upright：是否右上方向。
- retval：创建的 SURF 对象（返回值）。

使用 SURF 算法进行特征点检测的案例代码如下：

```python
import cv2
import numpy as np

#输入图像读取
src = cv2.imread('src.jpg', cv2.IMREAD_GRAYSCALE)
if src is None:
    print('Could not open or find the image! ')
    exit(0)

#SURF 算法特征点检测
minHessian = 400
detector = cv2.xfeatures2d_SURF.create(hessianThreshold=minHessian)
keypoints = detector.detect(src)

#特征点绘制
img_keypoints = np.empty((src.shape[0], src.shape[1], 3), dtype=np.uint8)
cv2.drawKeypoints(src, keypoints, img_keypoints)

#保存绘制结果
cv2.imwrite('SURF Keypoints.jpg', img_keypoints)
```

SURF 算法特征点检测绘制结果如图 10.2 所示。

图 10.2

提示：opencv-contrib-python 4.x 版本中没有发现 SURF 算法的调用，SURF 算法案例使用的 Python 版本为 Python 3.7，使用的 opencv-python 与 opencv-contrib-python 版本为 OpenCV 3.4.2.17，本案例在 Ubuntu 系统中运行完成。

10.2.3 案例 103：BRISK 特征点检测

BRISK（Binary Robust Invariant Scalable Keypoints，二进制鲁棒不变可伸缩关键点）算法是一种重要的特征提取算法，具有较好的尺度不变性、鲁棒性。OpenCV 中提供了 BRISK 算法的封装类 BRISK（见 10.1 节），在 Python 中调用该算法需要通过 BRISK_create 函数创建对象。BRISK_create 函数的定义如下：

```
retval = BRISK_create(thresh=None, octaves=None, patternScale=None)
```

参数说明如下。
- thresh：AGAST 检测得分阈值。
- octaves：检测层。
- patternScale：对特征点的邻域进行采样的模式的比例。
- retval：创建的 BRISK 对象（返回值）。

使用 BRISK 算法进行特征点检测的案例代码如下：

```
import cv2
import numpy as np

#图像读取
src = cv2.imread("src.jpg")
if src is None:
    print('Could not open or find the images!')
    exit(0)
gray = cv2.cvtColor(src, cv2.COLOR_BGR2GRAY)

#BRISK 算法对象创建
brisk = cv2.BRISK_create()
kpts = brisk.detect(gray, None)

#绘制特征点
img_keypoints = np.empty((src.shape[0], src.shape[1], 3), dtype=np.uint8)
cv2.drawKeypoints(src, kpts, img_keypoints)

#显示检测到的特征点
cv2.imshow('BRISK Keypoints', img_keypoints)
cv2.waitKey(0)
cv2.destroyAllWindows()
```

BRISK 算法特征点检测绘制结果如图 10.3 所示。

图 10.3

10.2.4 案例 104：ORB 特征点检测

ORB（Oriented Fast and Rotated Brief）算法使用 FAST 算法做特征点检测，使用 BRIEF 算法计算特征点的描述符。OpenCV 中提供了 ORB 算法的封装类 ORB（见 10.1 节），在 Python 中调用该算法需要通过 ORB_create 函数创建对象。ORB_create 函数的定义如下：

```
retval = ORB_create(nfeatures=None, scaleFactor=None, nlevels=None, edgeThreshold=None, firstLevel=None, WTA_K=None, scoreType=None, patchSize=None, fastThreshold=None)
```

参数说明如下。

- nfeatures：保留特征的最大数量。
- scaleFactor：金字塔抽取比，该值大于 1，如果该值为 2，则表示经典金字塔。
- nlevels：金字塔级数。
- edgeThreshold：未检测到特征的边界大小。
- firstLevel：原始图像置于金字塔中的级数。
- WTA_K：生成定向 BRIEF 描述符的每个元素的点数。
- scoreType：默认值为 HARRIS_SCORE，意思是使用 Harris 算法进行特征排序。
- patchSize：定向 BRIEF 描述符使用的 patch 大小。
- fastThreshold：FAST 算法阈值。
- retval：创建的 ORB 对象（返回值）。

使用 ORB 算法进行特征点检测的案例代码如下：

```python
import cv2
import numpy as np

#图像读取
src = cv2.imread("src.jpg")
if src is None:
    print('Could not open or find the images!')
```

```
    exit(0)
gray = cv2.cvtColor(src, cv2.COLOR_BGR2GRAY)

#ORB 算法对象创建
orb = cv2.ORB_create()
kpts = orb.detect(gray, None)

#绘制特征点
img_keypoints = np.empty((src.shape[0], src.shape[1], 3), dtype=np.uint8)
cv2.drawKeypoints(src, kpts , img_keypoints, color=(0, 0, 255))

#显示检测到的特征点
cv2.imshow('ORB Keypoints', img_keypoints)
cv2.waitKey(0)
cv2.destroyAllWindows()
```

ORB 算法特征点检测绘制结果如图 10.4 所示。

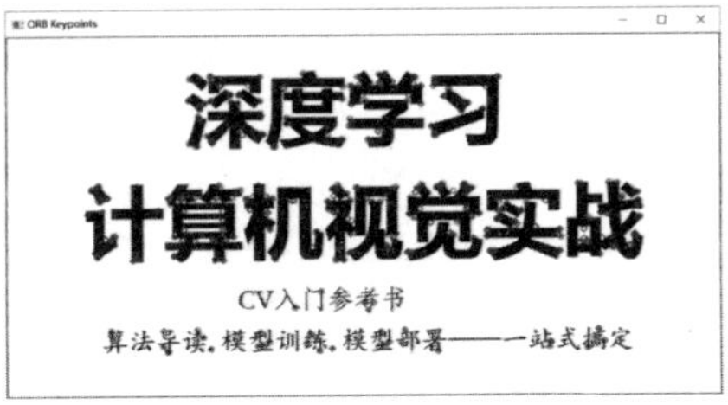

图 10.4

10.2.5　案例 105：KAZE 特征点检测

KAZE 是一种比 SIFT 更稳定的特征点检测算法，OpenCV 中提供了 KAZE 算法的封装类 KAZE（见 10.1 节），在 Python 中调用该算法需要通过 KAZE_create 函数创建对象。KAZE_create 函数的定义如下：

```
retval = KAZE_create(extended=None, upright=None, threshold=None, nOctaves=None,
nOctaveLayers=None, diffusivity=None)
```

参数说明如下。
- extended：设置为启用扩展（128 字节）描述符的提取。
- upright：设置为启用垂直描述符。
- threshold：检测器响应阈值，用来确定是否接受一个点为特征点。
- nOctaves：图像的最大层演变。
- nOctaveLayers：每个级别默认的子级数量。
- diffusivity：由 DiffusivityType 定义的类型（见 10.1 节）。

- retval：创建的 KAZE 对象（返回值）。

使用 KAZE 算法进行特征点检测的案例代码如下：

```python
import cv2
import numpy as np

#输入图像读取
src = cv2.imread("src.jpg")
if src is None:
    print('Could not open or find the images!')
    exit(0)
gray = cv2.cvtColor(src, cv2.COLOR_BGR2GRAY)

#KAZE 算法对象创建
kaze = cv2.KAZE_create()
kpts = kaze.detect(gray, None)

#绘制特征点
img_keypoints = np.empty((src.shape[0], src.shape[1], 3), dtype=np.uint8)
cv2.drawKeypoints(src, kpts, img_keypoints, color=(0, 0, 255))

#显示检测到的特征点
cv2.imshow('KAZE Keypoints', img_keypoints)
cv2.waitKey(0)
cv2.destroyAllWindows()
```

KAZE 算法特征点检测绘制结果如图 10.5 所示。

图 10.5

10.2.6 案例 106：AKAZE 特征点检测

AKAZE 是对 KAZE 算法的改进，OpenCV 中提供了 AKAZE 算法的封装类 AKAZE（见 10.1 节），在 Python 中调用该算法需要通过 AKAZE_create 函数创建对象。AKAZE_create 函数的定义如下：

```
retval = AKAZE_create(descriptor_type=None, descriptor_size=None, descriptor_channels=None,
threshold=None, nOctaves=None, nOctaveLayers=None, diffusivity=None)
```

参数说明如下：

- descriptor_type：提取的描述符类型，由 DescriptorType 定义（见 10.1 节）。
- descriptor_size：描述符的大小。
- descriptor_channels：描述符通道数。
- threshold：检测器响应阈值，用来确定是否接受一个点为特征点。
- nOctaves：图像的最大层演变。
- nOctaveLayers：每个级别默认的子级数量。
- diffusivity：由 DiffusivityType 定义的类型（见 10.1 节）。
- retval：创建的 AKAZE 对象（返回值）。

使用 AKAZE 算法进行特征点检测的案例代码如下：

```python
import cv2
import numpy as np

#输入图像读取
src = cv2.imread("src.jpg")
if src is None:
    print('Could not open or find the images!')
    exit(0)
gray = cv2.cvtColor(src, cv2.COLOR_BGR2GRAY)

#AKAZE 算法对象创建
akaze = cv2.AKAZE_create()
kpts, desc = akaze.detectAndCompute(gray, None)

#绘制特征点
img_keypoints = np.empty((src.shape[0], src.shape[1], 3), dtype=np.uint8)
cv2.drawKeypoints(src, kpts, img_keypoints, color=(0, 0, 255))

#显示检测到的特征点
cv2.imshow('AKAZE Keypoints', img_keypoints)
cv2.waitKey(0)
cv2.destroyAllWindows()
```

AKAZE 算法特征点检测绘制结果如图 10.6 所示。

图 10.6

10.2.7 案例107：AGAST 特征点检测

OpenCV 中提供了 AGAST 算法的封装类 AgastFeatureDetector（见 10.1 节），在 Python 中调用该算法需要通过 AgastFeatureDetector_create 函数创建对象。AgastFeatureDetector_create 函数的定义如下：

```
retval = AgastFeatureDetector_create(threshold=None, nonmaxSuppression=None, type=None)
```

参数说明如下。

- threshold：中心像素和该像素周围圆区域内的像素强度差阈值。
- nonmaxSuppression：若该值为 true，则非极大值抑制被应用于检测角点。
- type：3 种邻域之一，由 DetectorType 定义（见 10.1 节）。
- retval：创建的 AGAST 对象（返回值）。

使用 AGAST 算法进行特征点检测的案例代码如下：

```python
import cv2
import numpy as np

#图像读取
src = cv2.imread("src.jpg")
if src is None:
    print('Could not open or find the images!')
    exit(0)
gray = cv2.cvtColor(src, cv2.COLOR_BGR2GRAY)

#AGAST 算法对象创建
agast = cv2.AgastFeatureDetector_create()
kpts1 = agast.detect(gray, None)

#绘制特征点
img_keypoints = np.empty((src.shape[0], src.shape[1], 3), dtype=np.uint8)
cv2.drawKeypoints(src, kpts1, img_keypoints, color=(0, 0, 255))

#显示检测到的特征点
cv2.imshow('AGAST Keypoints', img_keypoints)
cv2.waitKey(0)
cv2.destroyAllWindows()
```

AGAST 算法特征点检测绘制结果如图 10.7 所示。

图 10.7

10.2.8　案例 108：FAST 特征点检测

OpenCV 中提供了 FAST 算法的封装类 FastFeatureDetector（见 10.1 节），在 Python 中调用该算法需要通过 FastFeatureDetector_create 函数创建对象，其定义如下：

```
retval = FastFeatureDetector_create(threshold=None, nonmaxSuppression=None, type=None)
```

参数说明如下。

- threshold：中心像素和该像素周围圆区域内的像素强度差阈值。
- nonmaxSuppression：若该值为 true，则非极大值抑制被应用于检测角点。
- type：3 种邻域之一，由 DetectorType 定义（见 10.1 节）。
- retval：创建的 FAST 对象（返回值）。

使用 FAST 算法进行特征点检测的案例代码如下：

```
import cv2
import numpy as np

#输入图像读取
src = cv2.imread("src.jpg")
if src is None:
    print('Could not open or find the images!')
    exit(0)
gray = cv2.cvtColor(src, cv2.COLOR_BGR2GRAY)

#FAST 算法对象创建
fast = cv2.FastFeatureDetector_create()
kpts1 = fast.detect(src, None)

#绘制特征点
img_keypoints = np.empty((src.shape[0], src.shape[1], 3), dtype=np.uint8)
cv2.drawKeypoints(src, kpts1, img_keypoints, color=(0,0,255))

#显示检测到的特征点
```

```
cv2.imshow('FAST Keypoints', img_keypoints)
cv2.waitKey(0)
cv2.destroyAllWindows()
```

FAST 算法特征点检测绘制结果如图 10.8 所示。

图 10.8

10.3 特征点匹配

10.3.1 案例 109：Brute-Force 特征点匹配

Brute-Force 匹配算法的原理较简单，该算法选取第一个描述符集合中的一个描述子，并将第二个描述符集合中所有的特征描述符与其进行距离计算，选取距离最近的进行匹配。

OpenCV 中提供了描述符匹配的封装类 DescriptorMatcher（见 10.1 节），在 Python 中可通过调用 DescriptorMatcher_create 函数创建匹配器对象，该函数的定义如下：

```
retval = DescriptorMatcher_create(descriptorMatcherType)
```

参数说明如下。

- descriptorMatcherType：描述符匹配器类型，由 MatcherType 定义（见 10.1 节）。
- retval：创建的匹配器对象（返回值）。

将 descriptorMatcherType 设置为 DescriptorMatcher_BRUTEFORCE，表示选用 Brute-Force 特征点匹配算法，案例代码如下：

```
import cv2
import numpy as np

img1 = cv2.imread('book1.jpg', cv2.IMREAD_GRAYSCALE)
img2 = cv2.imread('book2.jpg', cv2.IMREAD_GRAYSCALE)
if img1 is None or img2 is None:
    print('Could not open or find the images!')
    exit(0)
```

```
#使用SURF算法检测特征点并计算描述符
minHessian = 1500
detector = cv2.xfeatures2d.SURF_create(hessianThreshold=minHessian)
keypoints1, descriptors1 = detector.detectAndCompute(img1, None)
keypoints2, descriptors2 = detector.detectAndCompute(img2, None)

#使用Brute-Force匹配算法进行描述符匹配
matcher = cv2.DescriptorMatcher_create(cv2.DescriptorMatcher_BRUTEFORCE)
matches = matcher.match(descriptors1, descriptors2)

#匹配绘制
img_matches = np.empty((max(img1.shape[0], img2.shape[0]), img1.shape[1]+img2.shape[1], 3), dtype=np.uint8)
cv2.drawMatches(img1, keypoints1, img2, keypoints2, matches, img_matches)
#绘制结果保存
cv2.imwrite('BruteForce.jpg', img_matches)
```

本案例中参与匹配的两幅图像如图10.9所示。

图 10.9

本案例使用 SURF 算法进行特征点检测并计算描述符，使用 Brute-Force 匹配算法进行描述符匹配，匹配结果如图 10.10 所示。

图 10.10

10.3.2 案例 110：FLANN 特征点匹配

FLANN（Fast Library for Approximate Nearest Neighbors，快速近似最近邻搜索库）匹配算法实现了比 Brute-Force 算法更加快速高效的匹配，在 DescriptorMatcher_create 算法中传入参数 DescriptorMatcher_FLANNBASED 即可创建该算法的匹配器对象，案例代码如下：

```
import cv2
import numpy as np

img1 = cv2.imread('book2.jpg', cv2.IMREAD_GRAYSCALE)
img2 = cv2.imread('book1.jpg', cv2.IMREAD_GRAYSCALE)
if img1 is None or img2 is None:
    print('Could not open or find the images!')
    exit(0)

#使用 SURF 检测器检测特征点，计算描述符
minHessian = 400
detector = cv2.xfeatures2d_SURF.create(hessianThreshold=minHessian)
keypoints1, descriptors1 = detector.detectAndCompute(img1, None)
keypoints2, descriptors2 = detector.detectAndCompute(img2, None)

#使用 FLANN 匹配器匹配描述符
matcher = cv2.DescriptorMatcher_create(cv2.DescriptorMatcher_FLANNBASED)
knn_matches = matcher.knnMatch(descriptors1, descriptors2, 2)

#使用 Lowe 比率测试过滤匹配
ratio_thresh = 0.7
good_matches = []
for m,n in knn_matches:
    if m.distance < ratio_thresh * n.distance:
        good_matches.append(m)

#匹配绘制
img_matches = np.empty((max(img1.shape[0], img2.shape[0]), img1.shape[1]+img2.shape[1], 3), dtype=np.uint8)
cv2.drawMatches(img1, keypoints1, img2, keypoints2, good_matches, img_matches,
flags=cv2.DrawMatchesFlags_NOT_DRAW_SINGLE_POINTS)
#显示检测到的匹配
cv2.imwrite('FlannMatches.jpg', img_matches)
```

匹配结果如图 10.11 所示。

图 10.11

10.4 进阶必备：特征点检测算法概述

特征点（或称关键点）是图像的信息所在，包括角点、斑点等，它的研究在目标跟踪和图像识别领域有重要作用。目前有大量成熟的特征点检测算法，在 10.2 节中，介绍了它们的使用案例，本节对其中常用的算法做一个简单介绍。

1. SIFT

SIFT 由 David Lowe 在 2004 年提出，是一种局部特征检测算法，该算法具有良好的尺度不变性和方向不变性，旋转图像、改变图像亮度或拍摄视角依然能够得到良好的检测效果。

SIFT 算法进行特征检测主要包括以下 4 步。

（1）构建尺度空间，检测 DOG 尺度空间极值点。

（2）特征点过滤并精确定位。

（3）为特征点分配 128 个方向参数。

（4）生成特征点描述子。

读者可以参考论文 *Distinctive Image Features from Scale-Invariant Keypoints* 来深入了解 SIFT 特征检测算法。

2. SURF

SURF 由 Herbert Bay 等最初在 2006 年提出，该算法是 SIFT 算法的加速版，提升了算法的性能，为算法在实时场景中应用提供了可能。

SURF 算法进行特征检测主要包括以下 5 步。

（1）构建 Hessian 矩阵，生成所有的兴趣点，用于特征提取。

（2）构建尺度空间。

（3）特征点初步定位并过滤，生成稳定的特征点。

(4)基于 Harr 小波特征分配特征点主方向。

(5)生成特征点描述子。

读者可以参考论文 *SURF: Speeded Up Robust Features* 来深入研究 SURF 算法。

3. BRISK

BRISK 在 2011 年被提出,该算法主要基于 FAST 算法查找特征,BRISK 也是一种二进制的特征描述算子。BRISK 算法具有较好的旋转不变性、尺度不变性、鲁棒性。

读者可以阅读论文 *BRISK: Binary Robust Invariant Scalable Keypoints* 来深入了解 BRISK 算法。

4. ORB

ORB 算法具有比 SIFT 和 SURF 算法更快的速度,该算法主要基于 BRIEF 算法。ORB 算法通过方向计算增强了 BRIEF 描述符,方向计算是为了让 ORB 特征具有旋转不变性。

ORB 算法的深入学习可以参考论文 *ORB: an efficient alternative to SIFT or SURF*。

5. FAST

FAST(Features From Accelerated Segment Test,加速分段测试的特征)算法由 Rosten 和 Drummond 在 2006 年提出,该算法设置一个合理的阈值 t,将候选特征点 p 的亮度 I 与其包围圈内的点集亮度做比较,如果在包围圈内有 n(常用 9 或 12)个连续的像素点的像素值比 $I+t$ 大或比 $I+t$ 小,则认为该点是一个角点(特征点),如图 10.12 所示。

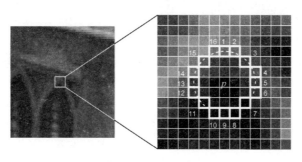

图 10.12

如图 10.12 所示,包围圈中有 16 个检测点,高亮显示的正方形是角点检测中使用的像素。p 处的像素是候选角点的中心。通过 12 个连续像素的虚线表示弧,这些检测点的像素与 p 点的像素相比超过阈值,因此可以确定 p 点为角点位置。

读者可以参考论文 *Machine learning for high-speed corner detection* 来深入了解 FAST 算法。

第 11 章

相机标定与三维重建模块 calib3d

相机在生活中发挥着重要的作用,由于镜头设计和镜片的光学性能,导致每个相机的镜头都存在不同程度的畸变,畸变影响照片的质量。因此,在机器视觉中,需要通过求出相机的内、外参数及畸变参数,用来对图像畸变进行矫正或对获得的图像重构三维场景,参数的求解过程称为相机标定。OpenCV 中的 calib3d 模块主要用于相机标定、三维重建等。

11.1 模块导读

OpenCV 中的 calib3d 模块封装了相机标定与三维重建算法,引入该模块需要包含头文件"opencv2/calib3d.hpp",通过该头文件,读者可以了解模块封装的算法。该头文件的定义如下:

```
#ifndef OPENCV_CALIB3D_HPP
#define OPENCV_CALIB3D_HPP
#include "opencv2/core.hpp"
#include "opencv2/features2d.hpp"
#include "opencv2/core/affine.hpp"

namespace cv
{
```

如下为与相机标定相关的标志的定义:

```
//! 鲁棒估计算法的类型
enum { LMEDS                   = 4,
       RANSAC                  = 8,
```

```
        RHO                  = 16,
        USAC_DEFAULT         = 32,
        USAC_PARALLEL        = 33,
        USAC_FM_8PTS         = 34,
        USAC_FAST            = 35,
        USAC_ACCURATE        = 36,
        USAC_PROSAC          = 37,
        USAC_MAGSAC          = 38
    };

enum SolvePnPMethod {
    SOLVEPNP_ITERATIVE   = 0,
    SOLVEPNP_EPNP        = 1,
    SOLVEPNP_P3P         = 2,
    SOLVEPNP_DLS         = 3,
    SOLVEPNP_UPNP        = 4,
    SOLVEPNP_AP3P        = 5,
    SOLVEPNP_IPPE        = 6,
    SOLVEPNP_IPPE_SQUARE = 7,
#ifndef CV_DOXYGEN
    SOLVEPNP_MAX_COUNT
#endif
};

enum { CALIB_CB_ADAPTIVE_THRESH = 1,
       CALIB_CB_NORMALIZE_IMAGE = 2,
       CALIB_CB_FILTER_QUADS    = 4,
       CALIB_CB_FAST_CHECK      = 8,
       CALIB_CB_EXHAUSTIVE      = 16,
       CALIB_CB_ACCURACY        = 32,
       CALIB_CB_LARGER          = 64,
       CALIB_CB_MARKER          = 128
     };

enum { CALIB_CB_SYMMETRIC_GRID  = 1,
       CALIB_CB_ASYMMETRIC_GRID = 2,
       CALIB_CB_CLUSTERING      = 4
     };

enum { CALIB_NINTRINSIC            = 18,
       CALIB_USE_INTRINSIC_GUESS   = 0x00001,
       CALIB_FIX_ASPECT_RATIO      = 0x00002,
       CALIB_FIX_PRINCIPAL_POINT   = 0x00004,
       CALIB_ZERO_TANGENT_DIST     = 0x00008,
       CALIB_FIX_FOCAL_LENGTH      = 0x00010,
       CALIB_FIX_K1                = 0x00020,
       CALIB_FIX_K2                = 0x00040,
```

```
    CALIB_FIX_K3                  = 0x00080,
    CALIB_FIX_K4                  = 0x00800,
    CALIB_FIX_K5                  = 0x01000,
    CALIB_FIX_K6                  = 0x02000,
    CALIB_RATIONAL_MODEL          = 0x04000,
    CALIB_THIN_PRISM_MODEL        = 0x08000,
    CALIB_FIX_S1_S2_S3_S4         = 0x10000,
    CALIB_TILTED_MODEL            = 0x40000,
    CALIB_FIX_TAUX_TAUY           = 0x80000,
    CALIB_USE_QR                  = 0x100000, //!< 用 QR 代替 SVD 分解求解，更快但可能不精确
    CALIB_FIX_TANGENT_DIST        = 0x200000,
    CALIB_FIX_INTRINSIC           = 0x00100,
    CALIB_SAME_FOCAL_LENGTH       = 0x00200,
    CALIB_ZERO_DISPARITY          = 0x00400,
    CALIB_USE_LU                  = (1 << 17), //!< 用 LU 代替 SVD 分解求解，速度快但可能不够精确
    CALIB_USE_EXTRINSIC_GUESS     = (1 << 22)
};

//! 求基矩阵的算法
enum { FM_7POINT = 1, //!< 7 点算法
       FM_8POINT = 2, //!< 8 点算法
       FM_LMEDS  = 4, //!< 最小中值算法
       FM_RANSAC = 8  //!< RANSAC 算法
     };

enum HandEyeCalibrationMethod
{
    CALIB_HAND_EYE_TSAI       = 0,
    CALIB_HAND_EYE_PARK       = 1,
    CALIB_HAND_EYE_HORAUD     = 2,
    CALIB_HAND_EYE_ANDREFF    = 3,
    CALIB_HAND_EYE_DANIILIDIS = 4
};

enum RobotWorldHandEyeCalibrationMethod
{
    CALIB_ROBOT_WORLD_HAND_EYE_SHAH = 0,
    CALIB_ROBOT_WORLD_HAND_EYE_LI   = 1
};

enum SamplingMethod {SAMPLING_UNIFORM,
                     SAMPLING_PROGRESSIVE_NAPSAC,
                     SAMPLING_NAPSAC,
                     SAMPLING_PROSAC };
enum LocalOptimMethod {LOCAL_OPTIM_NULL,
                       LOCAL_OPTIM_INNER_LO,
```

```
                LOCAL_OPTIM_INNER_AND_ITER_LO,
                LOCAL_OPTIM_GC, LOCAL_OPTIM_SIGMA};
enum ScoreMethod {SCORE_METHOD_RANSAC,
                SCORE_METHOD_MSAC,
                SCORE_METHOD_MAGSAC,
                SCORE_METHOD_LMEDS};
enum NeighborSearchMethod { NEIGH_FLANN_KNN,
                NEIGH_GRID,
                NEIGH_FLANN_RADIUS };

struct CV_EXPORTS_W_SIMPLE UsacParams
{
    double confidence          = 0.99;
    bool isParallel            = false;
    int loIterations           = 5;
    LocalOptimMethod loMethod  = LocalOptimMethod::LOCAL_OPTIM_INNER_LO;
    int loSampleSize           = 14;
    int maxIterations          = 5000;
    NeighborSearchMethod neighborsSearch  = NeighborSearchMethod::NEIGH_GRID;
    int randomGeneratorState   = 0;
    SamplingMethod sampler     = SamplingMethod::SAMPLING_UNIFORM;
    ScoreMethod score          = ScoreMethod::SCORE_METHOD_MSAC;
    double threshold           = 1.5;
};

/** 将旋转矩阵转换为旋转向量，反之亦然*/
CV_EXPORTS_W void Rodrigues( InputArray src, OutputArray dst, OutputArray jacobian = noArray() );
```

如下为 LMSolver 类的定义，该类为 Levenberg-Marquardt 分解器：

```
class CV_EXPORTS LMSolver : public Algorithm
{
public:
    class CV_EXPORTS Callback
    {
    public:
        virtual ~Callback() {}
        /** 计算指定参数向量的误差和雅可比矩阵*/
        virtual bool compute(InputArray param,
            OutputArray err,
            OutputArray J) const = 0;
    };

    /** 以传递的参数向量为起点运行 Levenberg-Marquardt 算法*/
    virtual int run(InputOutputArray param) const = 0;
    virtual void setMaxIters(int maxIters) = 0;
    virtual int getMaxIters() const = 0;
```

```cpp
        static Ptr<LMSolver> create(const Ptr<LMSolver::Callback>& cb,
                int maxIters);
        static Ptr<LMSolver> create(const Ptr<LMSolver::Callback>& cb,
                int maxIters, double eps);
};
```

findHomography 函数用于生成在两个平面之间查找透视变换的矩阵:

```cpp
CV_EXPORTS_W Mat findHomography( InputArray srcPoints,
        InputArray dstPoints,
        int method = 0,
        double ransacReprojThreshold = 3,
        OutputArray mask=noArray(),
        const int maxIters = 2000,
        const double confidence = 0.995);
CV_EXPORTS Mat findHomography( InputArray srcPoints,
        InputArray dstPoints,
        OutputArray mask,
        int method = 0,
        double ransacReprojThreshold = 3 );
CV_EXPORTS_W Mat findHomography(InputArray srcPoints,
        InputArray dstPoints,
        OutputArray mask,
        const UsacParams &params);
```

如下为与矩阵运算相关的函数定义:

```cpp
/** 计算 3×3 的 RQ 分解矩阵*/
CV_EXPORTS_W Vec3d RQDecomp3x3( InputArray src,
        OutputArray mtxR,
        OutputArray mtxQ,
        OutputArray Qx = noArray(),
        OutputArray Qy = noArray(),
        OutputArray Qz = noArray());


/** 将投影矩阵分解为旋转矩阵和相机内参矩阵*/
CV_EXPORTS_W void decomposeProjectionMatrix( InputArray projMatrix,
        OutputArray cameraMatrix,
        OutputArray rotMatrix,
        OutputArray transVect,
        OutputArray rotMatrixX = noArray(),
        OutputArray rotMatrixY = noArray(),
        OutputArray rotMatrixZ = noArray(),
        OutputArray eulerAngles =noArray() );

/** 计算每个相乘矩阵的矩阵积的偏导数*/
CV_EXPORTS_W void matMulDeriv( InputArray A,
        InputArray B,
        OutputArray dABdA,
```

```cpp
            OutputArray dABdB );

/** 合并两个旋转和平移变换*/
CV_EXPORTS_W void composeRT(InputArray rvec1,
            InputArray tvec1,
            InputArray rvec2,
            InputArray tvec2,
            OutputArray rvec3,
            OutputArray tvec3,
            OutputArray dr3dr1 = noArray(),
            OutputArray dr3dt1 = noArray(),
            OutputArray dr3dr2 = noArray(),
            OutputArray dr3dt2 = noArray(),
            OutputArray dt3dr1 = noArray(),
            OutputArray dt3dt1 = noArray(),
            OutputArray dt3dr2 = noArray(),
            OutputArray dt3dt2 = noArray());

/** 将三维点投影到图像平面上*/
CV_EXPORTS_W void projectPoints( InputArray objectPoints,
            InputArray rvec,
            InputArray tvec,
            InputArray cameraMatrix,
            InputArray distCoeffs,
            OutputArray imagePoints,
            OutputArray jacobian = noArray(),
            double aspectRatio = 0 );

/** 从三维-二维点对应中查找对象姿态*/
CV_EXPORTS_W bool solvePnP( InputArray objectPoints,
            InputArray imagePoints,
            InputArray cameraMatrix,
            InputArray distCoeffs,
            OutputArray rvec,
            OutputArray tvec,
            bool useExtrinsicGuess = false,
            int flags = SOLVEPNP_ITERATIVE );

/** 使用 RANSAC 方案从三维-二维点对应中查找对象姿态*/
CV_EXPORTS_W bool solvePnPRansac( InputArray objectPoints,
            InputArray imagePoints,
            InputArray cameraMatrix,
            InputArray distCoeffs,
            OutputArray rvec,
            OutputArray tvec,
            bool useExtrinsicGuess = false,
            int iterationsCount = 100,
```

```
                float reprojectionError = 8.0,
                double confidence = 0.99,
                OutputArray inliers = noArray(),
                int flags = SOLVEPNP_ITERATIVE );

/* 查找旋转和平移向量*/
CV_EXPORTS_W bool solvePnPRansac( InputArray objectPoints,
                InputArray imagePoints,
                InputOutputArray cameraMatrix,
                InputArray distCoeffs,
                OutputArray rvec,
                OutputArray tvec,
                OutputArray inliers,
                const UsacParams &params=UsacParams());

/** 从 3 个三维-二维点对应中查找对象姿态*/
CV_EXPORTS_W int solveP3P( InputArray objectPoints,
                InputArray imagePoints,
                InputArray cameraMatrix,
                InputArray distCoeffs,
                OutputArrayOfArrays rvecs,
                OutputArrayOfArrays tvecs,
                int flags );

/** 从三维-二维点对应和初始解开始细化姿态*/
CV_EXPORTS_W void solvePnPRefineLM( InputArray objectPoints,
                InputArray imagePoints,
                InputArray cameraMatrix,
                InputArray distCoeffs,
                InputOutputArray rvec,
                InputOutputArray tvec,
                TermCriteria criteria = TermCriteria(TermCriteria::EPS + TermCriteria::COUNT, 20,
FLT_EPSILON));

/** 从三维-二维点对应和初始解开始优化姿态*/
CV_EXPORTS_W void solvePnPRefineVVS( InputArray objectPoints,
                InputArray imagePoints,
                InputArray cameraMatrix,
                InputArray distCoeffs,
                InputOutputArray rvec,
                InputOutputArray tvec,
                TermCriteria criteria = TermCriteria(TermCriteria::EPS + TermCriteria::COUNT, 20,
FLT_EPSILON),
                double VVSlambda = 1);

/** 从三维-二维点对应中查找对象姿态*/
```

```cpp
CV_EXPORTS_W int solvePnPGeneric( InputArray objectPoints,
            InputArray imagePoints,
            InputArray cameraMatrix,
            InputArray distCoeffs,
            OutputArrayOfArrays rvecs,
            OutputArrayOfArrays tvecs,
            bool useExtrinsicGuess = false,
            SolvePnPMethod flags = SOLVEPNP_ITERATIVE,
            InputArray rvec = noArray(),
            InputArray tvec = noArray(),
            OutputArray reprojectionError = noArray() );

/** 从三维-二维点对应中寻找初始相机内参矩阵*/
CV_EXPORTS_W Mat initCameraMatrix2D( InputArrayOfArrays objectPoints,
            InputArrayOfArrays imagePoints,
            Size imageSize,
            double aspectRatio = 1.0 );
```

如下为与棋盘角点检测相关的函数的定义（见案例 113）：

```cpp
/** 查找棋盘内角的位置*/
CV_EXPORTS_W bool findChessboardCorners( InputArray image,
            Size patternSize,
            OutputArray corners,
            int flags = CALIB_CB_ADAPTIVE_THRESH + CALIB_CB_NORMALIZE_IMAGE );

CV_EXPORTS_W bool checkChessboard(InputArray img, Size size);

/** 使用基于扇区的方法查找棋盘内角的位置*/
CV_EXPORTS_AS(findChessboardCornersSBWithMeta)
bool findChessboardCornersSB(InputArray image,
            Size patternSize,
            OutputArray corners,
            int flags,
            OutputArray meta);
CV_EXPORTS_W inline
bool findChessboardCornersSB(InputArray image,
            Size patternSize,
            OutputArray corners,
            int flags = 0)
{
    return findChessboardCornersSB(image,
            patternSize,
            corners,
            flags,
            noArray());
}
```

```cpp
/** 估计检测到的棋盘的锐度*/
CV_EXPORTS_W Scalar estimateChessboardSharpness(InputArray image,
            Size patternSize,
            InputArray corners,
            float rise_distance=0.8F,
            bool vertical=false,
            OutputArray sharpness=noArray());

//! 找到棋盘转角的亚像素精确位置
CV_EXPORTS_W bool find4QuadCornerSubpix( InputArray img,
            InputOutputArray corners, Size region_size );

/** 渲染检测到的棋盘转角*/
CV_EXPORTS_W void drawChessboardCorners( InputOutputArray image,
            Size patternSize, InputArray corners, bool patternWasFound );

/** 根据姿态估计绘制世界/对象坐标系的轴*/
CV_EXPORTS_W void drawFrameAxes(InputOutputArray image,
            InputArray cameraMatrix,
            InputArray distCoeffs,
            InputArray rvec, InputArray tvec,
            float length, int thickness=3);

struct CV_EXPORTS_W_SIMPLE CirclesGridFinderParameters
{
    CV_WRAP CirclesGridFinderParameters();
    CV_PROP_RW cv::Size2f densityNeighborhoodSize;
    CV_PROP_RW float minDensity;
    CV_PROP_RW int kmeansAttempts;
    CV_PROP_RW int minDistanceToAddKeypoint;
    CV_PROP_RW int keypointScale;
    CV_PROP_RW float minGraphConfidence;
    CV_PROP_RW float vertexGain;
    CV_PROP_RW float vertexPenalty;
    CV_PROP_RW float existingVertexGain;
    CV_PROP_RW float edgeGain;
    CV_PROP_RW float edgePenalty;
    CV_PROP_RW float convexHullFactor;
    CV_PROP_RW float minRNGEdgeSwitchDist;

    enum GridType
    {
      SYMMETRIC_GRID, ASYMMETRIC_GRID
    };
    GridType gridType;

    CV_PROP_RW float squareSize;
```

```cpp
        CV_PROP_RW float maxRectifiedDistance;
};

#ifndef DISABLE_OPENCV_3_COMPATIBILITY
typedef CirclesGridFinderParameters CirclesGridFinderParameters2;
#endif

/** 在圆的网格中查找中心*/
CV_EXPORTS_W bool findCirclesGrid( InputArray image,
            Size patternSize,
            OutputArray centers, int flags,
            const Ptr<FeatureDetector> &blobDetector,
            const CirclesGridFinderParameters& parameters );
CV_EXPORTS_W bool findCirclesGrid( InputArray image,
            Size patternSize, OutputArray centers,
            int flags = CALIB_CB_SYMMETRIC_GRID,
            const Ptr<FeatureDetector> &blobDetector = SimpleBlobDetector::create());
```

如下为与相机标定相关的函数定义（见案例 114）：

```cpp
/** 从校准模式的多个视图中查找相机的内参和外参*/
CV_EXPORTS_AS(calibrateCameraExtended) double calibrateCamera(
            InputArrayOfArrays objectPoints,
            InputArrayOfArrays imagePoints, Size imageSize,
            InputOutputArray cameraMatrix,
            InputOutputArray distCoeffs,
            OutputArrayOfArrays rvecs,
            OutputArrayOfArrays tvecs,
            OutputArray stdDeviationsIntrinsics,
            OutputArray stdDeviationsExtrinsics,
            OutputArray perViewErrors, int flags = 0,
            TermCriteria criteria = TermCriteria( TermCriteria::COUNT + TermCriteria::EPS, 30,
DBL_EPSILON) );

CV_EXPORTS_W double calibrateCamera( InputArrayOfArrays objectPoints,
            InputArrayOfArrays imagePoints, Size imageSize,
            InputOutputArray cameraMatrix,
            InputOutputArray distCoeffs,
            OutputArrayOfArrays rvecs,
            OutputArrayOfArrays tvecs, int flags = 0,
            TermCriteria criteria = TermCriteria( TermCriteria::COUNT + TermCriteria::EPS, 30,
DBL_EPSILON) );

CV_EXPORTS_AS(calibrateCameraROExtended) double calibrateCameraRO(
            InputArrayOfArrays objectPoints,
            InputArrayOfArrays imagePoints, Size imageSize,
            int iFixedPoint, InputOutputArray cameraMatrix,
            InputOutputArray distCoeffs, OutputArrayOfArrays rvecs,
```

```
            OutputArrayOfArrays tvecs, OutputArray newObjPoints,
            OutputArray stdDeviationsIntrinsics,
            OutputArray stdDeviationsExtrinsics,
            OutputArray stdDeviationsObjPoints,
            OutputArray perViewErrors, int flags = 0,
            TermCriteria criteria = TermCriteria( TermCriteria::COUNT + TermCriteria::EPS, 30, DBL_EPSILON) );
CV_EXPORTS_W double calibrateCameraRO( InputArrayOfArrays objectPoints,
             InputArrayOfArrays imagePoints, Size imageSize,
            int iFixedPoint,
            InputOutputArray cameraMatrix,
            InputOutputArray distCoeffs,
            OutputArrayOfArrays rvecs, OutputArrayOfArrays tvecs,
            OutputArray newObjPoints, int flags = 0,
            TermCriteria criteria = TermCriteria(TermCriteria::COUNT + TermCriteria::EPS, 30, DBL_EPSILON) );

/** 根据相机固有矩阵计算有用的相机特性*/
CV_EXPORTS_W void calibrationMatrixValues( InputArray cameraMatrix,
            Size imageSize, double apertureWidth,
            double apertureHeight,
            CV_OUT double& fovx, CV_OUT double& fovy,
            CV_OUT double& focalLength,
            CV_OUT Point2d& principalPoint,
            CV_OUT double& aspectRatio );
```

如下函数用于标定立体摄像头，即可以同时标定两个摄像头：

```
CV_EXPORTS_AS(stereoCalibrateExtended) double stereoCalibrate(
            InputArrayOfArrays objectPoints,
            InputArrayOfArrays imagePoints1,
            InputArrayOfArrays imagePoints2,
            InputOutputArray cameraMatrix1,
            InputOutputArray distCoeffs1,
            InputOutputArray cameraMatrix2,
            InputOutputArray distCoeffs2,
            Size imageSize, InputOutputArray R,
            InputOutputArray T, OutputArray E, OutputArray F,
            OutputArray perViewErrors,
            int flags = CALIB_FIX_INTRINSIC,
            TermCriteria criteria = TermCriteria(TermCriteria::COUNT+TermCriteria::EPS, 30, 1e-6) );

CV_EXPORTS_W double stereoCalibrate( InputArrayOfArrays objectPoints,
            InputArrayOfArrays imagePoints1,
            InputArrayOfArrays imagePoints2,
            InputOutputArray cameraMatrix1,
            InputOutputArray distCoeffs1,
```

```cpp
                                InputOutputArray cameraMatrix2,
                                InputOutputArray distCoeffs2,
                                Size imageSize, OutputArray R,
                                OutputArray T, OutputArray E,
                                OutputArray F,
                                int flags = CALIB_FIX_INTRINSIC,
                                TermCriteria criteria = TermCriteria(TermCriteria::COUNT+TermCriteria::EPS, 30, 1e-6) );

/** 为校准立体相机的每个头部计算校正变换*/
CV_EXPORTS_W void stereoRectify( InputArray cameraMatrix1,
                                 InputArray distCoeffs1,
                                 InputArray cameraMatrix2,
                                 InputArray distCoeffs2,
                                 Size imageSize, InputArray R, InputArray T,
                                 OutputArray R1, OutputArray R2,
                                 OutputArray P1, OutputArray P2,
                                 OutputArray Q, int flags = CALIB_ZERO_DISPARITY,
                                 double alpha = -1, Size newImageSize = Size(),
                                 CV_OUT Rect* validPixROI1 = 0,
                                 CV_OUT Rect* validPixROI2 = 0 );

/** 计算未校准立体相机的校正变换*/
CV_EXPORTS_W bool stereoRectifyUncalibrated( InputArray points1,
                                             InputArray points2,
                                             InputArray F, Size imgSize,
                                             OutputArray H1, OutputArray H2,
                                             double threshold = 5 );

//! 计算三头相机的校正变换
CV_EXPORTS_W float rectify3Collinear( InputArray cameraMatrix1,
                                      InputArray distCoeffs1, InputArray cameraMatrix2,
                                      InputArray distCoeffs2, InputArray cameraMatrix3,
                                      InputArray distCoeffs3, InputArrayOfArrays imgpt1,
                                      InputArrayOfArrays imgpt3, Size imageSize,
                                      InputArray R12, InputArray T12,
                                      InputArray R13, InputArray T13,
                                      OutputArray R1, OutputArray R2, OutputArray R3,
                                      OutputArray P1, OutputArray P2, OutputArray P3,
                                      OutputArray Q, double alpha, Size newImgSize,
                                      CV_OUT Rect* roi1, CV_OUT Rect* roi2, int flags );

/** 基于自由缩放参数返回新的相机内参矩阵*/
CV_EXPORTS_W Mat getOptimalNewCameraMatrix( InputArray cameraMatrix,
                                            InputArray distCoeffs,
                                            Size imageSize, double alpha,
                                            Size newImgSize = Size(),
```

```
            CV_OUT Rect* validPixROI = 0,
            bool centerPrincipalPoint = false);
```

如下函数用于计算 Hand-Eye 校正：

```
CV_EXPORTS_W void calibrateHandEye( InputArrayOfArrays R_gripper2base,
            InputArrayOfArrays t_gripper2base,
            InputArrayOfArrays R_target2cam,
            InputArrayOfArrays t_target2cam,
            OutputArray R_cam2gripper, OutputArray t_cam2gripper,
            HandEyeCalibrationMethod method=CALIB_HAND_EYE_TSAI );

/** 计算 Robot-World/Hand-Eye 校正*/
CV_EXPORTS_W void calibrateRobotWorldHandEye(
            InputArrayOfArrays R_world2cam,
            InputArrayOfArrays t_world2cam,
            InputArrayOfArrays R_base2gripper,
            InputArrayOfArrays t_base2gripper,
            OutputArray R_base2world, OutputArray t_base2world,
            OutputArray R_gripper2cam, OutputArray t_gripper2cam,
            RobotWorldHandEyeCalibrationMethod method=CALIB_ROBOT_WORLD_HAND_EYE_SHAH );
```

如下函数用于空间转换：

```
/** 将点从欧氏空间转换为齐次空间*/
CV_EXPORTS_W void convertPointsToHomogeneous( InputArray src,
            OutputArray dst );

/** 将点从齐次空间转换为欧氏空间*/
CV_EXPORTS_W void convertPointsFromHomogeneous( InputArray src,
            OutputArray dst );

/** 将点转换为齐次坐标*/
CV_EXPORTS void convertPointsHomogeneous( InputArray src, OutputArray dst );
```

如下函数用于根据两幅图像中的对应点计算基础矩阵：

```
CV_EXPORTS_W Mat findFundamentalMat( InputArray points1, InputArray points2,
            int method, double ransacReprojThreshold, double confidence,
            int maxIters, OutputArray mask = noArray() );
CV_EXPORTS_W Mat findFundamentalMat( InputArray points1, InputArray points2,
            int method = FM_RANSAC, double ransacReprojThreshold = 3.,
            double confidence = 0.99, OutputArray mask = noArray() );
CV_EXPORTS Mat findFundamentalMat( InputArray points1, InputArray points2,
            OutputArray mask, int method = FM_RANSAC,
            double ransacReprojThreshold = 3., double confidence = 0.99 );

CV_EXPORTS_W Mat findFundamentalMat( InputArray points1, InputArray points2,
            OutputArray mask, const UsacParams &params);
```

如下函数用于根据两幅图像中的对应点计算本质矩阵：
```
CV_EXPORTS_W Mat findEssentialMat( InputArray points1, InputArray points2,
        InputArray cameraMatrix, int method = RANSAC,
        double prob = 0.999, double threshold = 1.0,
        OutputArray mask = noArray() );

CV_EXPORTS_W Mat findEssentialMat( InputArray points1, InputArray points2,
        double focal = 1.0, Point2d pp = Point2d(0, 0),
        int method = RANSAC, double prob = 0.999,
        double threshold = 1.0, OutputArray mask = noArray() );

CV_EXPORTS_W Mat findEssentialMat( InputArray points1, InputArray points2,
        InputArray cameraMatrix1, InputArray distCoeffs1,
        InputArray cameraMatrix2, InputArray distCoeffs2,
        int method = RANSAC, double prob = 0.999, double threshold = 1.0,
        OutputArray mask = noArray() );

CV_EXPORTS_W Mat findEssentialMat( InputArray points1, InputArray points2,
        InputArray cameraMatrix1, InputArray cameraMatrix2,
        InputArray dist_coeff1, InputArray dist_coeff2,
        OutputArray mask, const UsacParams &params);
```

如下为用于执行本质矩阵的相关操作的函数：
```
/** 将本质矩阵分解为可能的旋转和平移*/
CV_EXPORTS_W void decomposeEssentialMat( InputArray E, OutputArray R1,
        OutputArray R2, OutputArray t );

/** 根据估计的本质矩阵和两幅图像中的对应点恢复相机的相对旋转与平移*/
CV_EXPORTS_W int recoverPose( InputArray E, InputArray points1,
        InputArray points2, InputArray cameraMatrix,
        OutputArray R, OutputArray t,
        InputOutputArray mask = noArray() );
CV_EXPORTS_W int recoverPose( InputArray E, InputArray points1,
        InputArray points2, OutputArray R, OutputArray t,
        double focal = 1.0, Point2d pp = Point2d(0, 0),
        InputOutputArray mask = noArray() );
CV_EXPORTS_W int recoverPose( InputArray E, InputArray points1,
        InputArray points2, InputArray cameraMatrix,
        OutputArray R, OutputArray t, double distanceThresh,
        InputOutputArray mask = noArray(),
        OutputArray triangulatedPoints = noArray());
```

如下为与三维重建相关的函数定义：
```
/** 对于立体对图像中的点，计算另一个图像中相应的外线*/
CV_EXPORTS_W void computeCorrespondEpilines( InputArray points,
        int whichImage, InputArray F, OutputArray lines );
```

```cpp
/** 此函数通过使用立体相机的观察值重建三维点*/
CV_EXPORTS_W void triangulatePoints( InputArray projMatr1,
            InputArray projMatr2, InputArray projPoints1,
            InputArray projPoints2, OutputArray points4D );

/** 修正对应点的坐标*/
CV_EXPORTS_W void correctMatches( InputArray F, InputArray points1,
            InputArray points2, OutputArray newPoints1,
            OutputArray newPoints2 );

/** 滤除视差图中的小噪声点(斑点)*/
CV_EXPORTS_W void filterSpeckles( InputOutputArray img, double newVal,
            int maxSpeckleSize, double maxDiff,
            InputOutputArray buf = noArray() );

CV_EXPORTS_W Rect getValidDisparityROI( Rect roi1, Rect roi2,
            int minDisparity, int numberOfDisparities,
            int blockSize );
CV_EXPORTS_W void validateDisparity( InputOutputArray disparity,
            InputArray cost, int minDisparity, int numberOfDisparities,
            int disp12MaxDisp = 1 );

/** 将视差图像重新投影到三维空间*/
CV_EXPORTS_W void reprojectImageTo3D( InputArray disparity,
            OutputArray _3dImage, InputArray Q,
            bool handleMissingValues = false, int ddepth = -1 );

/** 计算两点之间的Sampson距离*/
CV_EXPORTS_W double sampsonDistance(InputArray pt1, InputArray pt2,
            InputArray F);

/** 计算两个三维点集之间的最佳仿射变换*/
CV_EXPORTS_W  int estimateAffine3D(InputArray src, InputArray dst,
            OutputArray out, OutputArray inliers,
            double ransacThreshold = 3, double confidence = 0.99);

/** 计算两个三维点集之间的最佳平移*/
CV_EXPORTS_W  int estimateTranslation3D(InputArray src, InputArray dst,
            OutputArray out, OutputArray inliers,
            double ransacThreshold = 3, double confidence = 0.99);

/** 计算两个二维点集之间的最佳仿射变换*/
CV_EXPORTS_W cv::Mat estimateAffine2D(InputArray from, InputArray to,
            OutputArray inliers = noArray(), int method = RANSAC,
            double ransacReprojThreshold = 3, size_t maxIters = 2000,
            double confidence = 0.99, size_t refineIters = 10);
```

```cpp
CV_EXPORTS_W cv::Mat estimateAffine2D(InputArray pts1, InputArray pts2,
        OutputArray inliers, const UsacParams &params);

/** 计算两个二维点集之间具有 4 个自由度的最优有限仿射变换*/
CV_EXPORTS_W cv::Mat estimateAffinePartial2D(InputArray from, InputArray to,
        OutputArray inliers = noArray(), int method = RANSAC,
        double ransacReprojThreshold = 3, size_t maxIters = 2000,
        double confidence = 0.99, size_t refineIters = 10);

/** 将单应矩阵分解为旋转、平移和平面法线*/
CV_EXPORTS_W int decomposeHomographyMat(InputArray H, InputArray K,
        OutputArrayOfArrays rotations, OutputArrayOfArrays translations,
        OutputArrayOfArrays normals);

/** 基于附加信息的滤波器单应分解*/
CV_EXPORTS_W void filterHomographyDecompByVisibleRefpoints(
        InputArrayOfArrays rotations,
        InputArrayOfArrays normals,
        InputArray beforePoints, InputArray afterPoints,
        OutputArray possibleSolutions,
        InputArray pointsMask = noArray());
```

如下为 StereoMatcher 类的定义，该类为立体匹配算法的基类：

```cpp
class CV_EXPORTS_W StereoMatcher : public Algorithm
{
public:
    enum { DISP_SHIFT = 4,
           DISP_SCALE = (1 << DISP_SHIFT)
        };

    /** 计算指定立体对的视差贴图*/
    CV_WRAP virtual void compute( InputArray left, InputArray right,
                                  OutputArray disparity ) = 0;

    CV_WRAP virtual int getMinDisparity() const = 0;
    CV_WRAP virtual void setMinDisparity(int minDisparity) = 0;

    CV_WRAP virtual int getNumDisparities() const = 0;
    CV_WRAP virtual void setNumDisparities(int numDisparities) = 0;

    CV_WRAP virtual int getBlockSize() const = 0;
    CV_WRAP virtual void setBlockSize(int blockSize) = 0;

    CV_WRAP virtual int getSpeckleWindowSize() const = 0;
    CV_WRAP virtual void setSpeckleWindowSize(int speckleWindowSize) = 0;
```

```cpp
    CV_WRAP virtual int getSpeckleRange() const = 0;
    CV_WRAP virtual void setSpeckleRange(int speckleRange) = 0;

    CV_WRAP virtual int getDisp12MaxDiff() const = 0;
    CV_WRAP virtual void setDisp12MaxDiff(int disp12MaxDiff) = 0;
};
```

如下为 StereoBM 类的定义，该类用于计算立体对应的块匹配算法：

```cpp
class CV_EXPORTS_W StereoBM : public StereoMatcher
{
public:
    enum { PREFILTER_NORMALIZED_RESPONSE = 0,
           PREFILTER_XSOBEL              = 1
         };

    CV_WRAP virtual int getPreFilterType() const = 0;
    CV_WRAP virtual void setPreFilterType(int preFilterType) = 0;

    CV_WRAP virtual int getPreFilterSize() const = 0;
    CV_WRAP virtual void setPreFilterSize(int preFilterSize) = 0;

    CV_WRAP virtual int getPreFilterCap() const = 0;
    CV_WRAP virtual void setPreFilterCap(int preFilterCap) = 0;

    CV_WRAP virtual int getTextureThreshold() const = 0;
    CV_WRAP virtual void setTextureThreshold(int textureThreshold) = 0;

    CV_WRAP virtual int getUniquenessRatio() const = 0;
    CV_WRAP virtual void setUniquenessRatio(int uniquenessRatio) = 0;

    CV_WRAP virtual int getSmallerBlockSize() const = 0;
    CV_WRAP virtual void setSmallerBlockSize(int blockSize) = 0;

    CV_WRAP virtual Rect getROI1() const = 0;
    CV_WRAP virtual void setROI1(Rect roi1) = 0;

    CV_WRAP virtual Rect getROI2() const = 0;
    CV_WRAP virtual void setROI2(Rect roi2) = 0;

    /** 创建 StereoBM 对象*/
    CV_WRAP static Ptr<StereoBM> create(int numDisparities = 0, int blockSize = 21);
};
```

如下为 StereoSGBM 类的定义，该类为 H. Hirschmuller 算法实现类：

```cpp
class CV_EXPORTS_W StereoSGBM : public StereoMatcher
{
```

```cpp
public:
    enum
    {
        MODE_SGBM      = 0,
        MODE_HH        = 1,
        MODE_SGBM_3WAY = 2,
        MODE_HH4       = 3
    };

    CV_WRAP virtual int getPreFilterCap() const = 0;
    CV_WRAP virtual void setPreFilterCap(int preFilterCap) = 0;

    CV_WRAP virtual int getUniquenessRatio() const = 0;
    CV_WRAP virtual void setUniquenessRatio(int uniquenessRatio) = 0;

    CV_WRAP virtual int getP1() const = 0;
    CV_WRAP virtual void setP1(int P1) = 0;

    CV_WRAP virtual int getP2() const = 0;
    CV_WRAP virtual void setP2(int P2) = 0;

    CV_WRAP virtual int getMode() const = 0;
    CV_WRAP virtual void setMode(int mode) = 0;

    /** 创建 StereoSGBM 对象*/
    CV_WRAP static Ptr<StereoSGBM> create(int minDisparity = 0,
            int numDisparities = 16, int blockSize = 3,
            int P1 = 0, int P2 = 0, int disp12MaxDiff = 0,
            int preFilterCap = 0, int uniquenessRatio = 0,
            int speckleWindowSize = 0, int speckleRange = 0,
            int mode = StereoSGBM::MODE_SGBM);
};

enum UndistortTypes
{
    PROJ_SPHERICAL_ORTHO  = 0,
    PROJ_SPHERICAL_EQRECT = 1
};
```

如下为与消除图像失真相关的函数定义（见案例 114）：

```cpp
/** 变换图像以补偿镜头失真*/
CV_EXPORTS_W void undistort( InputArray src, OutputArray dst,
                             InputArray cameraMatrix,
                             InputArray distCoeffs,
                             InputArray newCameraMatrix = noArray() );
```

```cpp
/** 计算不失真和校正变换映射*/
CV_EXPORTS_W
void initUndistortRectifyMap(InputArray cameraMatrix, InputArray distCoeffs,
            InputArray R, InputArray newCameraMatrix,
            Size size, int m1type, OutputArray map1,
            OutputArray map2);

CV_EXPORTS
float initWideAngleProjMap(InputArray cameraMatrix, InputArray distCoeffs,
            Size imageSize, int destImageWidth,
            int m1type, OutputArray map1, OutputArray map2,
            enum UndistortTypes projType = PROJ_SPHERICAL_EQRECT,
            double alpha = 0);
static inline float initWideAngleProjMap(InputArray cameraMatrix,
            InputArray distCoeffs, Size imageSize, int destImageWidth,
            int m1type, OutputArray map1, OutputArray map2,
            int projType, double alpha = 0)
{
    return initWideAngleProjMap(cameraMatrix, distCoeffs, imageSize,
            destImageWidth, m1type, map1, map2,
            (UndistortTypes)projType, alpha);
}

/** 返回默认的新相机矩阵*/
CV_EXPORTS_W Mat getDefaultNewCameraMatrix(
            InputArray cameraMatrix, Size imgsize = Size(),
            bool centerPrincipalPoint = false);

/** 从观测点坐标计算理想点坐标*/
CV_EXPORTS_W
void undistortPoints(InputArray src,
            OutputArray dst, InputArray cameraMatrix,
            InputArray distCoeffs, InputArray R = noArray(),
            InputArray P = noArray());
CV_EXPORTS_AS(undistortPointsIter) void undistortPoints(
            InputArray src, OutputArray dst,
            InputArray cameraMatrix, InputArray distCoeffs,
            InputArray R, InputArray P, TermCriteria criteria);
```

如下为与鱼眼相机模型相关的定义：

```cpp
namespace fisheye
{
    enum{
        CALIB_USE_INTRINSIC_GUESS     = 1 << 0,
        CALIB_RECOMPUTE_EXTRINSIC     = 1 << 1,
        CALIB_CHECK_COND              = 1 << 2,
```

```cpp
        CALIB_FIX_SKEW                  = 1 << 3,
        CALIB_FIX_K1                    = 1 << 4,
        CALIB_FIX_K2                    = 1 << 5,
        CALIB_FIX_K3                    = 1 << 6,
        CALIB_FIX_K4                    = 1 << 7,
        CALIB_FIX_INTRINSIC             = 1 << 8,
        CALIB_FIX_PRINCIPAL_POINT       = 1 << 9
    };

    /** 使用鱼眼模型投影点*/
    CV_EXPORTS void projectPoints(InputArray objectPoints,
            OutputArray imagePoints, const Affine3d& affine,
            InputArray K, InputArray D, double alpha = 0,
            OutputArray jacobian = noArray());
    CV_EXPORTS_W void projectPoints(InputArray objectPoints,
            OutputArray imagePoints, InputArray rvec, InputArray tvec,
            InputArray K, InputArray D, double alpha = 0,
            OutputArray jacobian = noArray());

    /** 使用鱼眼模型扭曲二维点*/
    CV_EXPORTS_W void distortPoints(InputArray undistorted,
            OutputArray distorted, InputArray K,
            InputArray D, double alpha = 0);

    /** 使用鱼眼模型不失真二维点*/
    CV_EXPORTS_W void undistortPoints(InputArray distorted,
            OutputArray undistorted, InputArray K, InputArray D,
            InputArray R = noArray(), InputArray P  = noArray());

    /** 计算图像变换的不失真和校正映射供 cv::remap()使用*/
    CV_EXPORTS_W void initUndistortRectifyMap(InputArray K, InputArray D,
            InputArray R, InputArray P,
            const cv::Size& size, int m1type, OutputArray map1,
            OutputArray map2);

    /** 变换图像以补偿鱼眼镜头失真*/
    CV_EXPORTS_W void undistortImage(InputArray distorted,
            OutputArray undistorted, InputArray K, InputArray D,
            InputArray Knew = cv::noArray(), const Size& new_size = Size());

    /** 为不失真或校正估计新的相机内参矩阵*/
    CV_EXPORTS_W void estimateNewCameraMatrixForUndistortRectify(
            InputArray K, InputArray D, const Size &image_size,
            InputArray R, OutputArray P, double balance = 0.0,
            const Size& new_size = Size(), double fov_scale = 1.0);
```

```
/** 执行相机校准*/
CV_EXPORTS_W double calibrate(InputArrayOfArrays objectPoints,
        InputArrayOfArrays imagePoints, const Size& image_size,
        InputOutputArray K, InputOutputArray D, OutputArrayOfArrays rvecs,
        OutputArrayOfArrays tvecs, int flags = 0,
        TermCriteria criteria = TermCriteria(TermCriteria::COUNT + TermCriteria::EPS, 100, DBL_EPSILON));

/** 鱼眼相机模型的立体校正*/
CV_EXPORTS_W void stereoRectify(InputArray K1, InputArray D1,
        InputArray K2, InputArray D2, const Size &imageSize, InputArray R,
        InputArray tvec, OutputArray R1, OutputArray R2,
        OutputArray P1, OutputArray P2, OutputArray Q, int flags,
        const Size &newImageSize = Size(),
        double balance = 0.0, double fov_scale = 1.0);

/** 执行立体校准*/
CV_EXPORTS_W double stereoCalibrate(InputArrayOfArrays objectPoints,
        InputArrayOfArrays imagePoints1,
        InputArrayOfArrays imagePoints2,
        InputOutputArray K1, InputOutputArray D1,
        InputOutputArray K2, InputOutputArray D2,
        Size imageSize, OutputArray R, OutputArray T,
        int flags = fisheye::CALIB_FIX_INTRINSIC,
        TermCriteria criteria = TermCriteria(TermCriteria::COUNT + TermCriteria::EPS, 100, DBL_EPSILON));
}
}
```

如下为类 CvLevMarq 的定义，该类为 Levenberg-Marquardt 非线性最小二乘算法的实现：

```
#if 0 //def __cplusplus
class CV_EXPORTS CvLevMarq
{
public:
    CvLevMarq();
    CvLevMarq( int nparams, int nerrs,
            CvTermCriteria criteria = cvTermCriteria(
                    CV_TERMCRIT_EPS+CV_TERMCRIT_ITER,30,DBL_EPSILON),
            bool completeSymmFlag=false );
    ~CvLevMarq();
    void init( int nparams, int nerrs,
            CvTermCriteria criteria= cvTermCriteria(
                    CV_TERMCRIT_EPS+CV_TERMCRIT_ITER,30,DBL_EPSILON),
            bool completeSymmFlag=false );
    bool update( const CvMat*& param, CvMat*& J, CvMat*& err );
```

```
bool updateAlt( const CvMat*& param, CvMat*& JtJ, CvMat*& JtErr,
                double*& errNorm );

void clear();
void step();
enum { DONE=0, STARTED=1, CALC_J=2, CHECK_ERR=3 };

cv::Ptr<CvMat> mask;
cv::Ptr<CvMat> prevParam;
cv::Ptr<CvMat> param;
cv::Ptr<CvMat> J;
cv::Ptr<CvMat> err;
cv::Ptr<CvMat> JtJ;
cv::Ptr<CvMat> JtJN;
cv::Ptr<CvMat> JtErr;
cv::Ptr<CvMat> JtJV;
cv::Ptr<CvMat> JtJW;
double prevErrNorm, errNorm;
int lambdaLg10;
CvTermCriteria criteria;
int state;
int iters;
bool completeSymmFlag;
int solveMethod;
};
#endif
#endif
```

11.2 单应性变换

11.2.1 案例 111：单应性变换矩阵

OpenCV 中提供了用于生成单应性变换矩阵的函数 findHomography，其定义如下：
```
retval, mask = findHomography(srcPoints, dstPoints, method=None, ransacReprojThreshold=None,
mask=None, maxIters=None, confidence=None)
```
参数说明如下。

- srcPoints：输入图像平面上点的坐标。
- dstPoints：输出图像平面上点的坐标。
- method：单应矩阵计算方式。
- ransacReprojThreshold：将点对视为 inlier 的最大允许重投影错误（仅用于 RANSAC 和 RHO）。

- mask：输出掩模（返回值）。
- maxIters：RANSAC 方法最大迭代次数。
- confidence：置信度等级，值在 0 到 1 之间。
- retval：计算的单应性矩阵（返回值）。

本案例使用如图 11.1 所示的输入图像。

单应性矩阵生成并变换的案例代码如下：

图 11.1

```python
import cv2
import numpy as np

def mouse_handler(event, x, y, flags, data):
    #选择4个点
    if event == cv2.EVENT_LBUTTONDOWN:
        cv2.circle(data['im'], (x, y), 3, (0, 0, 255), 5, 16);
        cv2.imshow("Image", data['im']);
        if len(data['points']) < 4:
            data['points'].append([x, y])

def get_four_points(im):
    #获取选择的4个点
    data = {}
    data['im'] = im.copy()
    data['points'] = []

    #为鼠标事件设置回调
    cv2.imshow("Image", im)
    cv2.setMouseCallback("Image", mouse_handler, data)
    cv2.waitKey(0)

    #数据类型转换
    points = np.vstack(data['points']).astype(float)
    return points

if __name__ == '__main__' :
    #输入图像读取
    im_src = cv2.imread("book1.jpg")
    #创建目标图像
    size = (300,400,3)
    im_dst = np.zeros(size, np.uint8)
    #选择待变换的4个点
    print('选择4个点，先左上，后右下，完成后按下 Enter 键')
    cv2.imshow("Image", im_src)
    pts_src = get_four_points(im_src);
    pts_dst = np.array( [[0,0], [size[0] - 1, 0], [size[0] - 1, size[1] -1], [0, size[1] - 1 ] ],
```

```
dtype=float )
    #计算单应性矩阵
    homography, mask = cv2.findHomography(pts_src, pts_dst)

    #透视变换
    im_dst = cv2.warpPerspective(im_src, homography, size[0:2])

    #结果展示
    cv2.imshow("Image", im_dst)
    cv2.waitKey(0)
```

在输入图像上选取 4 个点，4 个点的标注顺序如图 11.2 所示。

标注后，按下 Enter 键，生成单应性矩阵 homography，并进行透视变换，结果如图 11.3 所示。

图 11.2

图 11.3

11.2.2　案例 112：单应性应用之图像插入

本节介绍单应性变换的另外一个案例，即将一幅图像插入另一幅图像中的指定位置。

本案例的待插入图像如图 11.4 所示，本案例的被插入图像如图 11.5 所示。

图 11.4

图 11.5

图像插入案例代码如下：

```python
import cv2
import numpy as np

def mouse_handler(event, x, y, flags, data):
    #选择4个点
    if event == cv2.EVENT_LBUTTONDOWN:
        cv2.circle(data['im'], (x, y), 3, (0, 0, 255), 5, 16);
        cv2.imshow("Image", data['im']);
        if len(data['points']) < 4:
            data['points'].append([x, y])

def get_four_points(im):
    #获取选择的4个点
    data = {}
    data['im'] = im.copy()
    data['points'] = []

    #为鼠标事件设置回调
    cv2.imshow("Image", im)
    cv2.setMouseCallback("Image", mouse_handler, data)
    cv2.waitKey(0)

    #数据类型转换
    points = np.vstack(data['points']).astype(float)
    return points

if __name__ == '__main__' :
    #待插入图像读取
    im_src = cv2.imread('first-image.jpg');
    size = im_src.shape
    #原始图像中的4个点
    pts_src = np.array( [ [0,0], [size[1] - 1, 0], [size[1] - 1, size[0] -1], [0, size[0] - 1 ] ], dtype=float );
    #被插入图像（目标图像）读取
    im_dst = cv2.imread('times-square.jpg');
    #选取4个点作为待插入图像的插入位置
    pts_dst = get_four_points(im_dst)

    #计算单应性矩阵
    h, mask = cv2.findHomography(pts_src, pts_dst);
    #图像裁剪
    im_temp = cv2.warpPerspective(im_src, h, (im_dst.shape[1],im_dst.shape[0]))
    #在被插入图像中清除多边形区域
    cv2.fillConvexPoly(im_dst, pts_dst.astype(int), 0, 16);
```

```
#将裁剪的图像插入被插入图像中
im_dst = im_dst + im_temp;

#结果展示
cv2.imshow("Image", im_dst);
cv2.waitKey(0);
```

在被插入图像上选取 4 个点，4 个点的标注顺序如图 11.6 所示。

图像插入后的结果如图 11.7 所示。

图 11.6

图 11.7

11.3 相机标定

11.3.1 案例 113：棋盘角点检测并绘制

相机标定在目标定位跟踪中非常重要，OpenCV 中使用棋盘格模板进行相机标定。其中，现实世界中的点即三维点称为 object points，棋盘图像中的二维点称为 image points。

OpenCV 中提供了函数 findChessboardCorners，用于找到棋盘格模板，其定义如下：
`retval, corners = findChessboardCorners(image, patternSize, corners=None, flags=None)`

参数说明如下。

- image：输入的棋盘图像。
- patternSize：棋盘规格，棋盘内部角点的行列数。
- corners：检测到的角点（返回值）。
- flags：标定算法，标志可以组合使用。
- retval：函数调用状态，调用成功返回 true（返回值）。

建议使用 x 方向和 y 方向个数不相等的棋盘格模板。这个函数如果检测到模板，则会返回对应的角点，并返回 true。因为不一定所有的图像都能找到需要的模板，所以读者可以采集多幅图像进行标定。找到角点后，可以使用 cv2.cornerSubPix()得到更为精确的角点像素坐标。

OpenCV 中提供了函数 drawChessboardCorners，用于绘制检测到的棋盘角点，其定义如下：

```
image = drawChessboardCorners(image, patternSize, corners, patternWasFound)
```

参数说明如下。

- image：目标图像（返回值）。
- patternSize：棋盘规格，棋盘内部角点的行列数。
- corners：检测到的角点。
- patternWasFound：指示是否能找到整个棋盘。

棋盘角点检测案例代码如下：

```python
import cv2
import numpy as np

#定义棋盘维度
CHECKERBOARD = (6,9)
criteria = (cv2.TERM_CRITERIA_EPS + cv2.TERM_CRITERIA_MAX_ITER, 30, 0.001)

#存储每个棋盘图像的三维点
objpoints = []
#存储每个棋盘图像的二维点
imgpoints = []

#定义三维点的世界坐标
objp = np.zeros((1, CHECKERBOARD[0]*CHECKERBOARD[1], 3), np.float32)
objp[0,:,:2] = np.mgrid[0:CHECKERBOARD[0], 0:CHECKERBOARD[1]].T.reshape(-1, 2)
prev_img_shape = None

#读取其中一幅棋盘图像
img = cv2.imread("./images/image_2.jpg")
gray = cv2.cvtColor(img, cv2.COLOR_BGR2GRAY)
#寻找棋盘角点
ret, corners = cv2.findChessboardCorners(gray, CHECKERBOARD, cv2.CALIB_CB_ADAPTIVE_THRESH +
                                         cv2.CALIB_CB_FAST_CHECK +
cv2.CALIB_CB_NORMALIZE_IMAGE)

#如果检测到所需的角点个数，则微调像素坐标并将其显示在棋盘格图像上
if ret == True:
    objpoints.append(objp)
    #为给定的二维点微调像素坐标
    corners2 = cv2.cornerSubPix(gray, corners, (11, 11), (-1, -1), criteria)
```

```
    imgpoints.append(corners2)
    #绘制检测到的棋盘角点
    img = cv2.drawChessboardCorners(img, CHECKERBOARD, corners2, ret)

#结果展示
cv2.imshow('img', img)
cv2.waitKey(0)
cv2.destroyAllWindows()
```

检测的角点绘制结果如图 11.8 所示。

图 11.8

11.3.2　案例 114：消除图像失真

相机成像存在畸变（畸变原理可以参考相机模型），可以通过相机标定得到畸变参数等进行图像校正，相机标定并消除图像畸变主要包括以下 3 步。

（1）找到棋盘格模板，OpenCV 中提供了函数 cv2.findChessboardCorners()，用于找到棋盘格模板。

（2）通过三维点和对应的图像上的二维点对，使用 cv2.calibrateCamera() 函数进行标定，这个函数会返回标定结果、相机的内参矩阵、畸变系数、旋转矩阵和平移向量，其定义如下：

```
retval, cameraMatrix, distCoeffs, rvecs, tvecs = calibrateCamera(objectPoints, imagePoints,
imageSize, cameraMatrix, distCoeffs, rvecs=None, tvecs=None, flags=None, criteria=None)
```

参数说明如下。

- objectPoints：世界坐标系中的三维点。
- imagePoints：每一个内角点对应的图像坐标点。
- imageSize：图像尺寸。
- cameraMatrix：相机内参矩阵（返回值）。
- distCoeffs：畸变矩阵（返回值）。

- rvecs：旋转向量（返回值）。
- tvecs：位移向量（返回值）。
- flags：标定算法。
- criteria：迭代终止条件。
- retval：函数调用状态，调用成功返回 true（返回值）。

（3）使用新得到的内参矩阵和畸变系数对图像消除畸变，OpenCV 提供的直接消除畸变的函数为 cv2.undistort()，其定义如下：

```
dst = undistort(src, cameraMatrix, distCoeffs, dst=None, newCameraMatrix=None)
```

参数说明如下。

- src：输入图像。
- cameraMatrix：相机内参矩阵。
- distCoeffs：畸变矩阵。
- dst：输出图像（返回值）。
- newCameraMatrix：新的相机内参矩阵，在默认情况下，它与 cameraMatrix 相同，但读者可以通过使用不同的矩阵来缩放和移动结果。

OpenCV 中还提供了另外一种用于畸变校正的方法，即通过 cv2.initUndistortRectifyMap() 和 cv2.remap() 的组合来处理。

initUndistortRectifyMap 函数的定义如下：

```
map1, map2 = initUndistortRectifyMap(cameraMatrix, distCoeffs, R, newCameraMatrix, size, m1type, map1=None, map2=None)
```

参数说明如下。

- cameraMatrix：相机内参矩阵。
- distCoeffs：畸变矩阵。
- R：可选的修正变换矩阵。
- newCameraMatrix：新的相机内参矩阵。
- size：未畸变的图像尺寸。
- m1type：第一个输出的映射类型。
- map1：第一个输出映射（返回值）。
- map2：第二个输出映射（返回值）。

remap 函数执行重映射操作，函数定义及案例见 5.4 节案例。

使用相机标定生成畸变参数，并消除图像失真的案例代码如下：

```
import cv2
import numpy as np
```

```python
import glob

#定义棋盘维度
CHECKERBOARD = (6,9)
criteria = (cv2.TERM_CRITERIA_EPS + cv2.TERM_CRITERIA_MAX_ITER, 30, 0.001)

#存储每个棋盘图像的三维点
objpoints = []
#存储每个棋盘图像的二维点
imgpoints = []

#定义三维点的世界坐标
objp = np.zeros((1, CHECKERBOARD[0]*CHECKERBOARD[1], 3), np.float32)
objp[0,:,:2] = np.mgrid[0:CHECKERBOARD[0], 0:CHECKERBOARD[1]].T.reshape(-1, 2)
prev_img_shape = None

#提取给定目录下的图像路径
images = glob.glob('./images/*.jpg')
for fname in images:
    img = cv2.imread(fname)
    gray = cv2.cvtColor(img,cv2.COLOR_BGR2GRAY)
    #寻找棋盘角点
    ret, corners = cv2.findChessboardCorners(gray, CHECKERBOARD, cv2.CALIB_CB_ADAPTIVE_THRESH+
        cv2.CALIB_CB_FAST_CHECK+cv2.CALIB_CB_NORMALIZE_IMAGE)

    #如果检测到所需的角点个数,则微调像素坐标并将其显示在棋盘格图像上
    if ret == True:
        objpoints.append(objp)
        #为给定的二位点微调像素坐标
        corners2 = cv2.cornerSubPix(gray,corners,(11,11),(-1,-1),criteria)
        imgpoints.append(corners2)
        #绘制并显示角点
        img = cv2.drawChessboardCorners(img, CHECKERBOARD, corners2,ret)
h,w = img.shape[:2]

#通过传输已知三维点(objpoints)和检测角点对应的像素坐标(imgpoints)执行相机标定操作
ret, mtx, dist, rvecs, tvecs = cv2.calibrateCamera(objpoints, imgpoints,
    gray.shape[::-1],None,None)

#使用导出的相机参数消除图像失真
img = cv2.imread(images[0])
#利用标定得到的参数优化相机矩阵
newcameramtx, roi = cv2.getOptimalNewCameraMatrix(mtx, dist, (w,h), 1, (w,h))

#图像消除失真方法1
```

```
dst1 = cv2.undistort(img, mtx, dist, None, newcameramtx)

#图像消除失真方法 2
mapx,mapy=cv2.initUndistortRectifyMap(mtx,dist,None,newcameramtx,(w,h),5)
dst2 = cv2.remap(img,mapx,mapy,cv2.INTER_LINEAR)

#结果展示
cv2.imshow("original image", img)
cv2.imshow("undistorted method1 image",dst1)
cv2.imshow("undistorted method2 image",dst2)
cv2.waitKey(0)
```

本案例使用的输入图像如图 11.9 所示。

方法 1 消除失真的结果如图 11.10 所示。

 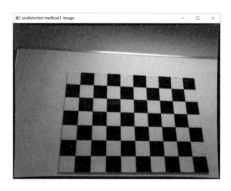

图 11.9　　　　　　　　　　　　　图 11.10

方法 2 消除失真的结果如图 11.11 所示。

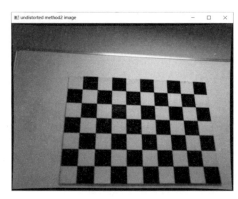

图 11.11

11.4 进阶必备：聊聊镜头失真

针孔相机模型是最简单的相机模型，该模型利用小孔成像的原理，将物体发出的光线经过小孔后在胶片上生成倒立的实像。成像小孔的孔径较小，通过针孔的光线就很少，因此，胶片需要很长时间才能获得充分曝光。镜头能聚焦光束，可以在胶片上产生清晰的影像，可以让胶片在很短的时间内获得适当的曝光，从而获得更高质量的图像。

但镜头会引入图像失真，这种失真是因为镜头制造和安装导致的，由此引入了两种主要类型的失真：径向畸变和切向畸变。

径向畸变是指沿着透镜半径方向分布的畸变，是由于透镜的形状导致了光传输过程中的不均匀弯曲所致的，光线在靠近透镜边缘的地方比靠近中心的地方弯曲得更多。因此，实际中直线传输的光线在到达图像传感器之前，从其理想位置径向向内或向外做了位移，在成像的图像上看起来是弯曲的，这种畸变在普通廉价的镜头中会表现得更为厉害，而高端镜头因为做了很多最小化径向畸变方面的工作则不是那么明显。

径向畸变分为桶形畸变（Barrel Distortions）和枕形畸变（Pincushion Distortions），如图 11.12 所示，其中，桶形畸变对应于负的径向位移，枕形畸变对应于正的径向位移。

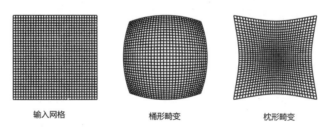

图 11.12

切向畸变是由镜头安装原因所致的，透镜与成像平面在安装后不平行，让图像看起来似乎被倾斜和拉伸了。

OpenCV 中提供了用于消除畸变的函数，根据畸变表示的数学模型，需要 5 个参数，分别为 k1、k2、p1、p2、k3（按顺序排列），它们被放在一个 5×1 的矩阵中作为函数的参数。而成像中还存在一些其他类型的畸变，因为这些畸变对成像的影响比径向畸变和切向畸变小，因此，OpenCV 中并没有提供更多的算法对其进行校正。

第 12 章
传统目标检测模块 objdetect

深度学习计算机视觉的惊人成绩让计算机视觉的传统算法在目标检测领域逐渐淡出人们的视野,但是在许多应用程序中,这些传统算法依旧发挥着重要的作用。它们在出现伊始也产生了轰动的效果,如 HOG 算法在行人检测方面的巨大优势,因此,若想深入学习图像处理,还是很有必要重温这些传统算法的。

12.1 模块导读

OpenCV 中的 objdetect 模块封装了传统计算机视觉的目标检测算法,引入该模块需要包含头文件"opencv2/objdetect.hpp",通过该头文件,读者可以了解模块封装的算法。该头文件的定义如下:

```
#ifndef OPENCV_OBJDETECT_HPP
#define OPENCV_OBJDETECT_HPP

#include "opencv2/core.hpp"

/** 基于 Haar 特征的级联目标检测分类器*/
typedef struct CvHaarClassifierCascade CvHaarClassifierCascade;

namespace cv
{
```

如下为 SimilarRects 类的定义,该类用于对候选对象进行分组,由级联分类器、HOG 等进行检测：
```cpp
class CV_EXPORTS SimilarRects
{
public:
    SimilarRects(double _eps) : eps(_eps) {}
    inline bool operator()(const Rect& r1, const Rect& r2) const
    {
        double delta = eps * ((std::min)(r1.width, r2.width) + (std::min)(r1.height, r2.height)) * 0.5;
        return std::abs(r1.x - r2.x) <= delta &&
            std::abs(r1.y - r2.y) <= delta &&
            std::abs(r1.x + r1.width - r2.x - r2.width) <= delta &&
            std::abs(r1.y + r1.height - r2.y - r2.height) <= delta;
    }
    double eps;
};
```

如下为 groupRectangles 函数的定义，该函数用于将目标候选矩形进行分组：
```cpp
CV_EXPORTS void groupRectangles(std::vector<Rect>& rectList, int groupThreshold,
                    double eps = 0.2);
CV_EXPORTS_W void groupRectangles(CV_IN_OUT std::vector<Rect>& rectList,
                    CV_OUT std::vector<int>& weights, int groupThreshold, double eps = 0.2);
CV_EXPORTS void groupRectangles(std::vector<Rect>& rectList, int groupThreshold,
                    double eps, std::vector<int>* weights, std::vector<double>* levelWeights );
CV_EXPORTS void groupRectangles(std::vector<Rect>& rectList, std::vector<int>& rejectLevels,
                    std::vector<double>& levelWeights, int groupThreshold, double eps = 0.2);
CV_EXPORTS void groupRectangles_meanshift(std::vector<Rect>& rectList,
                    std::vector<double>& foundWeights, std::vector<double>& foundScales,
                    double detectThreshold = 0.0, Size winDetSize = Size(64, 128));

template<> struct DefaultDeleter<CvHaarClassifierCascade>{
    CV_EXPORTS void operator ()(CvHaarClassifierCascade* obj) const;
};

enum { CASCADE_DO_CANNY_PRUNING      = 1,
       CASCADE_SCALE_IMAGE           = 2,
       CASCADE_FIND_BIGGEST_OBJECT   = 4,
       CASCADE_DO_ROUGH_SEARCH       = 8
};
```

如下为与级联分类器相关的类的定义：
```cpp
#基本级联分类器
class CV_EXPORTS_W BaseCascadeClassifier : public Algorithm
{
public:
    virtual ~BaseCascadeClassifier();
```

```cpp
    virtual bool empty() const CV_OVERRIDE = 0;
    virtual bool load( const String& filename ) = 0;
    virtual void detectMultiScale( InputArray image,
                     CV_OUT std::vector<Rect>& objects,
                     double scaleFactor,
                     int minNeighbors, int flags,
                     Size minSize, Size maxSize ) = 0;

    virtual void detectMultiScale( InputArray image,
                     CV_OUT std::vector<Rect>& objects,
                     CV_OUT std::vector<int>& numDetections,
                     double scaleFactor,
                     int minNeighbors, int flags,
                     Size minSize, Size maxSize ) = 0;

    virtual void detectMultiScale( InputArray image,
                         CV_OUT std::vector<Rect>& objects,
                         CV_OUT std::vector<int>& rejectLevels,
                         CV_OUT std::vector<double>& levelWeights,
                         double scaleFactor,
                         int minNeighbors, int flags,
                         Size minSize, Size maxSize,
                         bool outputRejectLevels ) = 0;

    virtual bool isOldFormatCascade() const = 0;
    virtual Size getOriginalWindowSize() const = 0;
    virtual int getFeatureType() const = 0;
    virtual void* getOldCascade() = 0;

    class CV_EXPORTS MaskGenerator
    {
    public:
        virtual ~MaskGenerator() {}
        virtual Mat generateMask(const Mat& src)=0;
        virtual void initializeMask(const Mat& /*src*/) { }
    };
    virtual void setMaskGenerator(const Ptr<MaskGenerator>& maskGenerator) = 0;
    virtual Ptr<MaskGenerator> getMaskGenerator() = 0;
};

/** 目标检测级联分类器类*/
class CV_EXPORTS_W CascadeClassifier
{
public:
    CV_WRAP CascadeClassifier();
    /** 从文件创建分类器*/
    CV_WRAP CascadeClassifier(const String& filename);
```

```cpp
    ~CascadeClassifier();
    /** 检查分类器加载是否成功*/
    CV_WRAP bool empty() const;
    /** 从文件加载分类器*/
    CV_WRAP bool load( const String& filename );
    /** 从 FileStorage 节点读取分类器*/
    CV_WRAP bool read( const FileNode& node );

    /** 检测输入图像中不同大小的对象，返回包围检测目标的矩形列表*/
    CV_WRAP void detectMultiScale( InputArray image,
                      CV_OUT std::vector<Rect>& objects,
                      double scaleFactor = 1.1,
                      int minNeighbors = 3, int flags = 0,
                      Size minSize = Size(),
                      Size maxSize = Size() );

    CV_WRAP_AS(detectMultiScale2) void detectMultiScale( InputArray image,
                      CV_OUT std::vector<Rect>& objects,
                      CV_OUT std::vector<int>& numDetections,
                      double scaleFactor=1.1,
                      int minNeighbors=3, int flags=0,
                      Size minSize=Size(),
                      Size maxSize=Size() );

    CV_WRAP_AS(detectMultiScale3) void detectMultiScale( InputArray image,
                          CV_OUT std::vector<Rect>& objects,
                          CV_OUT std::vector<int>& rejectLevels,
                          CV_OUT std::vector<double>& levelWeights,
                          double scaleFactor = 1.1,
                          int minNeighbors = 3, int flags = 0,
                          Size minSize = Size(),
                          Size maxSize = Size(),
                          bool outputRejectLevels = false );

    CV_WRAP bool isOldFormatCascade() const;
    CV_WRAP Size getOriginalWindowSize() const;
    CV_WRAP int getFeatureType() const;
    void* getOldCascade();

    CV_WRAP static bool convert(const String& oldcascade, const String& newcascade);
    void setMaskGenerator(const Ptr<BaseCascadeClassifier::MaskGenerator>& maskGenerator);
    Ptr<BaseCascadeClassifier::MaskGenerator> getMaskGenerator();
    Ptr<BaseCascadeClassifier> cc;
};

CV_EXPORTS Ptr<BaseCascadeClassifier::MaskGenerator> createFaceDetectionMaskGenerator();
```

```cpp
//! 感兴趣区域 ROI 检测
struct DetectionROI
{
    double scale;
    std::vector<cv::Point> locations;
    std::vector<double> confidences;
};
```

如下为 HOGDescriptor 类的定义，该类为 HOG 描述符和目标检测器的实现类：

```cpp
struct CV_EXPORTS_W HOGDescriptor
{
public:
    enum HistogramNormType { L2Hys = 0     //!< 默认的 histogramNormType
        };
    enum { DEFAULT_NLEVELS = 64            //!< 默认的 nlevels 值
        };
    enum DescriptorStorageFormat { DESCR_FORMAT_COL_BY_COL, DESCR_FORMAT_ROW_BY_ROW };

    /** 使用默认参数创建 HOGDescriptor 对象*/
    CV_WRAP HOGDescriptor() : winSize(64,128), blockSize(16,16),
    blockStride(8,8),cellSize(8,8), nbins(9), derivAperture(1), winSigma(-1),
    histogramNormType(HOGDescriptor::L2Hys), L2HysThreshold(0.2), gammaCorrection(true),
    free_coef(-1.f), nlevels(HOGDescriptor::DEFAULT_NLEVELS), signedGradient(false)
    {}
    /** 使用传递的参数创建 HOGDescriptor 对象*/
    CV_WRAP HOGDescriptor(Size _winSize, Size _blockSize, Size _blockStride,
            Size _cellSize, int _nbins, int _derivAperture=1,
            double _winSigma=-1,
            HOGDescriptor::HistogramNormType _histogramNormType=HOGDescriptor::L2Hys,
            double _L2HysThreshold=0.2, bool _gammaCorrection=false,
            int _nlevels=HOGDescriptor::DEFAULT_NLEVELS, bool _signedGradient=false)
    : winSize(_winSize), blockSize(_blockSize), blockStride(_blockStride), cellSize(_cellSize),
    nbins(_nbins), derivAperture(_derivAperture), winSigma(_winSigma),
    histogramNormType(_histogramNormType), L2HysThreshold(_L2HysThreshold),
    gammaCorrection(_gammaCorrection), free_coef(-1.f), nlevels(_nlevels),
    signedGradient(_signedGradient)
    {}

    CV_WRAP HOGDescriptor(const String& filename)
    {
        load(filename);
    }

    HOGDescriptor(const HOGDescriptor& d)
    {
```

```cpp
        d.copyTo(*this);
}

virtual ~HOGDescriptor() {}

/** 返回分类所需的系数*/
CV_WRAP size_t getDescriptorSize() const;

/** 检查探测器大小是否等于描述符大小*/
CV_WRAP bool checkDetectorSize() const;

/** 返回 winSigma 的值*/
CV_WRAP double getWinSigma() const;

/** 为线性 SVM 分类器设置系数*/
CV_WRAP virtual void setSVMDetector(InputArray svmdetector);

/** 从 FileNode 中读取 HOGDescriptor 参数*/
virtual bool read(FileNode& fn);

/** 保存 HOGDescriptor 参数*/
virtual void write(FileStorage& fs, const String& objname) const;

/** 为线性 SVM 分类器从文件中加载 HOGDescriptor 参数和系数*/
CV_WRAP virtual bool load(const String& filename, const String& objname = String());

/** 为线性 SVM 分类器保存 HOGDescriptor 参数和系数*/
CV_WRAP virtual void save(const String& filename, const String& objname = String()) const;

/** 克隆 HOGDescriptor 对象*/
virtual void copyTo(HOGDescriptor& c) const;

/** 计算给定图像的 HOG 描述符*/
CV_WRAP virtual void compute(InputArray img,
              CV_OUT std::vector<float>& descriptors,
              Size winStride = Size(), Size padding = Size(),
              const std::vector<Point>& locations = std::vector<Point>()) const;

/** 在没有多尺度窗口的情况下执行目标检测操作*/
CV_WRAP virtual void detect(InputArray img, CV_OUT std::vector<Point>& foundLocations,
              CV_OUT std::vector<double>& weights,
              double hitThreshold = 0, Size winStride = Size(),
              Size padding = Size(),
              const std::vector<Point>& searchLocations = std::vector<Point>()) const;

virtual void detect(InputArray img, CV_OUT std::vector<Point>& foundLocations,
```

```cpp
                double hitThreshold = 0, Size winStride = Size(),
                Size padding = Size(),
                const std::vector<Point>& searchLocations=std::vector<Point>()) const;

/** 多尺度目标检测*/
CV_WRAP virtual void detectMultiScale(InputArray img,
        CV_OUT std::vector<Rect>& foundLocations,
        CV_OUT std::vector<double>& foundWeights, double hitThreshold = 0,
        Size winStride = Size(), Size padding = Size(), double scale = 1.05,
        double finalThreshold = 2.0,bool useMeanshiftGrouping = false) const;
virtual void detectMultiScale(InputArray img, CV_OUT std::vector<Rect>& foundLocations,
        double hitThreshold = 0, Size winStride = Size(),
        Size padding = Size(), double scale = 1.05,
        double finalThreshold = 2.0, bool useMeanshiftGrouping = false) const;

/** 计算梯度和量化梯度方向*/
CV_WRAP virtual void computeGradient(InputArray img, InputOutputArray grad,
                InputOutputArray angleOfs,
                Size paddingTL = Size(), Size paddingBR = Size()) const;

CV_WRAP static std::vector<float> getDefaultPeopleDetector();
CV_WRAP static std::vector<float> getDaimlerPeopleDetector();
CV_PROP Size winSize;
CV_PROP Size blockSize;
 CV_PROP Size blockStride;
CV_PROP Size cellSize;
CV_PROP int nbins;
CV_PROP int derivAperture;
CV_PROP double winSigma;
CV_PROP HOGDescriptor::HistogramNormType histogramNormType;
CV_PROP double L2HysThreshold;
CV_PROP bool gammaCorrection;
CV_PROP std::vector<float> svmDetector;
UMat oclSvmDetector;
float free_coef;
CV_PROP int nlevels;
CV_PROP bool signedGradient;

/** 评估指定的 ROI 并返回每个位置的置信值*/
virtual void detectROI(InputArray img, const std::vector<cv::Point> &locations,
        CV_OUT std::vector<cv::Point>& foundLocations,
        CV_OUT std::vector<double>& confidences,
        double hitThreshold = 0, cv::Size winStride = Size(),
        cv::Size padding = Size()) const;

/** 在多尺度情况下评估指定的 ROI 并返回每个位置的置信值*/
virtual void detectMultiScaleROI(InputArray img,
```

```
                                    CV_OUT std::vector<cv::Rect>& foundLocations,
                                    std::vector<DetectionROI>& locations,
                                    double hitThreshold = 0,
                                    int groupThreshold = 0) const;

    /** 将对象候选矩形分组*/
    void groupRectangles(std::vector<cv::Rect>& rectList, std::vector<double>& weights,
                         int groupThreshold, double eps) const;
};
```

如下为二维码检测器类 QRCodeDetector 的定义：

```
class CV_EXPORTS_W QRCodeDetector
{
public:
    CV_WRAP QRCodeDetector();
    ~QRCodeDetector();

    /** 设置二维码水平扫描期间停止标记检测使用的 epsilon*/
    CV_WRAP void setEpsX(double epsX);
    /** 设置二维码垂直扫描期间停止标记检测使用的 epsilon*/
    CV_WRAP void setEpsY(double epsY);

    /** 图像中的二维码检测并返回二维码检测包围框矩形*/
    CV_WRAP bool detect(InputArray img, OutputArray points) const;

    /** 二维码解码*/
    CV_WRAP std::string decode(InputArray img, InputArray points,
                                    OutputArray straight_qrcode = noArray());

    /** 曲面二维码解码*/
    CV_WRAP cv::String decodeCurved(InputArray img, InputArray points,
                                    OutputArray straight_qrcode = noArray());

    /** 二维码检测并解码*/
    CV_WRAP std::string detectAndDecode(InputArray img, OutputArray points=noArray(),
                                    OutputArray straight_qrcode = noArray());

    /** 曲面二维码检测并解码*/
    CV_WRAP std::string detectAndDecodeCurved(InputArray img, OutputArray points=noArray(),
                                    OutputArray straight_qrcode = noArray());

    /** 检测图像中的多个二维码*/
    CV_WRAP
    bool detectMulti(InputArray img, OutputArray points) const;

    /** 多个二维码解码*/
```

```
    CV_WRAP
    bool decodeMulti(
            InputArray img, InputArray points,
            CV_OUT std::vector<std::string>& decoded_info,
            OutputArrayOfArrays straight_qrcode = noArray()
    ) const;

    /** 多个二维码检测并解码*/
    CV_WRAP
    bool detectAndDecodeMulti(
            InputArray img, CV_OUT std::vector<std::string>& decoded_info,
            OutputArray points = noArray(),
            OutputArrayOfArrays straight_qrcode = noArray()
    ) const;

protected:
    struct Impl;
    Ptr<Impl> p;
};
}

#include "opencv2/objdetect/detection_based_tracker.hpp"
#endif
```

> 提示：深度学习计算机视觉算法在目标检测等领域有着巨大的优势，如果读者没有尝试过深度学习的方法，则可能错过获得更好结果的机会。如果读者想尝试深度学习的方法，则可以参考本书笔者另外一本图书《深度学习计算机视觉实战》。

12.2 级联分类器的应用

12.2.1 案例 115：人脸检测

级联分类器是指将多个弱分类器（或简单分类器）级联形成一个强分类器（或复杂分类器），OpenCV 中封装了级联分类器的类 CascadeClassifier（见 12.1 节），该类提供了 load 成员函数，用于加载特征；提供了 detectMultiScale 成员函数，用于检测输入图像中的所有检测目标。

load 函数的定义如下：

retval = load(self, filename)

参数说明如下。

- filename：加载的分类器文件名。
- retval：函数调用状态，调用成功返回 true（返回值）。

detectMultiScale 函数的定义如下：

```
objects = detectMultiScale(self, image, scaleFactor=None, minNeighbors=None, flags=None,
minSize=None, maxSize=None)
```

参数说明如下。

- image：输入图像。
- scaleFactor：前后两次扫描搜索窗口的缩放比例。
- minNeighbors：指定认定检测目标存在需要的标记数量，默认值为 3，即有 3 个标记存在才能确定检测到目标（如人脸）。
- flags：与旧级联函数兼容而保留的相同参数，在新版本中可忽略。
- minSize：目标最小尺寸。
- maxSize：目标最大尺寸。
- objects：检测目标的包围矩形框向量（返回值）。

本节使用的人脸检测级联分类器为 Haar 级联分类器，OpenCV 提供了 Haar 级联分类器（在 OpenCV 源码路径下的 data/haarcascades 文件夹中），可用于人脸、人眼等的检测。

OpenCV 级联分类器进行人脸检测的案例代码如下：

```python
import cv2

#检测人脸并显示检测结果
def detectAndDisplay(image):
    gray = cv2.cvtColor(image, cv2.COLOR_BGR2GRAY)
    gray = cv2.equalizeHist(gray)

    #检测人脸
    faces = face_cascade.detectMultiScale(gray)
    for (x,y,w,h) in faces:
        image = cv2.rectangle(image, (x,y,w,h), color=(0,0,255), thickness=3)
    cv2.imshow('Face detection', image)
    cv2.waitKey(0)

face_cascade_name = "./haarcascades/haarcascade_frontalface_alt.xml"
#创建级联分类器
face_cascade = cv2.CascadeClassifier()
#加载特征文件
if not face_cascade.load(face_cascade_name):
    print('--(!)Error loading face cascade')
    exit(0)
```

```
#读取图像并进行人脸检测
img = cv2.imread("src.jpg")
detectAndDisplay(img)
```

人脸检测的显示结果如图 12.1 所示,多人脸检测的结果如图 12.2 所示。

图 12.1

图 12.2

OpenCV 中还提供了另外一种级联分类器——LBP 级联分类器,也可以用于人脸检测:

```
face_cascade_name = "./lbpcascades/lbpcascade_frontalface_improved.xml"
```

检测结果如图 12.3 所示。

图 12.3

如果读者想实时检测人脸,则可以调用摄像头设备,实时采集人脸图像并进行检测:

```
#视频帧捕获
cap = cv2.VideoCapture(camera_device)
if not cap.isOpened:
    print('--(!)Error opening video capture')
```

```
        exit(0)

while True:
    ret, frame = cap.read()
    if frame is None:
        print('--(!) No captured frame -- Break!')
        break
    #人脸检测并显示
    detectAndDisplay(frame)
    
    if cv2.waitKey(10) == 27:
        break
```

12.2.2 案例116：人眼检测

OpenCV 中提供了 Haar 特征人眼检测级联分类器，使用该级联分类器进行人眼检测的案例代码如下：

```
import cv2

#人脸与人眼检测
def detectAndDisplay(image):
    gray = cv2.cvtColor(image, cv2.COLOR_BGR2GRAY)
    gray = cv2.equalizeHist(gray)
    
    #人脸检测
    faces = face_cascade.detectMultiScale(gray)
    for (x,y,w,h) in faces:
        image = cv2.rectangle(image, (x, y, w, h), color=(0, 0, 255), thickness=3)
        faceROI = gray[y:y+h,x:x+w]
        #在人脸中检测人眼
        eyes = eyes_cascade.detectMultiScale(faceROI)
        for (x2,y2,w2,h2) in eyes:
            eye_center = (x + x2 + w2//2, y + y2 + h2//2)
            radius = int(round((w2 + h2)*0.25))
            image = cv2.circle(image, eye_center, radius, (255, 0, 0 ), 4)
    cv2.imshow('Eyes detection', image)
    cv2.waitKey(0)

face_cascade_name = "./haarcascades/haarcascade_frontalface_alt.xml"
eyes_cascade_name = "./haarcascades/haarcascade_eye_tree_eyeglasses.xml"

face_cascade = cv2.CascadeClassifier()
eyes_cascade = cv2.CascadeClassifier()

#加载分类器
```

```
if not face_cascade.load(face_cascade_name):
    print('--(!)Error loading face cascade')
    exit(0)
if not eyes_cascade.load(eyes_cascade_name):
    print('--(!)Error loading eyes cascade')
    exit(0)

#图像读取并检测
img = cv2.imread("src.jpg")
detectAndDisplay(img)
```

检测结果如图 12.4 所示。

图 12.4

12.3　案例 117：HOG 描述符行人检测

HOG（Histogram of Oriented Gradient，方向梯度直方图）算法是一种常见的特征提取算法，在行人检测中有着较好的效果，OpenCV 封装了 HOG 描述符类 HOGDescriptor（见 12.1 节）。

使用 HOG 描述符进行行人检测的案例代码如下：

```
import cv2

#图像读取
img = cv2.imread("src.jpg")
gray = cv2.cvtColor(img, cv2.COLOR_BGR2GRAY)

#创建 HOG 描述符
```

```
hog = cv2.HOGDescriptor()
hog.setSVMDetector(hog.getDefaultPeopleDetector())
#行人检测
(rects, scores) = hog.detectMultiScale(gray, winStride=(4,4), padding=(8,8),
                                       scale=1.25, useMeanshiftGrouping=False)
#检测结果绘制并显示
for (x, y, w, h) in rects:
    center = (x + w // 2, y + h // 2)
    img = cv2.rectangle(img, (x, y, w, h), color=(0, 0, 255), thickness=3)
cv2.imshow('Pedestrians detection', img)
cv2.waitKey(0)
cv2.destroyAllWindows()
```

HOG 描述符行人检测结果如图 12.5 所示。

图 12.5

提示：OpenCV 提供的 HOG 级联分类器（hogcascades/hogcascade_pedestrians.xml）在 OpenCV 3.x 之后就不再支持了。

12.4 二维码应用

二维码在我们的工作生活中发挥了重要作用，广泛应用于扫码支付、网络链接分享等场景，具有信息量大、易识别、成本低的特点。二维码有多种编码方式，常用的二维码为 QR Code。

二维码如果需要被识别，则第一步是在图像中检测到二维码，第二步是对检测到的二维码进行解码。

12.4.1 案例 118：二维码检测

OpenCV 中封装了二维码检测类 QRCodeDetector（见 12.1 节），其中用于二维码检测的成员函数为 detect，其定义如下：

```
retval, points = detect(self, img, points=None)
```

参数说明如下。
- img：输入图像。
- points：返回检测到包围二维码的最小四边形（返回值）。
- retval：函数调用状态，调用成功返回 true。

本案例使用的输入图像如图 12.6 所示。

使用 OpenCV 进行二维码检测的案例代码如下：

图 12.6

```
import cv2

#输入图像读取
image = cv2.imread("src.png")
#创建 QRCodeDetector 对象
qrDecoder = cv2.QRCodeDetector()
#二维码检测
_, bbox = qrDecoder.detect(image)
bbox = bbox.astype(int)
#检测结果显示
cv2.line(image, tuple(bbox[0][0]), tuple(bbox[0][1]), (255, 0, 0), 3)
cv2.line(image, tuple(bbox[0][1]), tuple(bbox[0][2]), (255, 0, 0), 3)
cv2.line(image, tuple(bbox[0][2]), tuple(bbox[0][3]), (255, 0, 0), 3)
cv2.line(image, tuple(bbox[0][3]), tuple(bbox[0][0]), (255, 0, 0), 3)
#结果显示
cv2.imshow("Results", image)
cv2.waitKey(0)
```

二维码检测结果如图 12.7 所示。

图 12.7

12.4.2 案例 119：二维码解码

QRCodeDetector 类的成员函数 decode 负责二维码的解码工作，其定义如下：

`retval, straight_qrcode = decode(self, img, points, straight_qrcode=None)`

参数说明如下。
- img：输入图像。
- points：检测到的包围二维码的矩形。
- straight_qrcode：可选的输出图像，该图像中包含经过校正和二值化的二维码（返回值）。
- retval：解码的数据结果（返回值）。

二维码解码的案例代码如下：

```
import cv2

#输入图像读取
image = cv2.imread("src.png")
#创建 QRCodeDetector 对象
qrDecoder = cv2.QRCodeDetector()
#二维码检测
_, bbox = qrDecoder.detect(image)
#二维码解码
data, _ = qrDecoder.decode(image, bbox)
print("Decoded Data : {}".format(data))
```

对检测到的二维码（见图 12.7）进行解码，结果如下：

```
Decoded Data :
《深度学习计算机视觉实战》肖铃，刘东著：
本书共有近 50 个案例，全书内容共分为四个部分：一、深度学习和计算机视觉算法理论基础；二、OpenCV 图像处理讲解；三、计算机视觉实战项目；四、模型部署。
```

QRCodeDetector 类的成员函数 detectAndDecode 封装了二维码检测和解码的功能，其定义如下：

`retval, points, straight_qrcode = detectAndDecode (self, img, points, straight_qrcode=None)`

参数说明如下。
- img：输入图像。
- points：检测到的包围二维码的矩形（返回值）。
- straight_qrcode：可选的输出图像，包含校正和二值化的二维码（返回值）。
- retval：解码的结果数据（返回值）。

使用 detectAndDecode 函数进行二维码检测和解码的案例代码如下：

```
import cv2
```

```
import time

#输入图像读取
image = cv2.imread("src.png")
#创建 QRCodeDetector 对象
qrDecoder = cv2.QRCodeDetector()
#二维码检测
t = time.time()
data, bbox, _ = qrDecoder.detectAndDecode(image)
#检测耗时输出
print("Time Taken for Detect and Decode : {:.3f} seconds".format(time.time() - t))
#二维码解码结果输出
print("Decoded Data : {}".format(data))
#检测结果显示
bbox = bbox.astype(int)
cv2.line(image, tuple(bbox[0][0]), tuple(bbox[0][1]), (255, 0, 0), 3)
cv2.line(image, tuple(bbox[0][1]), tuple(bbox[0][2]), (255, 0, 0), 3)
cv2.line(image, tuple(bbox[0][2]), tuple(bbox[0][3]), (255, 0, 0), 3)
cv2.line(image, tuple(bbox[0][3]), tuple(bbox[0][0]), (255, 0, 0), 3)
#结果显示
cv2.imshow("QRCode Detect", image)
cv2.waitKey(0)
```

二维码检测并解码的时间和解码结果输出信息如下：

```
Time Taken for Detect and Decode : 0.839 seconds
Decoded Data : 《深度学习计算机视觉实战》肖铃,刘东著:
本书共有近 50 个案例,全书内容共分为四个部分：一、深度学习和计算机视觉算法理论基础；二、OpenCV 图像处理讲解；三、计算机视觉实战项目；四、模型部署。
```

提示：读者可以在网上找到二维码生成平台，如果想使用 OpenCV 正确解码，则在二维码生成时注意码制选择 "QR Code"。

12.5 进阶必备：聊聊条形码与二维码

条形码（或一维码）是一种信息自动识别技术，图案由反射率相差极大的宽度不等的黑白条平行排列而成。条形码技术实现了信息快速录入计算机，多见于超市商品和图书上，该技术实现了物品的自动管理，在零售业、图书管理、仓储物流等许多领域有着广泛应用。

常用的条形码编码方式包括 EAN 码、39 码、交叉 25 码、UPC 码、128 码、93 码、ISBN 码等，图 12.8 为图书中常用的 ISBN 码条形码。

二维码因为其信息量大、储存信息多样、成本低廉等优点在近几年开始渗入生活的方方面面，如付款码、信息分享的二维码等。二维码是在平面（或二维方向）上按一定规律分布的、黑白或彩色与白色相间的、记录数据符号信息的图形，如图 12.9 所示。

图 12.8

图 12.9

常用的二维码编码方式有 PDF417、Data Matrix、QR Code、Code 49、Code 16K、Code one 等，如 12.4 节中的案例使用的二维码编码方式为 QR Code。

条形码与二维码的对比如下。

- 二维码包含的信息量远大于条形码包含的信息量：条形码包含的内容只能是字母和数字，尺寸较大，因而空间利用率较低，容纳的数据量一般为 30 个字符左右。二维码包含的内容为字母、数字、汉字、字符、片假名等，最大可承载信息量为 1850 个字符，信息内容多样且信息量巨大。
- 两种条码的信息表达方式不同：条形码只能在水平方向上表达信息，在垂直方向上不表达信息，具有一定的高度（为了便于条码设备的对准，从而进行信息读取）。二维码在二维层面上都有信息表达，因此可以存储的信息量更大。

条形码纠错能力较差，若条码有破损，则不能被读取；二维码在有部分破损的情况下，信息也可以被正常读取，纠错率高达 30%。

第 13 章
机器学习模块 ml

机器学习是人工智能的重要学科之一，主要研究以往经验或数据优化计算机算法，以获取新的知识或对新数据进行预测。机器学习涵盖概率论、统计学、近似理论和复杂算法多个领域，是一个多学科交叉专业。传统机器学习的主要研究方向包括决策树、随机森林、人工神经网络、贝叶斯学习等；而在大数据环境下，机器学习采用分布式和并行计算，开始对复杂多样的数据进行深层次的分析。另外，人工神经网络演变的深度学习是当下人工智能最火热的方向之一。OpenCV 中的机器学习模块 ml 封装了常用的机器学习算法，包括统计分类、回归和支持向量机等。

13.1 模块导读

机器学习模块 ml 的引用可以通过包含头文件"opencv2/ml.hpp"来实现，该头文件中定义了 OpenCV 封装的机器学习算法。该头文件的内容如下：

```
#ifndef OPENCV_ML_HPP
#define OPENCV_ML_HPP

#ifdef __cplusplus
# include "opencv2/core.hpp"
#endif

#ifdef __cplusplus
#include <float.h>
#include <map>
```

```cpp
#include <iostream>

namespace cv
{
namespace ml
{
```

如下为机器学习模块的一些枚举类型标志的定义：

```cpp
/** 变量类型 */
enum VariableTypes
{
    VAR_NUMERICAL      =0, //!< 同 VAR_ORDERED
    VAR_ORDERED        =0, //!< 有序变量
    VAR_CATEGORICAL    =1  //!< 分类变量
};

/** 错误类型 */
enum ErrorTypes
{
    TEST_ERROR = 0,
    TRAIN_ERROR = 1
};

/** 样本类型 */
enum SampleTypes
{
    ROW_SAMPLE = 0, //!< 每个训练样本都是一行样本
    COL_SAMPLE = 1  //!< 每个训练样本都是一列样本
};
```

如下为 ParamGrid 类的定义，该类用于表示 statmodel 参数的对数网格范围：

```cpp
class CV_EXPORTS_W ParamGrid
{
public:
    ParamGrid();
    ParamGrid(double _minVal, double _maxVal, double _logStep);

    CV_PROP_RW double minVal; //!< statmodel 参数的最小值，默认值为 0
    CV_PROP_RW double maxVal; //!< statmodel 参数的最大值，默认值为 0
    /** 迭代 statmodel 参数的对数步长*/
    CV_PROP_RW double logStep;

    /** 创建 ParamGrid 指针对象*/
    CV_WRAP static Ptr<ParamGrid> create(double minVal=0., double maxVal=0., double logstep=1.);
};
```

如下为 TrainData 类的定义，该类为训练数据封装的类：

```
class CV_EXPORTS_W TrainData
{
public:
    static inline float missingValue() { return FLT_MAX; }
    virtual ~TrainData();

    CV_WRAP virtual int getLayout() const = 0;
    CV_WRAP virtual int getNTrainSamples() const = 0;
    CV_WRAP virtual int getNTestSamples() const = 0;
    CV_WRAP virtual int getNSamples() const = 0;
    CV_WRAP virtual int getNVars() const = 0;
    CV_WRAP virtual int getNAllVars() const = 0;

    CV_WRAP virtual void getSample(InputArray varIdx, int sidx, float* buf) const = 0;
    CV_WRAP virtual Mat getSamples() const = 0;
    CV_WRAP virtual Mat getMissing() const = 0;

    /** 返回训练样本矩阵*/
    CV_WRAP virtual Mat getTrainSamples(int layout=ROW_SAMPLE,
                                        bool compressSamples=true,
                                        bool compressVars=true) const = 0;

    CV_WRAP virtual Mat getTrainResponses() const = 0;

    CV_WRAP virtual Mat getTrainNormCatResponses() const = 0;
    CV_WRAP virtual Mat getTestResponses() const = 0;
    CV_WRAP virtual Mat getTestNormCatResponses() const = 0;
    CV_WRAP virtual Mat getResponses() const = 0;
    CV_WRAP virtual Mat getNormCatResponses() const = 0;
    CV_WRAP virtual Mat getSampleWeights() const = 0;
    CV_WRAP virtual Mat getTrainSampleWeights() const = 0;
    CV_WRAP virtual Mat getTestSampleWeights() const = 0;
    CV_WRAP virtual Mat getVarIdx() const = 0;
    CV_WRAP virtual Mat getVarType() const = 0;
    CV_WRAP virtual Mat getVarSymbolFlags() const = 0;
    CV_WRAP virtual int getResponseType() const = 0;
    CV_WRAP virtual Mat getTrainSampleIdx() const = 0;
    CV_WRAP virtual Mat getTestSampleIdx() const = 0;
    CV_WRAP virtual void getValues(int vi, InputArray sidx, float* values) const = 0;
    virtual void getNormCatValues(int vi, InputArray sidx, int* values) const = 0;
    CV_WRAP virtual Mat getDefaultSubstValues() const = 0;

    CV_WRAP virtual int getCatCount(int vi) const = 0;
```

```cpp
/** 返回类标签*/
CV_WRAP virtual Mat getClassLabels() const = 0;

CV_WRAP virtual Mat getCatOfs() const = 0;
CV_WRAP virtual Mat getCatMap() const = 0;

/** 将训练数据分为训练集和测试集*/
CV_WRAP virtual void setTrainTestSplit(int count, bool shuffle=true) = 0;

/** 设置训练集和测试集划分比率*/
CV_WRAP virtual void setTrainTestSplitRatio(double ratio, bool shuffle=true) = 0;
CV_WRAP virtual void shuffleTrainTest() = 0;

/** 返回测试样本 */
CV_WRAP virtual Mat getTestSamples() const = 0;
CV_WRAP virtual void getNames(std::vector<String>& names) const = 0;
static CV_WRAP Mat getSubVector(const Mat& vec, const Mat& idx);
static CV_WRAP Mat getSubMatrix(const Mat& matrix, const Mat& idx, int layout);

/** 从.csv 文件读取数据集并返回可用的 TrainData 对象*/
static Ptr<TrainData> loadFromCSV(const String& filename,
                                  int headerLineCount,
                                  int responseStartIdx=-1,
                                  int responseEndIdx=-1,
                                  const String& varTypeSpec=String(),
                                  char delimiter=',',
                                  char missch='?');

/** 从内存创建 TrainData 对象*/
CV_WRAP static Ptr<TrainData> create(InputArray samples, int layout,
            InputArray responses,InputArray varIdx=noArray(),
            InputArray sampleIdx=noArray(), InputArray sampleWeights=noArray(),
            InputArray varType=noArray());
};
```

如下为 StatModel 类的定义，该类为统计模型基类：

```cpp
class CV_EXPORTS_W StatModel : public Algorithm
{
public:
    enum Flags {
        UPDATE_MODEL = 1,
        RAW_OUTPUT=1,
        COMPRESSED_INPUT=2,
        PREPROCESSED_INPUT=4
    };
```

```cpp
/** 返回训练样本中的变量数量 */
CV_WRAP virtual int getVarCount() const = 0;
CV_WRAP virtual bool empty() const CV_OVERRIDE;
CV_WRAP virtual bool isTrained() const = 0;
CV_WRAP virtual bool isClassifier() const = 0;

/** 训练统计模型*/
CV_WRAP virtual bool train( const Ptr<TrainData>& trainData, int flags=0 );
CV_WRAP virtual bool train( InputArray samples, int layout, InputArray responses );

/** 计算训练集和测试集误差*/
CV_WRAP virtual float calcError( const Ptr<TrainData>& data, bool test,
                    OutputArray resp ) const;

/** 根据样本进行预测*/
CV_WRAP virtual float predict( InputArray samples, OutputArray results=noArray(),
                    int flags=0 ) const = 0;

/** 使用默认参数创建训练模型*/
template<typename _Tp> static Ptr<_Tp> train(const Ptr<TrainData>& data, int flags=0)
{
    Ptr<_Tp> model = _Tp::create();
    return !model.empty() && model->train(data, flags) ? model : Ptr<_Tp>();
}
};
```

如下为 NormalBayesClassifier 类的定义，该类为正态分布数据的贝叶斯分类器的实现类：

```cpp
class CV_EXPORTS_W NormalBayesClassifier : public StatModel
{
public:
    /** 预测样本响应*/
    CV_WRAP virtual float predictProb( InputArray inputs, OutputArray outputs,
                        OutputArray outputProbs, int flags=0 ) const = 0;
    CV_WRAP static Ptr<NormalBayesClassifier> create();
    /** 从文件加载并创建序列化的 NormalBayesClassifier 对象*/
    CV_WRAP static Ptr<NormalBayesClassifier> load(const String& filepath , const String& nodeName = String());
};
```

如下为 KNearest 类的定义，该类为 K 最近邻（KNN）算法的实现类：

```cpp
class CV_EXPORTS_W KNearest : public StatModel
{
public:
    /** 模型预测使用的默认邻居数量 */
```

```
CV_WRAP virtual int getDefaultK() const = 0;
CV_WRAP virtual void setDefaultK(int val) = 0;

CV_WRAP virtual bool getIsClassifier() const = 0;
CV_WRAP virtual void setIsClassifier(bool val) = 0;

/** KDTree 参数实现*/
CV_WRAP virtual int getEmax() const = 0;
CV_WRAP virtual void setEmax(int val) = 0;

/** 算法类型*/
CV_WRAP virtual int getAlgorithmType() const = 0;
CV_WRAP virtual void setAlgorithmType(int val) = 0;

/** 查找邻居并预测输入向量的响应*/
CV_WRAP virtual float findNearest( InputArray samples, int k,
                    OutputArray results,
                    OutputArray neighborResponses=noArray(),
                    OutputArray dist=noArray() ) const = 0;

enum Types
{
    BRUTE_FORCE=1,
    KDTREE=2
};

/** 创建空模型*/
CV_WRAP static Ptr<KNearest> create();
/** 从文件中加载模型并创建 KNearest 对象*/
CV_WRAP static Ptr<KNearest> load(const String& filepath);
};
```

如下为 SVM 类的定义，该类实现了支持向量机算法（见案例 121）：

```
class CV_EXPORTS_W SVM : public StatModel
{
public:
    class CV_EXPORTS Kernel : public Algorithm
    {
    public:
        virtual int getType() const = 0;
        virtual void calc( int vcount, int n, const float* vecs, const float* another, float* results ) = 0;
    };

    CV_WRAP virtual int getType() const = 0;
    CV_WRAP virtual void setType(int val) = 0;
```

```cpp
CV_WRAP virtual double getGamma() const = 0;
CV_WRAP virtual void setGamma(double val) = 0;

CV_WRAP virtual double getCoef0() const = 0;
CV_WRAP virtual void setCoef0(double val) = 0;

CV_WRAP virtual double getDegree() const = 0;
CV_WRAP virtual void setDegree(double val) = 0;

CV_WRAP virtual double getC() const = 0;
CV_WRAP virtual void setC(double val) = 0;

CV_WRAP virtual double getNu() const = 0;
CV_WRAP virtual void setNu(double val) = 0;

CV_WRAP virtual double getP() const = 0;
CV_WRAP virtual void setP(double val) = 0;

CV_WRAP virtual cv::Mat getClassWeights() const = 0;
CV_WRAP virtual void setClassWeights(const cv::Mat &val) = 0;

CV_WRAP virtual cv::TermCriteria getTermCriteria() const = 0;
CV_WRAP virtual void setTermCriteria(const cv::TermCriteria &val) = 0;

CV_WRAP virtual int getKernelType() const = 0;

CV_WRAP virtual void setKernel(int kernelType) = 0;

virtual void setCustomKernel(const Ptr<Kernel> &_kernel) = 0;

//! SVM 类型
enum Types {
    /** 带惩罚乘法器 C 的支持向量分类*/
    C_SVC=100,
    /** NU 类 SVC*/
    NU_SVC=101,
    /** 分布估计（单类 SVM）*/
    ONE_CLASS=102,
    /** 支持向量回归*/
    EPS_SVR=103,
    /** NU 支持向量回归*/
    NU_SVR=104
};

/** SVM Kernel 类型*/
enum KernelTypes {
```

```cpp
    /** 自定义内核*/
    CUSTOM=-1,
    /** 线性内核*/
    LINEAR=0,
    /** Polynomial 内核*/
    POLY=1,
    /** 径向基函数（RBF）内核*/
    RBF=2,
    /** Sigmoid 内核*/
    SIGMOID=3,
    /** Exponential Chi2 内核*/
    CHI2=4,
    /** 直方图相交内核*/
    INTER=5
};

//! SVM 参数类型
enum ParamTypes {
    C=0,
    GAMMA=1,
    P=2,
    NU=3,
    COEF=4,
    DEGREE=5
};

/** 用最优参数训练 SVM*/
virtual bool trainAuto( const Ptr<TrainData>& data, int kFold = 10,
            ParamGrid Cgrid = getDefaultGrid(C),
            ParamGrid gammaGrid     = getDefaultGrid(GAMMA),
            ParamGrid pGrid         = getDefaultGrid(P),
            ParamGrid nuGrid        = getDefaultGrid(NU),
            ParamGrid coeffGrid     = getDefaultGrid(COEF),
            ParamGrid degreeGrid    = getDefaultGrid(DEGREE),
            bool balanced=false)    = 0;

CV_WRAP virtual bool trainAuto(InputArray samples,
        int layout,
        InputArray responses,
        int kFold = 10,
        Ptr<ParamGrid> Cgrid = SVM::getDefaultGridPtr(SVM::C),
        Ptr<ParamGrid> gammaGrid    = SVM::getDefaultGridPtr(SVM::GAMMA),
        Ptr<ParamGrid> pGrid        = SVM::getDefaultGridPtr(SVM::P),
        Ptr<ParamGrid> nuGrid       = SVM::getDefaultGridPtr(SVM::NU),
        Ptr<ParamGrid> coeffGrid    = SVM::getDefaultGridPtr(SVM::COEF),
        Ptr<ParamGrid> degreeGrid   = SVM::getDefaultGridPtr(SVM::DEGREE),
```

```cpp
            bool balanced=false) = 0;

    /** 检索所有支持向量*/
    CV_WRAP virtual Mat getSupportVectors() const = 0;
    /** 检索线性 SVM 的所有未压缩支持向量*/
    CV_WRAP virtual Mat getUncompressedSupportVectors() const = 0;
    /** 检索决策函数*/
    CV_WRAP virtual double getDecisionFunction(int i, OutputArray alpha,
                        OutputArray svidx) const = 0;
    static ParamGrid getDefaultGrid( int param_id );
    CV_WRAP static Ptr<ParamGrid> getDefaultGridPtr( int param_id );
    /** 创建空模型*/
    CV_WRAP static Ptr<SVM> create();
    /** 从文件加载模型并创建 SVM 对象*/
    CV_WRAP static Ptr<SVM> load(const String& filepath);
};
```

如下为 EM 类的定义,该类为期望最大化算法 EM 的实现类:

```cpp
class CV_EXPORTS_W EM : public StatModel
{
public:
    //! 协变矩阵类型
    enum Types {
        COV_MAT_SPHERICAL=0,
        COV_MAT_DIAGONAL=1,
        COV_MAT_GENERIC=2,
        COV_MAT_DEFAULT=COV_MAT_DIAGONAL
    };
    enum {DEFAULT_NCLUSTERS=5, DEFAULT_MAX_ITERS=100};
    enum {START_E_STEP=1, START_M_STEP=2, START_AUTO_STEP=0};

    /** 高斯混合模型中混合成分的个数*/
    CV_WRAP virtual int getClustersNumber() const = 0;
    CV_WRAP virtual void setClustersNumber(int val) = 0;

    /** 定义矩阵类型的协方差矩阵的约束*/
    CV_WRAP virtual int getCovarianceMatrixType() const = 0;
    CV_WRAP virtual void setCovarianceMatrixType(int val) = 0;

    /** EM 算法的终止条件*/
    CV_WRAP virtual TermCriteria getTermCriteria() const = 0;
    CV_WRAP virtual void setTermCriteria(const TermCriteria &val) = 0;

    CV_WRAP virtual Mat getWeights() const = 0;
    CV_WRAP virtual Mat getMeans() const = 0;
```

```cpp
    CV_WRAP virtual void getCovs(CV_OUT std::vector<Mat>& covs) const = 0;

    /** 返回所提供样本的后验概率*/
    CV_WRAP virtual float predict( InputArray samples, OutputArray results=noArray(),
                    int flags=0 ) const CV_OVERRIDE = 0;

    /** 返回给定样本的最大可能混合成分的似然对数值和索引*/
    CV_WRAP virtual Vec2d predict2(InputArray sample, OutputArray probs) const = 0;

    /** 根据样本集估计高斯混合参数*/
    CV_WRAP virtual bool trainEM(InputArray samples,
                        OutputArray logLikelihoods=noArray(),
                        OutputArray labels=noArray(),
                        OutputArray probs=noArray()) = 0;

    CV_WRAP virtual bool trainE(InputArray samples, InputArray means0,
                        InputArray covs0=noArray(),
                        InputArray weights0=noArray(),
                        OutputArray logLikelihoods=noArray(),
                        OutputArray labels=noArray(),
                        OutputArray probs=noArray()) = 0;

    CV_WRAP virtual bool trainM(InputArray samples, InputArray probs0,
                        OutputArray logLikelihoods=noArray(),
                        OutputArray labels=noArray(),
                        OutputArray probs=noArray()) = 0;

    /** 创建空的 EM 对象*/
    CV_WRAP static Ptr<EM> create();

    /** 从文件中加载模型并创建 EM 对象*/*/
    CV_WRAP static Ptr<EM> load(const String& filepath , const String& nodeName = String());
};
```

如下为 DTrees 类的定义，该类实现了决策树算法：

```cpp
class CV_EXPORTS_W DTrees : public StatModel
{
public:
    /** Predict 选项*/
    enum Flags {
                PREDICT_AUTO=0,
                PREDICT_SUM=(1<<8),
                PREDICT_MAX_VOTE=(2<<8),
                PREDICT_MASK=(3<<8)
    };

    /** 将分类变量的可能值聚类到 K<=maxCategories 聚类中，以找到次优分割*/
```

```cpp
CV_WRAP virtual int getMaxCategories() const = 0;
CV_WRAP virtual void setMaxCategories(int val) = 0;

/** 树的最大可能深度*/
CV_WRAP virtual int getMaxDepth() const = 0;
CV_WRAP virtual void setMaxDepth(int val) = 0;

/** 如果节点中的样本数小于此参数，则不会拆分节点，默认值为 10*/
CV_WRAP virtual int getMinSampleCount() const = 0;
CV_WRAP virtual void setMinSampleCount(int val) = 0;

CV_WRAP virtual int getCVFolds() const = 0;
CV_WRAP virtual void setCVFolds(int val) = 0;

CV_WRAP virtual bool getUseSurrogates() const = 0;
CV_WRAP virtual void setUseSurrogates(bool val) = 0;

CV_WRAP virtual bool getUse1SERule() const = 0;
CV_WRAP virtual void setUse1SERule(bool val) = 0;

CV_WRAP virtual bool getTruncatePrunedTree() const = 0;
CV_WRAP virtual void setTruncatePrunedTree(bool val) = 0;

/** 回归树的终止准则*/
CV_WRAP virtual float getRegressionAccuracy() const = 0;
CV_WRAP virtual void setRegressionAccuracy(float val) = 0;

/** 按类标签值排序的先验类概率数组*/
CV_WRAP virtual cv::Mat getPriors() const = 0;
CV_WRAP virtual void setPriors(const cv::Mat &val) = 0;

/** 决策树节点类*/
class CV_EXPORTS Node
{
public:
    Node();
    double value;
    int classIdx;
    int parent;
    int left;
    int right;
    int defaultDir;
    int split;
};

/** 决策树拆分类*/
```

```cpp
    class CV_EXPORTS Split
    {
    public:
        Split();
        int varIdx;
        bool inversed;
        float quality;
        int next;
        float c;
        int subsetOfs;
    };

    virtual const std::vector<int>& getRoots() const = 0;
    virtual const std::vector<Node>& getNodes() const = 0;
    virtual const std::vector<Split>& getSplits() const = 0;
    virtual const std::vector<int>& getSubsets() const = 0;
    CV_WRAP static Ptr<DTrees> create();
    CV_WRAP static Ptr<DTrees> load(const String& filepath , const String& nodeName = String());
};
```

如下为 RTrees 类的定义，该类实现了随机树分类器：

```cpp
class CV_EXPORTS_W RTrees : public DTrees
{
public:
    CV_WRAP virtual bool getCalculateVarImportance() const = 0;
    CV_WRAP virtual void setCalculateVarImportance(bool val) = 0;

    CV_WRAP virtual int getActiveVarCount() const = 0;
    CV_WRAP virtual void setActiveVarCount(int val) = 0;

    /** 指定训练算法何时停止的条件*/
    CV_WRAP virtual TermCriteria getTermCriteria() const = 0;
    CV_WRAP virtual void setTermCriteria(const TermCriteria &val) = 0;

    CV_WRAP virtual Mat getVarImportance() const = 0;
    CV_WRAP virtual void getVotes(InputArray samples, OutputArray results, int flags) const = 0;
#if CV_VERSION_MAJOR == 4
    CV_WRAP virtual double getOOBError() const { return 0; }
#else
    virtual double getOOBError() const = 0;
#endif
    /** 创建空的 RTrees 对象*/
    CV_WRAP static Ptr<RTrees> create();
    /** 从文件中加载模型并创建 RTrees 对象*/
    CV_WRAP static Ptr<RTrees> load(const String& filepath , const String& nodeName = String());
};
```

如下为 Boost 类的定义，该类实现了增强树分类器：

```cpp
class CV_EXPORTS_W Boost : public DTrees
{
public:
    CV_WRAP virtual int getBoostType() const = 0;
    CV_WRAP virtual void setBoostType(int val) = 0;

    CV_WRAP virtual int getWeakCount() const = 0;
    CV_WRAP virtual void setWeakCount(int val) = 0;

    /** 用于节省计算时间的介于 0 和 1 之间的阈值*/
    CV_WRAP virtual double getWeightTrimRate() const = 0;
    CV_WRAP virtual void setWeightTrimRate(double val) = 0;

    /** Boosting 类型*/
    enum Types {
        DISCRETE=0, //!< 离散 AdaBoost
        REAL=1, //!< 实数 AdaBoost
        LOGIT=2, //!< LogitBoost
        GENTLE=3 //!< Gentle AdaBoost
    };

    /** 创建空模型*/
    CV_WRAP static Ptr<Boost> create();

    /** 从文件中加载模型并创建 Boost 对象*/
    CV_WRAP static Ptr<Boost> load(const String& filepath , const String& nodeName = String());
};

/** 提示：源码头文件中梯度增强树 GBTrees : public DTrees 被注释掉了，此处就不做展示了*/
```

如下为 ANN_MLP 类的定义，该类实现了人工神经网络（ANN）、多层感知机（MLP）算法：

```cpp
class CV_EXPORTS_W ANN_MLP : public StatModel
{
public:
    /** 可用训练方法*/
    enum TrainingMethods {
        BACKPROP=0, //!< 反向传播算法
        RPROP = 1, //!< RPROP 算法
        ANNEAL = 2 //!< 模拟退火算法
    };

    /** 设置训练方法和通用参数*/
    CV_WRAP virtual void setTrainMethod(int method, double param1 = 0, double param2 = 0) = 0;
```

```cpp
    CV_WRAP virtual int getTrainMethod() const = 0;

    /** 初始化每个神经元的激活功能*/
    CV_WRAP virtual void setActivationFunction(int type, double param1 = 0, double param2 = 0) = 0;

    /** 整数向量，指定每层（包括输入层和输出层）中的神经元数量*/
    CV_WRAP virtual void setLayerSizes(InputArray _layer_sizes) = 0;
    CV_WRAP virtual cv::Mat getLayerSizes() const = 0;

    /** 训练算法终止条件*/
    CV_WRAP virtual TermCriteria getTermCriteria() const = 0;
    CV_WRAP virtual void setTermCriteria(TermCriteria val) = 0;

    CV_WRAP virtual double getBackpropWeightScale() const = 0;
    CV_WRAP virtual void setBackpropWeightScale(double val) = 0;

    CV_WRAP virtual double getBackpropMomentumScale() const = 0;
    CV_WRAP virtual void setBackpropMomentumScale(double val) = 0;

    CV_WRAP virtual double getRpropDW0() const = 0;
    CV_WRAP virtual void setRpropDW0(double val) = 0;

    CV_WRAP virtual double getRpropDWPlus() const = 0;
    CV_WRAP virtual void setRpropDWPlus(double val) = 0;

    CV_WRAP virtual double getRpropDWMinus() const = 0;
    CV_WRAP virtual void setRpropDWMinus(double val) = 0;

    CV_WRAP virtual double getRpropDWMin() const = 0;
    CV_WRAP virtual void setRpropDWMin(double val) = 0;

    CV_WRAP virtual double getRpropDWMax() const = 0;
    CV_WRAP virtual void setRpropDWMax(double val) = 0;

    CV_WRAP virtual double getAnnealInitialT() const = 0;
    CV_WRAP virtual void setAnnealInitialT(double val) = 0;

    CV_WRAP virtual double getAnnealFinalT() const = 0;
    CV_WRAP virtual void setAnnealFinalT(double val) = 0;

    CV_WRAP virtual double getAnnealCoolingRatio() const = 0;
    CV_WRAP virtual void setAnnealCoolingRatio(double val) = 0;

    CV_WRAP virtual int getAnnealItePerStep() const = 0;
    CV_WRAP virtual void setAnnealItePerStep(int val) = 0;
```

```cpp
    virtual void setAnnealEnergyRNG(const RNG& rng) = 0;

    /** 可选的激活函数 */
    enum ActivationFunctions {
        IDENTITY = 0,
        SIGMOID_SYM = 1,
        GAUSSIAN = 2,
        RELU = 3,
        LEAKYRELU= 4
    };
    /** 训练选项 */
    enum TrainFlags {
        UPDATE_WEIGHTS = 1,
        NO_INPUT_SCALE = 2,
        NO_OUTPUT_SCALE = 4
    };
    CV_WRAP virtual Mat getWeights(int layerIdx) const = 0;
    /** 创建空模型*/
    CV_WRAP static Ptr<ANN_MLP> create();
    /** 从文件中加载模型并创建 ANN_MLP 对象*/
    CV_WRAP static Ptr<ANN_MLP> load(const String& filepath);
};

#ifndef DISABLE_OPENCV_3_COMPATIBILITY
typedef ANN_MLP ANN_MLP_ANNEAL;
#endif
```

如下为 LogisticRegression 类的定义，该类实现了 Logistic 回归（见案例 120）：

```cpp
class CV_EXPORTS_W LogisticRegression : public StatModel
{
public:
    /** 学习率*/
    CV_WRAP virtual double getLearningRate() const = 0;
    CV_WRAP virtual void setLearningRate(double val) = 0;

    /** 迭代次数*/
    CV_WRAP virtual int getIterations() const = 0;
    CV_WRAP virtual void setIterations(int val) = 0;

    /** 正则化类型*/
    CV_WRAP virtual int getRegularization() const = 0;
    CV_WRAP virtual void setRegularization(int val) = 0;

    /** 训练方法*/
    CV_WRAP virtual int getTrainMethod() const = 0;
    CV_WRAP virtual void setTrainMethod(int val) = 0;
```

```cpp
/** 指定在小批量梯度下降的每个步骤中获取的训练样本数*/
CV_WRAP virtual int getMiniBatchSize() const = 0;
CV_WRAP virtual void setMiniBatchSize(int val) = 0;

/** 算法终止条件*/
CV_WRAP virtual TermCriteria getTermCriteria() const = 0;
CV_WRAP virtual void setTermCriteria(TermCriteria val) = 0;

//! 正则化类型
enum RegKinds {
    REG_DISABLE = -1,    //!< 禁用正则化
    REG_L1 = 0,          //!< L1 正则化
    REG_L2 = 1           //!< L2 正则化
};
//! 训练方法
enum Methods {
    BATCH = 0,
    MINI_BATCH = 1
};
/** 预测输入样本的响应并返回浮点类型结果*/
CV_WRAP virtual float predict( InputArray samples, OutputArray results=noArray(),
                    int flags=0 ) const CV_OVERRIDE = 0;
/** 此函数返回跨行排列的经过训练的参数*/
CV_WRAP virtual Mat get_learnt_thetas() const = 0;
/** 创建空模型*/
CV_WRAP static Ptr<LogisticRegression> create();
/** 从文件中加载模型并创建 LogisticRegression 对象*/
CV_WRAP static Ptr<LogisticRegression> load(const String& filepath ,
                const String& nodeName = String());
};
```

如下为 SVMSGD 类的定义,该类实现了随机梯度下降 SVM 分类器:

```cpp
class CV_EXPORTS_W SVMSGD : public cv::ml::StatModel
{
public:
    /** SVM SGD 类型*/
    enum SvmsgdType
    {
        SGD,
        ASGD
    };
    /** Margin 类型*/
    enum MarginType
    {
        SOFT_MARGIN,
```

```cpp
        HARD_MARGIN
    };

    CV_WRAP virtual Mat getWeights() = 0;
    CV_WRAP virtual float getShift() = 0;

    /** 创建空模型*/
    CV_WRAP static Ptr<SVMSGD> create();
    /** 从文件中加载模型并创建 SVMSGD 对象*/
    CV_WRAP static Ptr<SVMSGD> load(const String& filepath , const String& nodeName = String());
    CV_WRAP virtual void setOptimalParameters(int svmsgdType = SVMSGD::ASGD,
                    int marginType = SVMSGD::SOFT_MARGIN) = 0;

    CV_WRAP virtual int getSvmsgdType() const = 0;
    CV_WRAP virtual void setSvmsgdType(int svmsgdType) = 0;

    CV_WRAP virtual int getMarginType() const = 0;
    CV_WRAP virtual void setMarginType(int marginType) = 0;

    CV_WRAP virtual float getMarginRegularization() const = 0;
    CV_WRAP virtual void setMarginRegularization(float marginRegularization) = 0;

    CV_WRAP virtual float getInitialStepSize() const = 0;
    CV_WRAP virtual void setInitialStepSize(float InitialStepSize) = 0;

    CV_WRAP virtual float getStepDecreasingPower() const = 0;
    CV_WRAP virtual void setStepDecreasingPower(float stepDecreasingPower) = 0;

    /** 算法终止条件*/
    CV_WRAP virtual TermCriteria getTermCriteria() const = 0;
    CV_WRAP virtual void setTermCriteria(const cv::TermCriteria &val) = 0;
};
```

如下函数用于从多元正态分布生成样本:

```cpp
CV_EXPORTS void randMVNormal( InputArray mean, InputArray cov, int nsamples, OutputArray samples);
```

如下函数用于创建测试集:

```cpp
CV_EXPORTS void createConcentricSpheresTestSet( int nsamples, int nfeatures,
            int nclasses, OutputArray samples, OutputArray responses);
```

如下为与模拟退火算法相关的函数定义:

```cpp
/** 模拟退火求解器*/
#ifdef CV_DOXYGEN
/** 模拟退火优化算法中使用的系统状态声明示例函数*/
struct SimulatedAnnealingSolverSystem
{
```

```
        double energy() const;
        void changeState();
        void reverseState();
};
#endif // CV_DOXYGEN

/** 该类实现模拟退火优化*/
template<class SimulatedAnnealingSolverSystem>
int simulatedAnnealingSolver(SimulatedAnnealingSolverSystem& solverSystem,
     double initialTemperature, double finalTemperature, double coolingRatio,
     size_t iterationsPerStep,
     CV_OUT double* lastTemperature = NULL,
     cv::RNG& rngEnergy = cv::theRNG()
);
}
}
#include <opencv2/ml/ml.inl.hpp>
#endif // __cplusplus
#endif // OPENCV_ML_HPP
```

ml 模块中封装了贝叶斯分类器算法、K 最近邻算法、支持向量机算法、期望最大化算法、决策树算法、随机树算法、增强树算法、人工神经网络（感知机）算法、Logistic 回归算法等机器学习中的常用算法,本章会选取 Logistic 回归与支持向量机两个案例做讲解。另外,在 core 模块中定义的主成分分析（PCACompute，见 3.1 节）在机器学习中也较常用,本章也会提供案例进行讲解。

13.2　案例 120：基于 OpenCV 的 Logistic 回归

Logistic 回归（逻辑回归）是常用分类算法，OpenCV 中封装了逻辑回归算法类 LogisticRegression（见 13.1 节）。

鸢尾花数据集是学习 Logistic 回归时常用的数据集，可以使用 sklearn 库导入，该数据集一共包含 4 个特征（花萼长度、花萼宽度、花瓣长度、花瓣宽度）；1 个类别变量；3 种种类（setosa、versicolor、virginica），每类 50 个样本，共 150 个样本。

```
from sklearn import datasets

#加载鸢尾花数据集
iris = datasets.load_iris()

#输出数据集对象信息
print(dir(iris))
#输出特征名称
```

```python
print(iris['feature_names'])
#输出分类目标名称
print(iris.target_names)
#输出数据集样本数量
print(len(iris.data))
```

上述打印信息结果如下：

```
['DESCR', 'data', 'feature_names', 'filename', 'frame', 'target', 'target_names']
['sepal length (cm)', 'sepal width (cm)', 'petal length (cm)', 'petal width (cm)']
['setosa' 'versicolor' 'virginica']
150
```

使用鸢尾花数据集进行 Logistic 回归，其中有一个类别可分，另两个类别不可分（如分类结果是 setosa 或 versicolor+virginica 中的一种）。也可以选取其中的两个类别和对应的样本进行分类，本案例选取 setosa 和 versicolor 两个类别，即分类结果为 setosa 或 versicolor。案例代码如下：

```python
import numpy as np
import cv2
from sklearn import datasets
from sklearn import model_selection
from sklearn import metrics
import matplotlib.pyplot as plt

#加载鸢尾花数据集
iris = datasets.load_iris()
#提取第一类
index0 = iris.target == 1
data0 = iris.data[index0].astype(np.float32)
target0 = iris.target[index0].astype(np.float32)
#提取第二类
index1 = iris.target == 0
data1 = iris.data[index1].astype(np.float32)
target1 = iris.target[index1].astype(np.float32)

#绘制两个类别不同花的花萼长宽数据散点图
plt.scatter(data0[:, 0], data0[:, 1], c='r', marker="*")
plt.scatter(data1[:, 0], data1[:, 1], c='b', marker="x")
plt.xlabel(iris.feature_names[0])
plt.ylabel(iris.feature_names[1])
plt.show()

#取出 3 类中的 2 类，取前面 2 类
index = iris.target!=2
data = iris.data[index].astype(np.float32)
target = iris.target[index].astype(np.float32)
```

```python
#划分训练集和测试集，测试集比例为 0.1
x_train, x_test, y_train, y_test = model_selection.train_test_split(data, target, test_size=0.1, random_state=42)
#训练 Logistic 回归分类器
lr = cv2.ml.LogisticRegression_create()
#设置训练方法
lr.setTrainMethod(cv2.ml.LogisticRegression_MINI_BATCH)
lr.setMiniBatchSize(1)
#设置迭代次数
lr.setIterations(100)
#开始训练
lr.train(x_train, cv2.ml.ROW_SAMPLE, y_train)
#打印训练参数
print(lr.get_learnt_thetas())
#测试集准确率测试
_, y_pred=lr.predict(x_test)
print("test accuracy: ", metrics.accuracy_score(y_test, y_pred))
```

绘制的两个特征的散点图如图 13.1 所示。

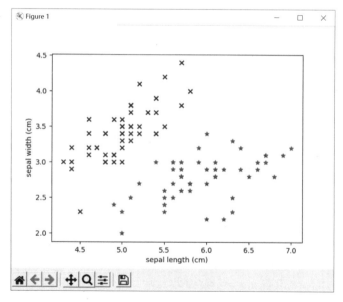

图 13.1

训练参数的训练结果及测试集准确率测试结果输出如下：

```
[[-0.04090133 -0.01910263 -0.16340333  0.28743777  0.11909772]]
test accuracy:  1.0
```

13.3 案例 121：基于 OpenCV 的支持向量机

支持向量机（SVM）是机器学习中最常用的分类算法之一，OpenCV 中封装了支持向量机算法的类 SVM（见 13.1 节）。

本案例来源于 OpenCV 官方示例（samples/python/tutorial_code/ml/non_linear_svms），对示例代码做了部分调整，案例代码如下：

```
import cv2
import numpy as np
import random as rng

#每个类别训练样本数
NTRAINING_SAMPLES = 100
#线性可分样本比例
FRAC_LINEAR_SEP = 0.9

#可视化的图像创建
WIDTH = 512
HEIGHT = 512
I = np.zeros((HEIGHT, WIDTH, 3), dtype=np.uint8)

#-----------1. 随机创建训练数据 -------------
trainData = np.empty((2*NTRAINING_SAMPLES, 2), dtype=np.float32)
labels = np.empty((2*NTRAINING_SAMPLES, 1), dtype=np.int32)
rng.seed(100) #Random value generation class

#设置线性可分部分数据
nLinearSamples = int(FRAC_LINEAR_SEP * NTRAINING_SAMPLES)

#生成第一个类别的随机点
trainClass = trainData[0:nLinearSamples,:]
#点的 x 坐标在[0, 0.4)区间
c = trainClass[:,0:1]
c[:] = np.random.uniform(0.0, 0.4 * WIDTH, c.shape)
#点的 y 坐标在[0, 1)区间
c = trainClass[:,1:2]
c[:] = np.random.uniform(0.0, HEIGHT, c.shape)

#生成第二个类别的随机点
trainClass = trainData[2*NTRAINING_SAMPLES-nLinearSamples:2*NTRAINING_SAMPLES,:]
#点的 x 坐标在[0.6, 1]区间
c = trainClass[:,0:1]
```

```python
c[:] = np.random.uniform(0.6*WIDTH, WIDTH, c.shape)
#点的y坐标在[0, 1)区间
c = trainClass[:,1:2]
c[:] = np.random.uniform(0.0, HEIGHT, c.shape)

#-------------创建非线性可分部分数据 --------
#为类别1和2生成随机点
trainClass = trainData[nLinearSamples:2*NTRAINING_SAMPLES-nLinearSamples,:]
#x坐标在[0.4, 0.6)区间
c = trainClass[:,0:1]
c[:] = np.random.uniform(0.4*WIDTH, 0.6*WIDTH, c.shape)
#y坐标在[0, 1)区间
c = trainClass[:,1:2]
c[:] = np.random.uniform(0.0, HEIGHT, c.shape)

#-------------设置类别标签 -----------------
labels[0:NTRAINING_SAMPLES,:] = 1                      #类别1
labels[NTRAINING_SAMPLES:2*NTRAINING_SAMPLES,:] = 2 #类别2

#-------------2. 设置SVM参数 -------------
print('Starting training process')
svm = cv2.ml.SVM_create()
svm.setType(cv2.ml.SVM_C_SVC)
svm.setC(0.1)
svm.setKernel(cv2.ml.SVM_LINEAR)
svm.setTermCriteria((cv2.TERM_CRITERIA_MAX_ITER, int(1e7), 1e-6))

#-----------3. 训练SVM -------------------
svm.train(trainData, cv2.ml.ROW_SAMPLE, labels)
print('Finished training process')

#-----------4. 展示决策边界 --------------
white = (255,255,255)
gray = (230,230,230)
for i in range(I.shape[0]):
    for j in range(I.shape[1]):
        sampleMat = np.matrix([[j,i]], dtype=np.float32)
        response = svm.predict(sampleMat)[1]
        
        if response == 1:
            I[i,j] = white
        elif response == 2:
            I[i,j] = gray

#------------5. 展示训练数据----------------
thick = -1
```

```python
#类别1
for i in range(NTRAINING_SAMPLES):
    px = trainData[i,0]
    py = trainData[i,1]
    cv2.circle(I, (int(px), int(py)), 3, (0, 255, 0), thick)

#类别2
for i in range(NTRAINING_SAMPLES, 2*NTRAINING_SAMPLES):
    px = trainData[i,0]
    py = trainData[i,1]
    cv2.circle(I, (int(px), int(py)), 3, (255, 0, 0), thick)

#----------------6. 显示支持向量 --------------
thick = 2
sv = svm.getUncompressedSupportVectors()

for i in range(sv.shape[0]):
    cv2.circle(I, (int(sv[i,0]), int(sv[i,1])), 6, (128, 128, 128), thick)

#结果展示
cv2.imshow('SVM for Non-Linear Training Data', I)
cv2.waitKey(0)
cv2.destroyAllWindows()
```

案例中创建了两个类别，并为两个类别生成了随机数据，线性可分的数据比例为 0.9；然后使用生成的数据训练 SVM 分类器，训练结果如图 13.2 所示。

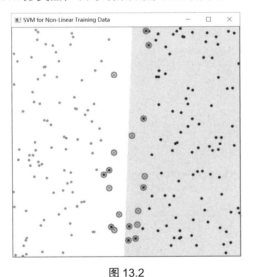

图 13.2

在图 13.2 中，绘制的圆圈所在的点称为支持向量。

13.4 案例 122：基于 OpenCV 的主成分分析

主成分分析在机器学习中有重要作用，是一种分析多维并从中提取带有最多信息量的维度子集的方法，主要用于数据降维。

本案例来源于 OpenCV 官方示例（samples/python/tutorial_code/ml/introduction_to_pca），对示例代码做了部分调整，案例代码如下：

```python
import cv2
import numpy as np
from math import atan2, cos, sin, sqrt, pi

#可视化
def drawAxis(img, p_, q_, colour, scale):
    p = list(p_)
    q = list(q_)
    #角度（单位：弧度）
    angle = atan2(p[1] - q[1], p[0] - q[0])
    hypotenuse = sqrt((p[1] - q[1]) * (p[1] - q[1]) + (p[0] - q[0]) * (p[0] - q[0]))

    #在这里将箭头按比例延长为原来的 2 倍
    q[0] = p[0] - scale * hypotenuse * cos(angle)
    q[1] = p[1] - scale * hypotenuse * sin(angle)
    cv2.line(img, (int(p[0]), int(p[1])), (int(q[0]), int(q[1])), colour, 1, cv2.LINE_AA)

    #创建箭头钩子
    p[0] = q[0] + 9 * cos(angle + pi / 4)
    p[1] = q[1] + 9 * sin(angle + pi / 4)
    cv2.line(img, (int(p[0]), int(p[1])), (int(q[0]), int(q[1])), colour, 1, cv2.LINE_AA)

    p[0] = q[0] + 9 * cos(angle - pi / 4)
    p[1] = q[1] + 9 * sin(angle - pi / 4)
    cv2.line(img, (int(p[0]), int(p[1])), (int(q[0]), int(q[1])), colour, 1, cv2.LINE_AA)

#PCA
def getOrientation(pts, img):
    #构造用于主成分分析的缓冲区
    sz = len(pts)
    data_pts = np.empty((sz, 2), dtype=np.float64)
    for i in range(data_pts.shape[0]):
        data_pts[i,0] = pts[i,0,0]
        data_pts[i,1] = pts[i,0,1]
```

```python
    #执行主成分分析
    mean = np.empty((0))
    mean, eigenvectors, eigenvalues = cv2.PCACompute2(data_pts, mean)

    #存储对象中心
    cntr = (int(mean[0,0]), int(mean[0,1]))

    #绘制主成分
    cv2.circle(img, cntr, 3, (255, 0, 255), 2)
    p1 = (cntr[0] + 0.02 * eigenvectors[0,0] * eigenvalues[0,0], cntr[1] + 0.02 * eigenvectors[0,1] * eigenvalues[0,0])
    p2 = (cntr[0] - 0.02 * eigenvectors[1,0] * eigenvalues[1,0], cntr[1] - 0.02 * eigenvectors[1,1] * eigenvalues[1,0])
    drawAxis(img, cntr, p1, (0, 255, 0), 1)
    drawAxis(img, cntr, p2, (255, 255, 0), 5)

    angle = atan2(eigenvectors[0,1], eigenvectors[0,0])
    return angle

#载入图像,图像为OpenCV提供的图像资源
src = cv2.imread('pca_test.jpg')
#检查图像载入是否成功
if src is None:
    print('Could not open or find the image! ')
    exit(0)
#显示输入图像
cv2.imshow('src', src)

#转为灰度图像
gray = cv2.cvtColor(src, cv2.COLOR_BGR2GRAY)

#转为二值图像
_, bw = cv2.threshold(gray, 50, 255, cv2.THRESH_BINARY | cv2.THRESH_OTSU)

#查找轮廓
contours, _ = cv2.findContours(bw, cv2.RETR_LIST, cv2.CHAIN_APPROX_NONE)

for i, c in enumerate(contours):
    #计算每个轮廓的面积
    area = cv2.contourArea(c)
    #忽略太小或太大的轮廓
    if area < 1e2 or 1e5 < area:
        continue

    #绘制所有轮廓
    cv2.drawContours(src, contours, i, (0, 0, 255), 2)
```

```
#找到每个形状的方向
getOrientation(c, src)

#结果显示
cv2.imshow('output', src)
cv2.waitKey(0)
cv2.destroyAllWindows()
```

案例中使用的图像为 OpenCV 官方提供的图像资源，输入图像如图 13.3 所示。

主成分分析后，绘制的每个形状的方向如图 13.4 所示。

图 13.3

图 13.4

13.5 进阶必备：机器学习算法概述

常用的机器学习算法总结如下。

1. 线性回归

线性回归是指用一条直线的线性关系表示输入值 x 与输出值 y 之间的关系，线性回归的建模即找出直线的斜率和截距参数。

线性回归的建模算法包括最小二乘法、梯度下降优化算法等。

2. Logistic 回归

Logistic 回归采用逻辑函数表示输入与输出之间的关系，将逻辑函数的输出结果增加一个阈值，就可以得到分类结果。

3. 支持向量机

支持向量机（SVM）是最受欢迎的机器学习算法，该算法是一种快速可靠的分类算法，可以在数据量有限的情况下很好地完成任务。

支持向量机算法的原理是求解一个几何间隔最大的超平面（或分割线），将数据进行正确的划分，如图 13.5 所示。

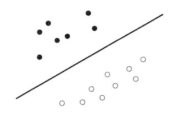

图 13.5

距离分割超平面最近的点称为支持向量，训练完成后的支持向量机模型仅与支持向量有关。支持向量机中引入了核技巧，以解决非线性可分问题。

4. 决策树

决策树（Decision Tree）是以特征属性构建的树结构（可以是二叉树或非二叉树），非叶子节点表示一个特征属性上的测试，每个分支代表该特征属性在某个属性值上的输出，树的叶子节点为一个分类类别。筛选西瓜的二叉树的构建如图 13.6 所示。

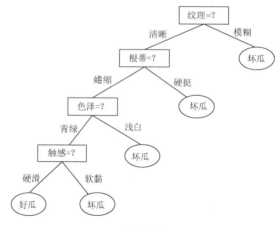

图 13.6

决策树进行决策的过程是：从根节点开始，测试待分类项中相应的特征属性，并按照其值选择输出分支，直到到达叶子节点，将叶子节点存放的类别作为决策结果。

根据属性划分方法，常用的决策树有 ID3 决策树（选择信息增益最大的属性来划分）、C4.5

决策树（选择增益率大的属性来划分）和 CART 决策树（选择基尼指数最小的属性来划分）。

决策树决策中常见的过拟合问题可以通过剪枝的方法缓解，剪枝分为预剪枝和后剪枝。其中，预剪枝是在划分前对每个结点进行估计，如果结点划分不能带来决策树泛化性能的提升，则停止划分；后剪枝是对已经构建完成的决策树采用自底向上的方法对非叶子结点进行考查，如果将该结点替换为叶子结点能带来决策树泛化性能的提升，则将该非叶子节点替换为叶子结点。

5. 随机森林

决策树可以选取特征对数据集进行划分并把数据贴上标签进行分类，随机森林是一个包含多个决策树的分类器，其输出的类别是由决策树分类结果投票产生的。随机森林是一种常见的集成学习算法。

随机森林由多个决策树组成，决策树之间没有关联。针对新的输入数据，每一棵决策树分别进行判断分类，随机森林以决策树分类结果中居多的类别作为最终分类结果。随机森林实现简单、不容易过拟合、稳定性高，但是计算成本较高。

6. 贝叶斯分类器

贝叶斯分类器是一种重要的分类算法，以贝叶斯理论为基础，通过求解后验概率分布来计算样本属于某类别的概率。

贝叶斯分类器包括朴素贝叶斯分类器（假设所有属性相互独立）、半朴素贝叶斯分类器（考虑部分属性之间的依赖关系）、贝叶斯网等。

7. KNN 算法

KNN（K-NearestNeighbor）算法即 K 最近邻算法，是一种简单的分类算法。在 KNN 算法分类中，每个样本都可以用它最邻近的 K 个邻近值来代表。KNN 算法简单、理论成熟，但是要计算每个样本到全体已知样本的距离，计算量较大。

8. 线性判别分析

线性判别分析（LDA）是将训练样例投影到一条直线上，使同类样例的投影点尽可能接近，使异类样例的投影点尽可能远，该算法是一种经典的监督线性降维方法。

9. 神经网络

神经网络也称人工神经网络，是目前人工智能研究中最热门的技术之一。神经网络是由多层网络构建的复杂结构，由训练数据对算法进行不断的优化以得到一组合理的参数（w 和 b），即训练得到模型。对于新的数据，输入模型计算就可以得到输出。神经网络的典型结构为 CNN（卷积神经网络）和 RNN（循环神经网络）。

第 14 章
深度学习模块 dnn

深度学习是目前的热点课题,国家层面对人工智能非常重视,企业或高校也都积极开展相关课题的研究。在深度学习的研究中,算法的研发和模型部署是两个最重要的方向。目前,有很多开源的深度学习框架(如 Tensorflow、Torch、Caffe 等),这些框架提供了模型训练的平台,是算法研发的重要支撑。而另外一个方向,即模型部署只需要模型推理框架,Tensorflow 提供了 Lite 支持模型推理部署,相较于模型训练,推理框架支持的算子相对较少,框架开发难度相对低一些,因此有很多的软件库只提供推理功能,OpenCV 4 中的 dnn 模块就实现了模型推理功能。

14.1 模块导读

OpenCV 中引入 dnn 模块需要包含头文件"opencv2/dnn.hpp",为了保持未来的扩展性,该头文件只对内部头文件做了一个包装。因而,该头文件中引用了内部模块头文件#include <opencv2/dnn/dnn.hpp>。本节通过解读头文件 opencv2/dnn/dnn.hpp 来帮助读者了解 dnn 模块封装的模型推理函数。该头文件的定义如下:

```
#ifndef OPENCV_DNN_DNN_HPP
#define OPENCV_DNN_DNN_HPP

#include <vector>
#include <opencv2/core.hpp>
#include "opencv2/core/async.hpp"
#include "../dnn/version.hpp"
```

```cpp
#include <opencv2/dnn/dict.hpp>

namespace cv {
namespace dnn {
CV__DNN_INLINE_NS_BEGIN
    typedef std::vector<int> MatShape;
```
如下为 dnn 模块函数使用的枚举类型标志的定义：
```cpp
    /** 枚举可支持的计算后端*/
    enum Backend
    {
        DNN_BACKEND_DEFAULT = 0,
        DNN_BACKEND_HALIDE,
        DNN_BACKEND_INFERENCE_ENGINE,                    //!< Intel 的推理计算后端
        DNN_BACKEND_OPENCV,
        DNN_BACKEND_VKCOM,
        DNN_BACKEND_CUDA,
#ifdef __OPENCV_BUILD
        DNN_BACKEND_INFERENCE_ENGINE_NGRAPH = 1000000,   // 内部使用
        DNN_BACKEND_INFERENCE_ENGINE_NN_BUILDER_2019,    // 内部使用
#endif
    };

    /** 枚举计算的目标设备*/
    enum Target
    {
        DNN_TARGET_CPU = 0,
        DNN_TARGET_OPENCL,
        DNN_TARGET_OPENCL_FP16,
        DNN_TARGET_MYRIAD,
        DNN_TARGET_VULKAN,
        DNN_TARGET_FPGA,
        DNN_TARGET_CUDA,
        DNN_TARGET_CUDA_FP16,
        DNN_TARGET_HDDL
    };
```
如下为目标设备相关函数的定义：
```cpp
    CV_EXPORTS std::vector< std::pair<Backend, Target> > getAvailableBackends();
    CV_EXPORTS_W std::vector<Target> getAvailableTargets(dnn::Backend be);

    /** 启用使用 CV DNN API 加载 DNN 模型的详细打印信息记录*/
    CV_EXPORTS void enableModelDiagnostics(bool isDiagnosticsMode);

    /** 该类提供层初始化所需的所有数据*/
    class CV_EXPORTS LayerParams : public Dict
    {
```

```cpp
public:
    std::vector<Mat> blobs; //!< 存储为 blob 的学习参数列表
    String name; //!< 层实例名（可选参数）
    String type; //!< 层类型名（可选参数）
};
```

如下为与框架后端相关的类的定义：

```cpp
/** 该类封装了某些后端的函数导出*/
class BackendNode
{
public:
    BackendNode(int backendId);
    virtual ~BackendNode();
    int backendId; //!< 后端标识符
};


/** 该类为不同的后端和目标设备封装 cv::Mat*/
class BackendWrapper
{
public:
    BackendWrapper(int backendId, int targetId);
    /** 为指定后端或目标设备封装*/
    BackendWrapper(int targetId, const cv::Mat& m);
    BackendWrapper(const Ptr<BackendWrapper>& base, const MatShape& shape);

    virtual ~BackendWrapper(); //!< Virtual destructor to make polymorphism.

    /** 向 CPU 主机存储传输数据*/
    virtual void copyToHost() = 0;
    /** 设置 CPU 上的真实数据*/
    virtual void setHostDirty() = 0;

    int backendId;   //!< 后端标识 ID
    int targetId;    //!< 目标设备标识 ID
};
```

如下为模型层相关类的定义：

```cpp
class CV_EXPORTS ActivationLayer;
/** 层创建*/
class CV_EXPORTS_W Layer : public Algorithm
{
public:
    //! 列举存储的学习参数以便被 Net::getParam()使用
    CV_PROP_RW std::vector<Mat> blobs;

    /** 根据输入/输出计算和设置内部参数*/
    CV_DEPRECATED_EXTERNAL
```

```cpp
virtual void finalize(const std::vector<Mat*> &input, std::vector<Mat> &output);

CV_WRAP virtual void finalize(InputArrayOfArrays inputs, OutputArrayOfArrays outputs);

/** 前向推理*/
CV_DEPRECATED_EXTERNAL virtual void forward(std::vector<Mat*> &input,
                std::vector<Mat> &output, std::vector<Mat> &internals);

virtual void forward(InputArrayOfArrays inputs, OutputArrayOfArrays outputs,
                OutputArrayOfArrays internals);

void forward_fallback(InputArrayOfArrays inputs, OutputArrayOfArrays outputs,
                OutputArrayOfArrays internals);

CV_DEPRECATED_EXTERNAL void finalize(const std::vector<Mat> &inputs,
                CV_OUT std::vector<Mat> &outputs);

CV_DEPRECATED std::vector<Mat> finalize(const std::vector<Mat> &inputs);
/** 分配层并计算输出,该函数在未来版本中会被抛弃*/
CV_DEPRECATED CV_WRAP void run(const std::vector<Mat> &inputs,
            CV_OUT std::vector<Mat> &outputs, CV_IN_OUT std::vector<Mat> &internals);

/** 将输入 blob 的索引返回输入数组*/
virtual int inputNameToIndex(String inputName);
/** 将输出 blob 的索引返回输出数组*/
CV_WRAP virtual int outputNameToIndex(const String& outputName);

/** 判断是否支持指定后端*/
virtual bool supportBackend(int backendId);

/** 返回 Halide 后端节点*/
virtual Ptr<BackendNode> initHalide(const std::vector<Ptr<BackendWrapper> > &inputs);
virtual Ptr<BackendNode> initInfEngine(const std::vector<Ptr<BackendWrapper> > &inputs);
virtual Ptr<BackendNode> initNgraph(const std::vector<Ptr<BackendWrapper> > &inputs,
                        const std::vector<Ptr<BackendNode> >& nodes);
virtual Ptr<BackendNode> initVkCom(const std::vector<Ptr<BackendWrapper> > &inputs);

/** 返回 CUDA 后端节点*/
virtual Ptr<BackendNode> initCUDA( void *context,
                        const std::vector<Ptr<BackendWrapper>>& inputs,
                        const std::vector<Ptr<BackendWrapper>>& outputs
);

/** 基于层超参数的 Halide 自动调度 */
virtual void applyHalideScheduler(Ptr<BackendNode>& node,
            const std::vector<Mat*> &inputs, const std::vector<Mat> &outputs,
```

```cpp
                    int targetId) const;
    /** 实现层融合*/
    virtual Ptr<BackendNode> tryAttach(const Ptr<BackendNode>& node);
    /** 尝试将后续激活层附加到该层*/
    virtual bool setActivation(const Ptr<ActivationLayer>& layer);
    /** 尝试将当前层和下一层融合*/
    virtual bool tryFuse(Ptr<Layer>& top);
    /** 返回具有按通道进行乘法和加法的层的参数*/
    virtual void getScaleShift(Mat& scale, Mat& shift) const;
    /** 附加到特定层*/
    virtual void unsetAttached();
    virtual bool getMemoryShapes(const std::vector<MatShape> &inputs,
                    const int requiredOutputs, std::vector<MatShape> &outputs,
                    std::vector<MatShape> &internals) const;

    virtual int64 getFLOPS(const std::vector<MatShape> &inputs,
                    const std::vector<MatShape> &outputs) const {
        CV_UNUSED(inputs); CV_UNUSED(outputs); return 0;
    }

    virtual bool updateMemoryShapes(const std::vector<MatShape> &inputs);
    CV_PROP String name;              //!< 层实例名
    CV_PROP String type;              //!< 层类型名
    CV_PROP int preferableTarget;     //!< 推理时的偏好目标设备

    Layer();
    explicit Layer(const LayerParams &params);
    void setParamsFrom(const LayerParams &params);
    virtual ~Layer();
};
```

如下为 Net 类的定义，该类允许创建和操作复杂的人工神经网络：

```cpp
class CV_EXPORTS_W_SIMPLE Net
{
public:
    CV_WRAP Net();
    CV_WRAP ~Net();

    /** 从 Intel 的模型优化器中间表示（IR）创建层*/
    CV_WRAP static Net readFromModelOptimizer(const String& xml, const String& bin);
    /** 从 Intel 的模型优化器在内存缓冲区中的中间表示（IR）创建网络*/
    CV_WRAP static Net readFromModelOptimizer(const std::vector<uchar>& bufferModelConfig,
                        const std::vector<uchar>& bufferWeights);

    static Net readFromModelOptimizer(const uchar* bufferModelConfigPtr,
                        size_t bufferModelConfigSize, const uchar* bufferWeightsPtr,
```

```cpp
                        size_t bufferWeightsSize);

CV_WRAP bool empty() const;
CV_WRAP String dump();
CV_WRAP void dumpToFile(const String& path);
int addLayer(const String &name, const String &type, LayerParams &params);
int addLayerToPrev(const String &name, const String &type, LayerParams &params);
CV_WRAP int getLayerId(const String &layer);
CV_WRAP std::vector<String> getLayerNames() const;
typedef DictValue LayerId;
CV_WRAP Ptr<Layer> getLayer(LayerId layerId);
std::vector<Ptr<Layer> > getLayerInputs(LayerId layerId);

/** 将第一层的输出连接到第二层的输入*/
CV_WRAP void connect(String outPin, String inpPin);
void connect(int outLayerId, int outNum, int inpLayerId, int inpNum);

CV_WRAP void setInputsNames(const std::vector<String> &inputBlobNames);
CV_WRAP void setInputShape(const String &inputName, const MatShape& shape);

/** 前向推理*/
CV_WRAP Mat forward(const String& outputName = String());
CV_WRAP AsyncArray forwardAsync(const String& outputName = String());
CV_WRAP void forward(OutputArrayOfArrays outputBlobs,
                     const String& outputName = String());
CV_WRAP void forward(OutputArrayOfArrays outputBlobs,
                     const std::vector<String>& outBlobNames);
CV_WRAP_AS(forwardAndRetrieve) void forward(
                     CV_OUT std::vector<std::vector<Mat> >& outputBlobs,
                     const std::vector<String>& outBlobNames);

CV_WRAP void setHalideScheduler(const String& scheduler);
CV_WRAP void setPreferableBackend(int backendId);
CV_WRAP void setPreferableTarget(int targetId);

/** 设置网络输入*/
CV_WRAP void setInput(InputArray blob, const String& name = "",
                      double scalefactor = 1.0, const Scalar& mean = Scalar());

CV_WRAP void setParam(LayerId layer, int numParam, const Mat &blob);
CV_WRAP Mat getParam(LayerId layer, int numParam = 0);

CV_WRAP std::vector<int> getUnconnectedOutLayers() const;
CV_WRAP std::vector<String> getUnconnectedOutLayersNames() const;

CV_WRAP void getLayersShapes(const std::vector<MatShape>& netInputShapes,
             CV_OUT std::vector<int>& layersIds,
```

```
                    CV_OUT std::vector<std::vector<MatShape> >& inLayersShapes,
                    CV_OUT std::vector<std::vector<MatShape> >& outLayersShapes) const;
    CV_WRAP void getLayersShapes(const MatShape& netInputShape,
                    CV_OUT std::vector<int>& layersIds,
                    CV_OUT std::vector<std::vector<MatShape> >& inLayersShapes,
                    CV_OUT std::vector<std::vector<MatShape> >& outLayersShapes) const;
    void getLayerShapes(const MatShape& netInputShape, const int layerId,
                    CV_OUT std::vector<MatShape>& inLayerShapes,
                    CV_OUT std::vector<MatShape>& outLayerShapes) const;
    void getLayerShapes(const std::vector<MatShape>& netInputShapes, const int layerId,
                    CV_OUT std::vector<MatShape>& inLayerShapes,
                    CV_OUT std::vector<MatShape>& outLayerShapes) const;

    CV_WRAP int64 getFLOPS(const std::vector<MatShape>& netInputShapes) const;
    CV_WRAP int64 getFLOPS(const MatShape& netInputShape) const;
    CV_WRAP int64 getFLOPS(const int layerId,
                    const std::vector<MatShape>& netInputShapes) const;
    CV_WRAP int64 getFLOPS(const int layerId,
                           const MatShape& netInputShape) const;

    CV_WRAP void getLayerTypes(CV_OUT std::vector<String>& layersTypes) const;
    CV_WRAP int getLayersCount(const String& layerType) const;
    void getMemoryConsumption(const std::vector<MatShape>& netInputShapes,
                    CV_OUT size_t& weights, CV_OUT size_t& blobs) const;
    CV_WRAP void getMemoryConsumption(const MatShape& netInputShape,
                    CV_OUT size_t& weights, CV_OUT size_t& blobs) const;
    CV_WRAP void getMemoryConsumption(const int layerId,
                    const std::vector<MatShape>& netInputShapes,
                    CV_OUT size_t& weights, CV_OUT size_t& blobs) const;
    CV_WRAP void getMemoryConsumption(const int layerId, const MatShape& netInputShape,
                    CV_OUT size_t& weights, CV_OUT size_t& blobs) const;

    void getMemoryConsumption(const std::vector<MatShape>& netInputShapes,
                    CV_OUT std::vector<int>& layerIds, CV_OUT std::vector<size_t>& weights,
                    CV_OUT std::vector<size_t>& blobs) const;
    void getMemoryConsumption(const MatShape& netInputShape,
                    CV_OUT std::vector<int>& layerIds, CV_OUT std::vector<size_t>& weights,
                    CV_OUT std::vector<size_t>& blobs) const;

    CV_WRAP void enableFusion(bool fusion);
    CV_WRAP int64 getPerfProfile(CV_OUT std::vector<double>& timings);

private:
    struct Impl;
    Ptr<Impl> impl;
};
```

如下为与模型读取等操作相关的函数的定义:
```
/** 读取 Darknet 框架的模型*/
CV_EXPORTS_W Net readNetFromDarknet(const String &cfgFile,
                const String &darknetModel = String());
CV_EXPORTS_W Net readNetFromDarknet(const std::vector<uchar>& bufferCfg,
                const std::vector<uchar>& bufferModel = std::vector<uchar>());
CV_EXPORTS Net readNetFromDarknet(const char *bufferCfg, size_t lenCfg,
                const char *bufferModel = NULL, size_t lenModel = 0);

/** 读取 Caffe 模型*/
CV_EXPORTS_W Net readNetFromCaffe(const String &prototxt,
                const String &caffeModel = String());
CV_EXPORTS_W Net readNetFromCaffe(const std::vector<uchar>& bufferProto,
                const std::vector<uchar>& bufferModel = std::vector<uchar>());
CV_EXPORTS Net readNetFromCaffe(const char *bufferProto, size_t lenProto,
                const char *bufferModel = NULL, size_t lenModel = 0);

/** 读取 Tensorflow 模型*/
CV_EXPORTS_W Net readNetFromTensorflow(const String &model, const String &config = String());
CV_EXPORTS_W Net readNetFromTensorflow(const std::vector<uchar>& bufferModel,
                const std::vector<uchar>& bufferConfig = std::vector<uchar>());
CV_EXPORTS Net readNetFromTensorflow(const char *bufferModel, size_t lenModel,
                const char *bufferConfig = NULL, size_t lenConfig = 0);

/** 读取 Torch 模型*/
 CV_EXPORTS_W Net readNetFromTorch(const String &model, bool isBinary = true,
                bool evaluate = true);

 /** 读取.caffemodel、.pb、.t7、.weights、.bin、.onnx 等格式的网络模型*/
 CV_EXPORTS_W Net readNet(const String& model, const String& config = "",
                const String& framework = "");
 CV_EXPORTS_W Net readNet(const String& framework, const std::vector<uchar>& bufferModel,
                const std::vector<uchar>& bufferConfig = std::vector<uchar>());

/** 加载序列化为 Torch7 框架的 torch.Tensor 对象的 blob*/
CV_EXPORTS_W Mat readTorchBlob(const String &filename, bool isBinary = true);

/** 从 Intel 的 Model Optimizer 加载网络*/
CV_EXPORTS_W Net readNetFromModelOptimizer(const String &xml, const String &bin);
CV_EXPORTS_W Net readNetFromModelOptimizer(const std::vector<uchar>& bufferModelConfig,
                const std::vector<uchar>& bufferWeights);
CV_EXPORTS Net readNetFromModelOptimizer(const uchar* bufferModelConfigPtr,
                size_t bufferModelConfigSize, const uchar* bufferWeightsPtr,
                size_t bufferWeightsSize);

/** 读取 ONNX 模型*/
```

```cpp
CV_EXPORTS_W Net readNetFromONNX(const String &onnxFile);
CV_EXPORTS Net readNetFromONNX(const char* buffer, size_t sizeBuffer);
CV_EXPORTS_W Net readNetFromONNX(const std::vector<uchar>& buffer);

/** 从.pb 文件创建 blob*/
CV_EXPORTS_W Mat readTensorFromONNX(const String& path);

/** 从图像创建 4 维 blob*/
CV_EXPORTS_W Mat blobFromImage(InputArray image, double scalefactor=1.0,
                const Size& size = Size(), const Scalar& mean = Scalar(),
                bool swapRB=false, bool crop=false, int ddepth=CV_32F);

CV_EXPORTS void blobFromImage(InputArray image, OutputArray blob,
                double scalefactor=1.0, const Size& size = Size(),
                const Scalar& mean = Scalar(), bool swapRB=false,
                bool crop=false, int ddepth=CV_32F);

/** 从图像序列创建 4 维 blob*/
CV_EXPORTS_W Mat blobFromImages(InputArrayOfArrays images, double scalefactor=1.0,
                Size size = Size(), const Scalar& mean = Scalar(), bool swapRB=false,
                bool crop=false, int ddepth=CV_32F);
CV_EXPORTS void blobFromImages(InputArrayOfArrays images, OutputArray blob,
                double scalefactor=1.0, Size size = Size(),
                const Scalar& mean = Scalar(), bool swapRB=false, bool crop=false,
                int ddepth=CV_32F);

/** 将 4 维 blob 解析为图像*/
CV_EXPORTS_W void imagesFromBlob(const cv::Mat& blob_, OutputArrayOfArrays images_);

/** 将 Caffe 模型的权重全部转为半精度浮点型*/
CV_EXPORTS_W void shrinkCaffeModel(const String& src, const String& dst,
                const std::vector<String>& layersTypes = std::vector<String>());

/** 为 pb 格式存储的二进制网络创建文本表示*/
CV_EXPORTS_W void writeTextGraph(const String& model, const String& output);

/** 对给定的框和相应的得分执行非极大值抑制操作*/
CV_EXPORTS void NMSBoxes(const std::vector<Rect>& bboxes,
                const std::vector<float>& scores, const float score_threshold,
                const float nms_threshold, CV_OUT std::vector<int>& indices,
                const float eta = 1.f, const int top_k = 0);
CV_EXPORTS_W void NMSBoxes(const std::vector<Rect2d>& bboxes,
                const std::vector<float>& scores, const float score_threshold,
                const float nms_threshold, CV_OUT std::vector<int>& indices,
                const float eta = 1.f, const int top_k = 0);
CV_EXPORTS_AS(NMSBoxesRotated) void NMSBoxes(const std::vector<RotatedRect>& bboxes,
```

```
                const std::vector<float>& scores, const float score_threshold,
                const float nms_threshold, CV_OUT std::vector<int>& indices,
                const float eta = 1.f, const int top_k = 0);
```

如下为 Model 类的定义，该类为神经网络高级 API 类：

```
class CV_EXPORTS_W_SIMPLE Model
{
public:
    CV_DEPRECATED_EXTERNAL
    Model();

    Model(const Model&) = default;
    Model(Model&&) = default;
    Model& operator=(const Model&) = default;
    Model& operator=(Model&&) = default;

    /** 创建模型*/
    CV_WRAP Model(const String& model, const String& config = "");
    CV_WRAP Model(const Net& network);

    /** 设置输入尺寸*/
    CV_WRAP Model& setInputSize(const Size& size);
    CV_WRAP inline
    Model& setInputSize(int width, int height) { return setInputSize(Size(width, height)); }

    CV_WRAP Model& setInputMean(const Scalar& mean);
    CV_WRAP Model& setInputScale(double scale);
    CV_WRAP Model& setInputCrop(bool crop);
    CV_WRAP Model& setInputSwapRB(bool swapRB);
    CV_WRAP void setInputParams(double scale = 1.0, const Size& size = Size(),
                const Scalar& mean = Scalar(), bool swapRB = false, bool crop = false);

    /** 模型推理运行预测*/
    CV_WRAP void predict(InputArray frame, OutputArrayOfArrays outs) const;

    CV_WRAP Model& setPreferableBackend(dnn::Backend backendId);
    CV_WRAP Model& setPreferableTarget(dnn::Target targetId);

    CV_DEPRECATED_EXTERNAL
    operator Net&() const { return getNetwork_(); }

    Net& getNetwork_() const;
    inline Net& getNetwork_() { return const_cast<const Model*>(this)->getNetwork_(); }

    struct Impl;
    inline Impl* getImpl() const { return impl.get(); }
    inline Impl& getImplRef() const { CV_DbgAssert(impl); return *impl.get(); }
```

```
protected:
    Ptr<Impl> impl;
};
```

如下为 ClassificationModel 类的定义，该类为分类模型高级 API 类：

```
class CV_EXPORTS_W_SIMPLE ClassificationModel : public Model
{
public:
    /** 创建分类模型*/
    CV_WRAP ClassificationModel(const String& model, const String& config = "");
    CV_WRAP ClassificationModel(const Net& network);

    /** 执行分类*/
    std::pair<int, float> classify(InputArray frame);
    CV_WRAP void classify(InputArray frame, CV_OUT int& classId, CV_OUT float& conf);
};
```

如下为 KeypointsModel 类的定义，该类为特征点模型高级 API 类：

```
class CV_EXPORTS_W_SIMPLE KeypointsModel: public Model
{
public:
    /** 创建模型*/
    CV_WRAP KeypointsModel(const String& model, const String& config = "");
    CV_WRAP KeypointsModel(const Net& network);

    /** 运行网络*/
    CV_WRAP std::vector<Point2f> estimate(InputArray frame, float thresh=0.5);
};
```

如下为 SegmentationModel 类的定义，该类为分割模型高级 API 类：

```
class CV_EXPORTS_W_SIMPLE SegmentationModel: public Model
{
public:
    /** 创建模型*/
    CV_WRAP SegmentationModel(const String& model, const String& config = "");
    CV_WRAP SegmentationModel(const Net& network);

    /** 执行分割*/
    CV_WRAP void segment(InputArray frame, OutputArray mask);
};
```

如下为 DetectionModel 类的定义，该类为目标检测网络高级 API 类：

```
class CV_EXPORTS_W_SIMPLE DetectionModel : public Model
{
public:
    /** 模型创建*/
    CV_WRAP DetectionModel(const String& model, const String& config = "");
    CV_WRAP DetectionModel(const Net& network);
```

```
    CV_DEPRECATED_EXTERNAL
    DetectionModel();

    CV_WRAP DetectionModel& setNmsAcrossClasses(bool value);
    CV_WRAP bool getNmsAcrossClasses();

    /** 执行检测*/
    CV_WRAP void detect(InputArray frame, CV_OUT std::vector<int>& classIds,
                CV_OUT std::vector<float>& confidences,
                CV_OUT std::vector<Rect>& boxes, float confThreshold = 0.5f,
                float nmsThreshold = 0.0f);
};
```

如下为 TextRecognitionModel 类的定义，该类为文本识别网络高级 API 类：

```
class CV_EXPORTS_W_SIMPLE TextRecognitionModel : public Model
{
public:
    CV_DEPRECATED_EXTERNAL TextRecognitionModel();

    /** 创建模型*/
    CV_WRAP TextRecognitionModel(const Net& network);
    CV_WRAP inline
    TextRecognitionModel(const std::string& model, const std::string& config = "")
        : TextRecognitionModel(readNet(model, config)) { /* nothing */ }

    /** 网络输出转译为字符串时的解码方式，如 CTC 方法 CTC-greedy*/
    CV_WRAP TextRecognitionModel& setDecodeType(const std::string& decodeType);
    CV_WRAP const std::string& getDecodeType() const;

    /** 设置词表*/
    CV_WRAP TextRecognitionModel& setVocabulary(const std::vector<std::string>& vocabulary);
    CV_WRAP const std::vector<std::string>& getVocabulary() const;

    /** 执行文本识别操作*/
    CV_WRAP std::string recognize(InputArray frame) const;
    CV_WRAP void recognize(InputArray frame, InputArrayOfArrays roiRects,
                CV_OUT std::vector<std::string>& results) const;
};
```

如下为 TextDetectionModel 类的定义，该类为文本检测类：

```
class CV_EXPORTS_W_SIMPLE TextDetectionModel : public Model
{
protected:
    CV_DEPRECATED_EXTERNAL TextDetectionModel();

public:
```

```cpp
    /** 运行文本检测*/
    CV_WRAP void detect(InputArray frame,
                CV_OUT std::vector< std::vector<Point> >& detections,
                CV_OUT std::vector<float>& confidences) const;
    CV_WRAP void detect(InputArray frame,
                CV_OUT std::vector< std::vector<Point> >& detections) const;
    CV_WRAP void detectTextRectangles(InputArray frame,
                CV_OUT std::vector<cv::RotatedRect>& detections,
                CV_OUT std::vector<float>& confidences) const;
    CV_WRAP void detectTextRectangles(InputArray frame,
                CV_OUT std::vector<cv::RotatedRect>& detections) const;
};
```

如下为 TextDetectionModel_EAST 类的定义，该类为 EAST 文本检测算法类：

```cpp
class CV_EXPORTS_W_SIMPLE TextDetectionModel_EAST : public TextDetectionModel
{
public:
    CV_DEPRECATED_EXTERNAL TextDetectionModel_EAST();

    /** 创建模型*/
    CV_WRAP TextDetectionModel_EAST(const Net& network);
    CV_WRAP inline TextDetectionModel_EAST(const std::string& model,
                    const std::string& config = "")
        : TextDetectionModel_EAST(readNet(model, config)) {}

    /** 设置检测置信度阈值*/
    CV_WRAP TextDetectionModel_EAST& setConfidenceThreshold(float confThreshold);
    CV_WRAP float getConfidenceThreshold() const;

    CV_WRAP TextDetectionModel_EAST& setNMSThreshold(float nmsThreshold);
    CV_WRAP float getNMSThreshold() const;
};
```

如下为 TextDetectionModel_DB 类的定义，该类为 DB 文本检测算法类：

```cpp
class CV_EXPORTS_W_SIMPLE TextDetectionModel_DB : public TextDetectionModel
{
public:
    CV_DEPRECATED_EXTERNAL TextDetectionModel_DB();

    /** 创建模型*/
    CV_WRAP TextDetectionModel_DB(const Net& network);

    CV_WRAP inline TextDetectionModel_DB(const std::string& model,
                    const std::string& config = "")
        : TextDetectionModel_DB(readNet(model, config)) {}

    CV_WRAP TextDetectionModel_DB& setBinaryThreshold(float binaryThreshold);
```

```cpp
    CV_WRAP float getBinaryThreshold() const;

    CV_WRAP TextDetectionModel_DB& setPolygonThreshold(float polygonThreshold);
    CV_WRAP float getPolygonThreshold() const;

    CV_WRAP TextDetectionModel_DB& setUnclipRatio(double unclipRatio);
    CV_WRAP double getUnclipRatio() const;

    CV_WRAP TextDetectionModel_DB& setMaxCandidates(int maxCandidates);
    CV_WRAP int getMaxCandidates() const;
};
CV__DNN_INLINE_NS_END
}
}
#include <opencv2/dnn/layer.hpp>
#include <opencv2/dnn/dnn.inl.hpp>
#include <opencv2/dnn/utils/inference_engine.hpp>
#endif  /* OPENCV_DNN_DNN_HPP */
```

14.2 风格迁移

14.2.1 深度学习风格迁移

风格迁移是指将一种图像的风格应用到某一幅图像中去。风格迁移的实现有较多算法，如生成对抗网络 GAN 可以实现风格迁移，用这些方法生成迁移图像的过程也是一个模型训练的过程，因而迁移速度缓慢。

深度学习中有一种快速进行风格迁移的方法，该算法的网络结构如图 14.1 所示。

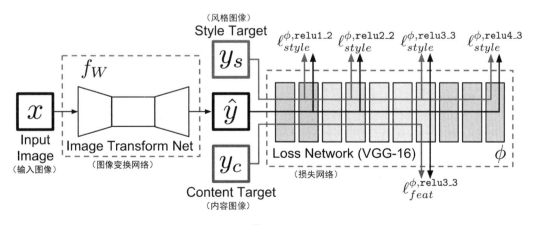

图 14.1

快速风格迁移网络主要包括两部分：图像变换网络和损失网络。将图像变换网络生成的图像与输入的内容图像（Content Target）和风格图像（Style Target）经过损失网络计算损失，不断优化损失得到最终的目标风格图像。

该算法主要包括两个阶段：模型训练与推理执行。模型训练阶段就是上述的选定一幅风格图像，训练得到输出目标风格图像的算法模型。推理执行阶段是输入一幅图像，送入网络模型，得到风格迁移之后的图像结果。对快速风格迁移有兴趣的读者可以参考论文 *Perceptual Losses for Real-Time Style Transfer and Super-Resolution*，以深入了解实现细节。

14.2.2 案例 123：OpenCV 实现风格迁移推理

本案例使用的模型来源于快速风格迁移算法，模型训练使用的是 Torch 框架，在 GitHub 的 jcjohnson 用户下的 fast-neural-style 仓库中，提供了模型下载的脚本 download_style_transfer_models.sh。

本案例在使用时做了部分修改，脚本内容如下：

```
BASE_URL="http://cs.stanford.edu/people/jcjohns/fast-neural-style/models/"   #下载的基本链接

mkdir -p models/instance_norm
cd models/instance_norm
curl -O "$BASE_URL/instance_norm/candy.t7"
curl -O "$BASE_URL/instance_norm/la_muse.t7"
curl -O "$BASE_URL/instance_norm/mosaic.t7"
curl -O "$BASE_URL/instance_norm/feathers.t7"
curl -O "$BASE_URL/instance_norm/the_scream.t7"
curl -O "$BASE_URL/instance_norm/udnie.t7"

mkdir -p ../eccv16
cd ../eccv16
curl -O "$BASE_URL/eccv16/the_wave.t7"
curl -O "$BASE_URL/eccv16/starry_night.t7"
curl -O "$BASE_URL/eccv16/la_muse.t7"
curl -O "$BASE_URL/eccv16/composition_vii.t7"
cd ../../
```

执行脚本 download_style_transfer_models.sh，就可以下载不同风格的模型。

提示：download_style_transfer_models.sh 是 Linux 系统中的执行脚本，在 Windows 系统中无法执行。用户可以在虚拟机中安装 Linux 系统，使用 Linux 系统下载模型；还有一个技巧就是使用 Git Bash 执行脚本进行下载。

本案例使用的原始图像如图 14.2 所示，来源于上述 fast-neural-style 论文代码开源仓库。

图 14.2

风格迁移模型推理执行包括 4 步。

第一步,输入模型加载。加载训练好的模型,本案例直接下载快速风格迁移算法开源的模型,读者可以训练自己的模型,并转换加载。

第二步,输入图像预处理。读取输入图像,并按照输入模型的要求转换维度和图像大小。另外,本案例的预处理还包括均值减法的运用,将输入图像的 RGB 3 个通道减去一组经验均值,作用是减小光照变换的影响,这是计算机视觉算法中图像预处理的常见方法之一。

第三步,模型推理执行。设置模型的输入,执行模型的推理计算,得到输出结果。

第四步,输出结果后处理。模型推理执行的结果并不是期望的最终结果,需要做一些变换才能得到最终有意义的结果。例如,在本案例中,需要调整输出的维度,将第二步预处理中均值减法减去的均值加回来,这样得到的就是最终期望的结果。

本案例的代码如下:

```python
import cv2
import time

model_base_dir = "./models/"                                          #基本路径

model_name = "candy"                                                  #模型名称
model_path = model_base_dir + model_name + ".t7"                      #模型所在路径
model = cv2.dnn.readNetFromTorch(model_path)                          #读取 Torch 模型

#读取模型有效性判断
if model.empty():
    print("load model error!")
else:
    print("load model sucess!")

src = cv2.imread("src.jpg")                                           #读取原始图像
(img_h, img_w) = src.shape[:2]                                        #获取图像宽高
blob_image = cv2.dnn.blobFromImage(src, 1.0, (img_w, img_h), (103.939, 116.779, 123.680),
swapRB=False, crop=False)                                             #图像维度变换
```

```
start = time.time()
model.setInput(blob_image)                                          #设置模型输入
output = model.forward()                                            #模型推理,得到输出

#对输出结果调整维度,将减去的平均值加回来,并交换各颜色空间
output = output.reshape((3, output.shape[2], output.shape[3]))      #输出结果维度变换
output[0] += 103.939
output[1] += 116.779
output[2] += 123.680
output = output.transpose(1, 2, 0)
end = time.time()

print("time for single image is:", end - start)
cv2.imwrite("candy.jpg", output)
```

上述案例使用的是 candy 风格,迁移后的结果如图 14.3 所示。

图 14.3

快速风格迁移的推理速度较快,单幅图像的推理时间在 0.48s 左右,该时间没有计算模型加载的时间。在应用部署时,模型只需要加载一次,读者可以测试预热(WarmUp)之后的单幅图像的推理时间。

```
time for single image is: 0.4827094078063965
```

快速风格迁移算法提供的开源模型中有多种风格,如 feathers 风格的迁移结果如图 14.4 所示。

图 14.4

对于其他风格结果,本案例就不做一一展示了,读者可以自己动手,选择不同的风格模型进行推理。

14.3 图像分类

14.3.1 深度学习图像分类

深度学习计算机视觉中有几类常见的任务，图像分类就是其中之一。图像分类是给定输入图像，通过图像分类模型推理计算，输出图像中的事物所属的类别。

> 提示：读者如果想详细了解计算机视觉的常见任务，则可以参考笔者图书《深度学习计算机视觉实战》，该书通过详细的案例展示了深度学习计算机视觉的常见任务及部署应用。

图像分类任务相对简单，读者可以选择该任务入门深度学习计算机视觉方向。图像分类任务模型训练包括以下 3 步。

第一步，数据预处理。

数据预处理包括数据集的准备，以及对数据集的预处理。图像分类任务有较多的开源数据集，如手写数字识别、CIFAR-10 和 CIFAR-100 等，这些数据集在很多框架中都有载入，不必下载，可以直接通过框架库调用。如下是在 Keras 和 TensorFlow 中引入数据集的方法：

```
#Keras 中 MNIST 数据集的引入
from keras.datasets import mnist
#数据导入
input = mnist.load_data()

#Keras 中 CIFAR-10 数据集的引入
from keras.datasets import cifar10
#Keras 中 CIFAR-100 数据集的引入
from keras.datasets import cifar100

#Tensorflow 1.x 中 MNIST 数据集的引入
from tensorflow.examples.tutorials import mnist

#Tensorflow 2.x 中 MNIST 数据集的引入
from tensorflow.keras.datasets import mnist
```

在 TensorFlow 1.15.0 版本之后，可以安装 tensorflow-datasets 库，该库提供了一系列可以在 Tensorflow 中使用的数据集：

```
import tensorflow_datasets as tfds
```

可以通过 tfds.list_builders() 列出所有可用的数据集（部分如下）：

```
['abstract_reasoning',
 'aeslc',
 'aflw2k3d',
 'amazon_us_reviews',
 ...
 'cifar10',
 'cifar100',
 'cifar10_1',
 ...
 'coco',
 ...
 'fashion_mnist',    #案例中会用到
 ...
 'imagenet2012',
 'imagenet2012_corrupted',
 'imagenet_resized',
 ...
 'iris',
 'kitti',
 'kmnist',
 ...
 'wider_face',
 'wikihow',
 'wikipedia',
 'wmt14_translate',
 'wmt15_translate',
 'wmt16_translate',
 'wmt17_translate',
 'wmt18_translate',
 'wmt19_translate',
 'wmt_t2t_translate',
 'wmt_translate',
 'xnli',
 'xsum',
 'yelp_polarity_reviews']
```

列表中有常见的数据集，如 coco 和 imagenet2012 等，可以通过 name 进行数据集的调用：

```
mnist_train = tfds.load(name="mnist", split="train")
assert isinstance(mnist_train, tf.data.Dataset)
print(mnist_train)
```

数据集加载会下载并准备好数据，除非读者显式指定 download=False，数据集一旦下载完成，就可以重复利用，如上述 MNIST 数据集的下载过程输出信息如下：

```
Downloading and preparing dataset mnist/3.0.0 (download: 14.06 MiB, generated: Unknown size,
```

```
total: 14.06 MiB) to /home/kbuilder/tensorflow_datasets/mnist/3.0.0...
WARNING:absl:Dataset mnist is hosted on GCS. It will automatically be downloaded to your
local data directory. If you'd instead prefer to read directly from our public
GCS bucket (recommended if you're running on GCP), you can instead set
data_dir=gs://tfds-data/datasets.
HBox(children=(FloatProgress(value=0.0, description='Dl Completed...', max=4.0,
style=ProgressStyle(descriptio…
Dataset mnist downloaded and prepared to /home/kbuilder/tensorflow_datasets/mnist/3.0.0.
Subsequent calls will reuse this data.
<DatasetV1Adapter shapes: {image: (28, 28, 1), label: ()}, types: {image: tf.uint8, label:
tf.int64}>
```

另外，读者还可以根据需要制作自己的数据集，图像分类数据集制作包括图像数据收集处理与标签制作。

第二步，网络搭建。

网络搭建是算法实现的过程，选择合适的网络对输入数据提取特征，然后使用分类器输出不同类别的得分。特征提取网络可以选取经典网络，如 VGG、ResNet 等；分类器常见的有 softmax 等。

在模型训练之前，需要为不同的任务制定损失函数、优化器及监控指标，分类任务中常见的损失函数为交叉熵损失，监控指标为计算准确率（图像正确分类的比率）。

第三步，模型训练。

模型训练是指对算法进行迭代，最小化损失函数，并计算评价函数，得到监控指标最好（如准确率最高）的模型。

14.3.2 案例 124：基于 TensorFlow 训练 Fashion-MNIST 算法模型

本案例参考 TensorFlow 官网示例，训练一个神经网络模型，对运动鞋和衬衫等图像进行分类，功能和实现原理与手写数字识别实现算法类似，步骤如下。

第一步，数据预处理。

常用的算法库导入如下：

```
#TensorFlow 和 tf.keras 导入
import tensorflow as tf
from tensorflow import keras

import numpy as np
import matplotlib.pyplot as plt

#本案例使用的是 2.4.1 版本的 TensorFlow
print(tf.__version__)
```

本案例使用的训练数据为 Fashion MNIST 数据集，直接导入数据集。从数据集名称上就可以看出，Fashion MNIST 数据集和 MNIST 数据集类似，该数据集包含 10 个类别的 70000 幅灰度图像，数据集中的图像以低分辨率（28 像素×28 像素，与 MNIST 数据集中的图像分辨率相同）展示单件衣物，如图 14.5 所示。

图 14.5

数据集分为训练集和测试集，其中 60000 幅图像作为训练集用于训练模型，剩余的 10000 幅图像作为测试集用于训练模型的评估。

```
#数据集导入，按训练集与测试集进行数据划分
fashion_mnist = keras.datasets.fashion_mnist
(train_images, train_labels), (test_images, test_labels) = fashion_mnist.load_data()
```

数据集对应的类别标签如表 14.1 所示，每幅图像都有一个对应的标签。

表 14.1

标　　签	类	标　　签	类
0	T恤/上衣	6	衬衫
1	裤子	7	运动鞋
2	套头衫	8	包
3	连衣裙	9	短靴
4	外套	—	—
5	凉鞋	—	—

数据集导入时会首先进行下载，读者也可以按照输出信息中的链接下载数据集并保存。

```
Downloading data from
https://storage.googleapis.com/tensorflow/tf-keras-datasets/train-labels-idx1-ubyte.gz
32768/29515 [==============================] - 0s 1us/step
Downloading data from
https://storage.googleapis.com/tensorflow/tf-keras-datasets/train-images-idx3-ubyte.gz
26427392/26421880 [==============================] - 22s 1us/step
Downloading data from
https://storage.googleapis.com/tensorflow/tf-keras-datasets/t10k-labels-idx1-ubyte.gz
8192/5148 [==============================] - 0s 0us/step
Downloading data from
https://storage.googleapis.com/tensorflow/tf-keras-datasets/t10k-images-idx3-ubyte.gz
4423680/4422102 [==============================] - 4s 1us/step
```

导入的数据集中的每一幅图像的像素值都在 0 到 255 之间，在送入网络之前，需要进行归一化处理，让像素值变换到 0 到 1 之间，归一化处理有利于网络的收敛，可以加快模型训练速度。

第二步，网络搭建。

本案例中使用的框架是 Keras，通过 tf.keras 引入（读者也可以直接使用 Keras），通过如下方式构建神经网络的层：

```
model = keras.Sequential([
    keras.layers.Flatten(input_shape=(28, 28)),
    keras.layers.Dense(128, activation='relu'),
    keras.layers.Dense(10)
    keras.layers.Softmax()
])
```

本案例中使用的网络结构很简单，将输入图像展平（Flatten），然后使用两层全连接网络（Dense），读者可以增加网络层以构建复杂的模型。

在模型训练之前，需要先编译模型，制定损失函数、优化器及评价指标，如本案例使用的优化器为'adam'、损失函数为 tf.keras.losses.SparseCategoricalCrossentropy、评价指标为准确率'accuracy'。

```
model.compile(optimizer='adam',
```

```
            loss=tf.keras.losses.SparseCategoricalCrossentropy(from_logits=True),
            metrics=['accuracy'])
```

第三步，模型训练。

模型训练调用 fit 函数，训练完成的模型通过 save 函数保存，evaluate 函数在测试集上评估模型训练的效果：

```
#模型训练
model.fit(train_images, train_labels, epochs=10)
model.save("fashion-mnist.h5")

test_loss, test_acc = model.evaluate(test_images,  test_labels, verbose=2)
print('\nTest accuracy:', test_acc)
```

本案例进行了 10 个轮次的训练，训练完成后，将模型保存，训练过程如下：

```
Epoch 1/10
1875/1875 [==============================] - 3s 1ms/step - loss: 0.6270 - accuracy: 0.7879
Epoch 2/10
1875/1875 [==============================] - 2s 1ms/step - loss: 0.3871 - accuracy: 0.8608
Epoch 3/10
1875/1875 [==============================] - 2s 1ms/step - loss: 0.3423 - accuracy: 0.8761
Epoch 4/10
1875/1875 [==============================] - 2s 1ms/step - loss: 0.3202 - accuracy: 0.8834
Epoch 5/10
1875/1875 [==============================] - 2s 1ms/step - loss: 0.2913 - accuracy: 0.8934
Epoch 6/10
1875/1875 [==============================] - 2s 1ms/step - loss: 0.2788 - accuracy: 0.8983
Epoch 7/10
1875/1875 [==============================] - 2s 1ms/step - loss: 0.2707 - accuracy: 0.8995
Epoch 8/10
1875/1875 [==============================] - 2s 1ms/step - loss: 0.2554 - accuracy: 0.9047
Epoch 9/10
1875/1875 [==============================] - 2s 1ms/step - loss: 0.2485 - accuracy: 0.9075
Epoch 10/10
1875/1875 [==============================] - 2s 1ms/step - loss: 0.2411 - accuracy: 0.9101
```

训练完成后，使用测试集数据测试，结果如下：

```
313/313 - 0s - loss: 0.3421 - accuracy: 0.8781

Test accuracy: 0.8780999779701233
```

对比结果表明，模型在测试集上的准确率略低于训练集，这种差距表明模型存在轻微过拟合。

保存的模型文件可以通过可视化工具 Netron 打开，本案例训练的模型结构读取如图 14.6 所示。

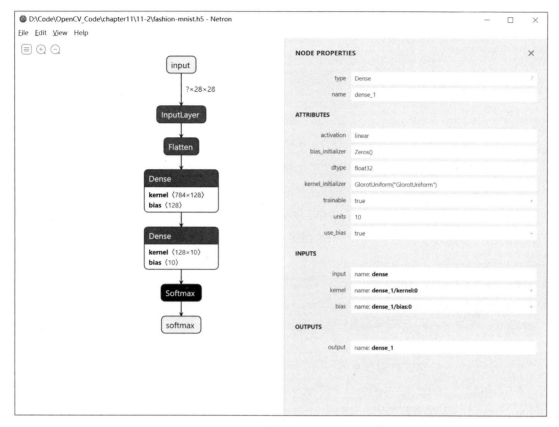

图 14.6

在图 14.6 中,左边为训练模型的结构图,图中的每一个节点代表一层,单击某个节点,就可以看到图中右边所示的模型节点信息,信息中展示了该节点的节点属性(NODE PROPERTIES)、输入(INPUTS)和输出(OUTPUTS)。

14.3.3 案例 125:OpenCV 实现图像分类推理

模型训练完成后,就可以将其部署到应用中,有很多推理框架可以提供部署服务,本节案例使用 OpenCV 进行该模型的部署。

模型部署包括以下 3 步。

第一步,模型转换。

在案例 124 中,训练的模型保存的是 HDF5(.h5)格式,也可以保存为 SavedModel 格式,这也是 tf.keras 模型保存的默认格式。

```
model.save("./model")          #模型保存在当前目录的 model 子文件夹下
```

model 文件夹中的内容如图 14.7 所示。

名称	修改日期	类型	大小
assets	2021/4/26 10:55	文件夹	
variables	2021/4/26 10:55	文件夹	
saved_model.pb	2021/4/26 10:55	PB 文件	75 KB

图 14.7

- saved_model.pb 文件用于存储模型图和训练配置参数。
- variables 目录包含一个标准训练检查点，保存每一次训练的权重。
- assets 目录包含 TensorFlow 计算图使用的文件，如用于初始化词汇表的文本文件。本案例中没有使用这种文件，因此目录为空。

有的项目可能还包括 assets.extra 目录，用于保存 TensorFlow 计算图未使用的任何文件。

模型转换需要转换为推理框架需要的模型格式，OpenCV 中提供了用于读取 TensorFlow 算法模型的函数 readNetFromTensorflow，其定义如下：

```
retval = readNetFromTensorflow(model, config=None):
```

参数说明如下。

- model：.pb 格式的模型文件。
- config：.pbtxt 文件路径。
- retval：模型对象（返回值）。

```
#重载函数
retval = readNetFromTensorflow(bufferModel[, bufferConfig])
```

参数说明如下。

- bufferModel：.pb 模型文件 buffer。
- bufferConfig：.pbtxt 文件 buffer。
- retval：模型对象（返回值）。

readNetFromTensorflow 函数传入的模型需要的模型格式为.pb 格式，因此，读者需要将训练的模型转换为.pb 格式。

知识点：转换为.pb 模型的过程称为计算图固化，即将计算图与权重数据固化到一个文件中。在 TensorFlow 1.x 中，提供了脚本 freeze_graph.py 以实现固化；2.x 不再支持，用户需要使用 1.x 中的脚本将模型转换为.pb 格式。

第二步，模型推理。

调用 readNetFromTensorflow 函数读取模型后，调用 forward 函数进行模型推理，得到模型输出。

第三步，输出后处理。

对于图像分类任务，模型推理输出的是每个类别对应的得分，如本案例中的输出结果是输入图像分类为 T 恤等 10 个类别的得分。但是这种输出不是最终的期望结果，有意义的输出是图像对应的类别，因此需要选取得分最高的类别，这个类别就是图像分类的类别，对应表 14.1，就可以找出最终的分类标签名称，这个将输出结果转换为有意义的输出的过程就是后处理。

图 14.8

本案例使用的测试图像如图 14.8 所示。

使用 OpenCV 进行图像分类推理的案例代码如下：

```python
#库导入
import cv2
import numpy as np

#模型读取
model = cv2.dnn.readNetFromTensorflow("./fashion-mnist.pb")
if model.empty():
    print("model error!")
else:
    print("load model sucess!")

#输入图像处理
src = cv2.imread("bag.jpg", 0)                          #以灰度图像方式读取输入图像
img_resize = cv2.resize(src, (28, 28))                  #调整图像大小为(28, 28)
img_resize = (np.expand_dims(img_resize, 0))/255        #输入图像维度调整并归一化，调整后为(1,28,28)

#模型推理
model.setInput(img_resize)          #设置模型输入数据
output = model.forward()            #执行模型推理操作

#后处理
result = np.argmax(output, axis=-1)
print("识别结果为:", result)
```

最终识别输出的结果如下：

识别结果为:[8]

对应表 14.1，index（index 是从 0 开始的）为 8 对应的标签为 bag，预测结果正确。

14.4 目标检测

14.4.1 深度学习目标检测

图像分类任务是输出给定图像中目标的类别，目标检测是分类与定位的结合，不仅需要给出目标的类别，还需要给出目标的位置，目标检测在人脸检测、行人检测、无人驾驶等领域有着重要的作用。

深度学习目标检测常用的算法分为两阶段目标检测（Two-Stage）算法和单阶段目标检测（One-Stage）算法两大类。

Two-Stage 算法分为候选框提取和分类两个阶段，代表算法如 R-CNN 系列（R-CNN、Fast R-CNN、Faster R-CNN）。

One-Stage 算法不需要生成候选框，故称为单阶段，代表算法如 SSD、YOLO 系列（YOLO v1~v5）。

提示：笔者图书《深度学习计算机视觉实战》的第 2 章对深度学习目标检测算法做了系统讲解，第 10 章和第 11 章的文本检测与识别（OCR）是目标检测中的经典案例，读者可以参考学习。

14.4.2 案例 126：OpenCV 实现目标检测推理

本案例使用的模型是 SSD_MobileNet_v2 网络训练 COCO 数据得到的，案例代码如下：

```python
import cv2
import numpy as np

#加载 COCO 分类类名
with open('object_detection_classes_coco.txt', 'r') as f:
    class_names = f.read().split('\n')

#为每个类创建随机颜色矩阵
COLORS = np.random.uniform(0, 255, size=(len(class_names), 3))

#加载深度学习模型
model = cv2.dnn.readNet(model='frozen_inference_graph.pb',
```

```python
                    config='ssd_mobilenet_v2_coco_2018_03_29.pbtxt.txt',
                    framework='TensorFlow')

#读取输入图像
image = cv2.imread('src.jpg')
image_height, image_width, _ = image.shape
#对图像做预处理
blob = cv2.dnn.blobFromImage(image=image, size=(300, 300), mean=(104, 117, 123),
                             swapRB=True)
#设置模型输入
model.setInput(blob)
#模型推理
output = model.forward()

#遍历所有的检测结果
for detection in output[0, 0, :, :]:
    #提取检测置信度
    confidence = detection[2]
    #当检测置信度大于 0.4 时,绘制检测框
    if confidence > 0.4:
        #获取类别 id
        class_id = detection[1]
        #与 class_names 映射得到检测的类别名称
        class_name = class_names[int(class_id)-1]
        color = COLORS[int(class_id)]
        #获取包围框的坐标
        box_x = detection[3] * image_width
        box_y = detection[4] * image_height
        #获取包围框的宽高
        box_width = detection[5] * image_width
        box_height = detection[6] * image_height
        #在每个检测目标周围绘制矩形框
        cv2.rectangle(image, (int(box_x), int(box_y)), (int(box_width), int(box_height)),
                      color, thickness=2)
        #在图像上标注检测的类别名称
        cv2.putText(image, class_name, (int(box_x), int(box_y - 5)),
                    cv2.FONT_HERSHEY_SIMPLEX, 1, color, 2)

#结果显示
cv2.imshow('image', image)
cv2.waitKey(0)
cv2.destroyAllWindows()
```

object_detection_classes_coco.txt 中存储的内容为检测目标的类别,部分类别如下:
person
bicycle

```
car
motorcycle
airplane
bus
...
```

案例中使用的输入图像需要按照模型输入尺寸（300 像素×300 像素）的要求做调整。另外，输入图像预处理还做了图像减均值、交换 RB 通道的操作，预处理后利用加载的深度学习模型做推理，得到检测结果并绘制其包围框，绘制结果如图 14.9 所示。

图 14.9

14.5　图像超分

14.5.1　深度学习图像超分算法

图像超分是指将输入图像的尺寸放大或改善输入图像细节的过程，即让图像看起来更清晰。OpenCV 中提供了 4 种深度学习算法对图像进行超分。

- EDSR：该算法使用 ResNet 网络架构，没有批量归一化层（BN 层），不同的尺度需要不同的模型。若想对该算法进行深入研究，则可以参考论文 *Enhanced Deep Residual Networks for Single Image Super-Resolution*。
- ESPCN：该算法的网络结构参考了 SRCNN 算法，使用亚像素卷积层代替常用的卷积层，亚像素卷积层的作用类似于反卷积层，网络最后一层使用亚像素卷积层来提高分辨率。若想对该算法进行深入研究，则可以参考论文 *Real-Time Single Image and Video Super-Resolution Using an Efficient Sub-Pixel Convolutional Neural*

Network。
- FSRCNN：与 ESPCN 算法类似，是对 SRCNN 算法的改进，针对计算量大的问题做了处理。若想对该算法进行深入研究，则可以参考论文 *Accelerating the Super-Resolution Convolutional Neural Network*。
- LapSRN：该算法结合传统图像处理算法拉普拉斯金字塔（见 5.6.2 节）与深度学习的方法，由低分辨率图像生成不同超分辨率的图像，实现多级超分辨率模型。若想对该算法进行深入研究，则可以参考论文 *Fast and Accurate Image Super-Resolution with Deep Laplacian Pyramid Networks*。

注意：本节案例需要用户安装 opencv_contrib 库，Python 环境安装方法为 pip install opencv-contrib-python（注意：需要安装 OpenCV 4.3 之后的版本）。

14.5.2 案例 127：OpenCV 实现图像超分推理

本节图像超分应用 EDSR 算法模型，进行 4 倍超分，案例代码如下：

```
import cv2

img=cv2.imread("pyrdown2.jpg")
cv2.imshow("image", img)
sr = cv2.dnn_superres.DnnSuperResImpl_create()
path = "EDSR_x4.pb"
#模型读取
sr.readModel(path)
#设置模型名称与超分倍数，注意：倍数与模型对应
sr.setModel("edsr", 4)
#图像超分
result = sr.upsample(img)
cv2.imshow("super_res", result)
#应用拉普拉斯金字塔上采样两次
pyrup = cv2.pyrUp(img)
pyrup = cv2.pyrUp(pyrup)
cv2.imshow("pyrup", pyrup)

cv2.waitKey(0)
cv2.destroyAllWindows()
```

案例中使用的输入图像为图 3.10 下采样两次的结果，即图 5.19 中的第三幅图像（从左向右数），如图 14.10 所示。

图 14.10

本案例将超分结果与上采样两次的结果进行对比，如图 14.11 所示。

图 14.11

在图 14.11 中，左图为超分结果，右图为上采样两次的结果，超分图像比上采样两次的图像清晰很多。

14.6 进阶必备：OpenCV 与计算机视觉

计算机视觉是深度学习的主要方向之一，在很多人工智能应用（如人脸识别、无人驾驶）中发挥着重要作用，是目前研究和求职的热点，前景巨大。OpenCV 是图像处理和计算机视觉中常用的工具库，掌握 OpenCV 常用的图像处理算法，对计算机视觉中的预处理和后处理有着重要的作用，因此，OpenCV 为计算机视觉算法的研发提供了强有力的支撑。

14.6.1 计算机视觉的发展

计算机视觉的发展由来已久，不过，以前的计算机视觉采用的都是传统图像处理的方法，进展比较缓慢。目前火热的计算机视觉方向主要是指利用深度学习的方法进行视觉任务的开发。深度学习计算机视觉的方法性能远优于传统图像处理的方法性能，且能够部署落地，是 AI 中目前为数不多的能够投入使用的方向之一。

计算机视觉的应用方向很多，如人脸识别、无人驾驶、智慧医疗等，但这些都是综合学科，作为初学者，可以选取计算机视觉中比较基础的内容（如图像分类、目标检测及图像分割等）作为入门的方向。掌握常用计算机视觉算法的原理，通过案例实现常用的功能，在此基础上进行相关任务的应用扩展，就可以逐步入门该方向了。

计算机视觉的入门目前有两大误区需要读者警惕。

第一，计算机视觉本质是图像处理，在深度学习算法成熟的当下，可以直接使用 CNN 网络提取图像特征，因此很多读者不会注意这些特征的差异，但是这种差异在处理不同任务时至关重要。

建议读者在入门计算机视觉的过程中，不仅要关注深度学习算法，还要学习传统图像处理知识，这样才能更好地满足视觉任务的开发。

第二，很多读者只关注计算机视觉算法模型的开发，而忽略了模型部署的细节。

对于企业，一切算法的开发的最终目的都是投入商用，因此，模型部署也是算法工程师需要考虑的问题。模型部署会面临很多问题，如平台硬件限制、推理框架的选择、模型的性能等，建议初学者在入门的同时留心模型部署。

提示：笔者图书《深度学习计算机视觉实战》对计算机视觉算法与模型部署做了大量的案例解读，有助于读者学习提升。

14.6.2 OpenCV 在计算机视觉中的应用

前面说过，计算机视觉是对图像信息的处理，属于图像处理的范畴，读者在学习计算机视觉的同时需要关注传统图像处理算法的研究，而 OpenCV 对传统图像处理算法做了系统的封装，本书就是按照 OpenCV 模块化结构进行讲解的。

OpenCV 在计算机视觉中的应用包括以下几方面。

- 数据集图像读取。

计算机视觉中的数据集主要是图像，在模型训练阶段，OpenCV 可以用于读取数据集图像文件，读取后的数据可以被打包成批次（batch）送入网络参与模型训练。

- 输入图像预处理。

在模型测试或部署阶段，图像被送入模型之前，需要进行一些预处理，如将图像尺寸、维度调整到满足模型输入的要求。

- 数据集增强。

利用常用的图像处理算法（如旋转、随意裁剪等）对原有数据集图像进行处理，得到新的图像并标注，扩充到原有数据集中，实现数据集的增强。

- 输出结果后处理。

对于有些计算机视觉任务，需要使用 OpenCV 对输出结果进行后处理，转换为有意义的输出。例如，在文本检测中，检测算法输出检测到的文本行的坐标，需要根据坐标裁剪出文本行，送入文本识别模型进行识别；在图像矫正算法中，模型输出的也是待矫正区域的坐标，需要使用透视变换算法将图像进行矫正。

作为开源的图像处理算法库，OpenCV 的功能是非常完备且强大的，这是其作为计算机视觉研究辅助工具的重要原因。

OpenCV 提供的 dnn 模块引入了模型推理功能，支持 TensorFlow、Caffe、Torch 等多种 AI 框架模型的推理，在某些场景中可以不用引入这些框架的推理库，而使用 OpenCV 进行模型部署。OpenCV 进行模型推理使用也有其局限性，如支持的算子有限、支持的模型格式有限等，读者可以根据开发环境选择使用。